OPERADS AND UNIVERSAL ALGEBRA

NANKAI SERIES IN PURE, APPLIED MATHEMATICS AND THEORETICAL PHYSICS

Editors: S. S. Chern, C. N. Yang, M. L. Ge, Y. Long

Published:

Vol. 1 Probability and Statistics
eds. Z. P. Jiang, S. J. Yan, P. Cheng and R. Wu

Vol. 2 Nonlinear Anaysis and Microlocal Analysis
eds. K. C. Chung, Y. M. Huang and T. T. Li

Vol. 3 Algebraic Geometry and Algebraic Number Theory
eds. K. Q. Feng and K. Z. Li

Vol. 4 Dynamical Systems
eds. S. T. Liao, Y. Q. Ye and T. R. Ding

Vol. 5 Computer Mathematics
eds. W.-T. Wu and G.-D. Hu

Vol. 6 Progress in Nonlinear Analysis
eds. K.-C. Chang and Y. Long

Vol. 7 Progress in Variational Methods
eds. C. Liu and Y. Long

Vol. 8 Quantized Algebra and Physics
Proceedings of International Workshop
eds. M.-L. Ge, C. Bai and N. Jing

Vol. 9 Operads and Universal Algebra
Proceedings of the International Conference
eds. C. Bai, L. Guo and J.-L. Loday

Proceedings of the International Conference on
Operads and Universal Algebra

Tianjin, China 5–9 July 2010

Nankai
Series in
Pure,
Applied
Mathematics
and
Theoretical
Physics

Vol. 9

OPERADS AND UNIVERSAL ALGEBRA

Edited by

Chengming Bai
Nankai University, China

Li Guo
Rutgers University at Newark, USA

Jean-Louis Loday
CNRS, France & Université de Strasbourg, France

NEW JERSEY • LONDON • SINGAPORE • BEIJING • SHANGHAI • HONG KONG • TAIPEI • CHENNAI

Published by

World Scientific Publishing Co. Pte. Ltd.

5 Toh Tuck Link, Singapore 596224

USA office: 27 Warren Street, Suite 401-402, Hackensack, NJ 07601

UK office: 57 Shelton Street, Covent Garden, London WC2H 9HE

British Library Cataloguing-in-Publication Data
A catalogue record for this book is available from the British Library.

Nankai Series in Pure, Applied Mathematics and Theoretical Physics — Vol. 9
OPERADS AND UNIVERSAL ALGEBRA
Proceedings of the International Conference

ISBN-13 978-981-4365-11-6
ISBN-10 981-4365-11-4

Printed in Singapore.

Preface

The Summer School and the International Conference "Operads and Universal Algebra" were held at the Chern Institute of Mathematics (Nankai University, Tianjin) from June 28 to July 9 of 2011. The present volume is the Proceedings of these events.

The goal of the Summer School was to introduce the non-specialists to the theory of algebraic operads and to the current problems in universal algebra. The aim of the International Conference was to expose the forefront research on these themes. These events successfully brought together teachers and students in mathematics from all around China (including Beijing, Tianjin, Guangzhou, Hangzhou, Harbin, Kunming and Lanzhou) and from France (including Paris, Lyon and Strasbourg), but the International Conference hosted also specialists from several other countries (see the list of participants).

During these two weeks we found a strong convergence of the methods: the Gröbner-Shirshov bases. They are a key tool in the work on rewriting systems of the mathematical school of South China Normal University and they turn out to be an efficient ingredient in the algorithmic proof of Koszul duality, both for algebras and for operads. It was also a first opportunity to compare the work of the French school and the Chern Institute of Mathematics school on the "higher algebra" types like dendriform algebras and the like.

These Proceedings show the strong relationship between topics coming from universal algebra, algebraic topology, computer science and mathematical physics. Also included in these proceedings are a list of problems in operad theory and an encyclopedia of types of algebras accumulated as of 2010.

The fruitfulness of this meeting can be judged not only by these Proceedings, but also by the productive collaborations which occurred during and after this meeting, and the ties forged through this meeting that result future visits and events, including the next meeting "Operads and rewriting" to be held in Lyon in November 2011.

The Summer School and the International Conference were preceded by a preparatory week which was held at the Capital Normal University, Beijing from June 21 to June 25, 2011.

Acknowledgements

The Preparatory School, the Summer School and the International Conference on "Operads and Universal Algebra" have been made possible through the financial support of the National Science Foundation of China (NSFC) and the Centre National de la Recherche Scientifique (CNRS), France, through the project "Franco-Chinese interactions in Mathematics" and also the Chinese Mathematical Society; the Mathematical Center of Ministry of Education, China; K.C. Wong Education Foundation, Hong Kong; NKBRPC, Ministry of Science and Technology, China; SRFDP, Ministry of Education, China.

We warmly thank all these organizations for their help and particularly the Capital Normal University, Beijing, and the Chern Institute of Mathematics, Nankai University, Tianjin, for hosting us during these events.

Chengming Bai (Chern Institute of Mathematics, Tianjin)
Li Guo (Rutgers University, Newark)
Jean-Louis Loday (CNRS, Strasbourg)

On behalf of all the participants I would like to warmly thank the local organizers: Professor Shanzhong Sun, Professor Chengming Bai and their teams, for the excellence of the organization both in Beijing and in Tianjin.

J.-L. Loday

Organizing Committees

Scientific Committee

- Jean-Louis Loday (chair), CNRS, University of Strasbourg, France,
- Chengming Bai (vice-chair), Nankai University, China,
- Marcelo Aguiar, Texas A & M University, USA,
- Yuqun Chen, South China Normal University, China,
- Pierre-Louis Curien, University of Paris VII, France,
- Li Guo, Rutgers University at Newark, USA,
- Maria Ronco, University of Valparaiso, Chile,
- Jim Stasheff, University of Pennsylvania, USA,
- Ke Wu, Capital Normal University, China.

Organization Committee

- Mo-Lin Ge (chair), Nankai University, China,
- Chengming Bai, Nankai University, China,
- Leonid A. Bokut, South China Normal University, China and Sobolev Institute of Mathematics, Russia,
- Yuqi Guo, Southwest University, China,
- Shanzhong Sun, Capital Normal University, China.

Primary Organizers

- Jean-Louis Loday (chair), CNRS, University of Strasbourg, France,
- Chengming Bai, Nankai University, China,
- Li Guo, Rutgers University at Newark, USA.

Speakers and Lectures

Preparatory School

- Li Guo, Rutgers University at Newark, USA. "Two lectures on Rota-Baxter algebras".
- Manfred Hartl, Valenciennes, France. "Two lectures on model categories".
- Yi-Zhi Huang, Rutgers Unversity at New Brunswick, USA. "Three lectures on vertex operator algebras".
- Jean-Louis Loday, CNRS and University of Strasbourg, France. "Four lectures on Associative algebras, Lie algebras, bialgebras, Lie bialgebras, dendriform and Zinbiel algebras".
- Shanzhong Sun, Capital Normal University, China. "Three lectures on A-infinity algebras".
- Lizhong Wang, Peking University, China. "Two lectures on representations of symmetric groups, Young diagrams and Frobenius characteristic".
- Zifeng Yang, Capital Normal University, China. "Three lectures on categories, functors, monads and homological algebra".

Summer School

- Chengming Bai, Nankai University, China. "Pre-Lie algebras and Lie and Jordan analogues of Loday algebras".
- Yuqun Chen, South China Normal University, China. "Universal algebra: Groebner-Shirshov bases (for some operads)".
- Pierre-Louis Curien, CNRS and University of Paris VII, France. "Higher category theory and programming languages".
- Li Guo, Rutgers University at Newark, USA. "Rota-Baxter algebras and regular quadratic operads with unit action".
- Yun Liu, Yuxi Normal University, China. "Universal algebra and some related topics in theoretical computer science".
- Muriel Livernet, University of Paris 13, France. "Topological operads".

- Jean-Louis Loday, CNRS and University of Strasbourg, France. "Algebraic operads".
- Bruno Vallette, University of Nice, France. "Koszul duality".

International Conference

- Emily Burgunder, Toulouse University, France. "Operads and Kontsevich graph complexes".
- Yongshan Chen, South China Normal University, China. "On Groebner-Shirshov bases".
- Yuqun Chen, South China Normal University, China. "Some new results on Groebner-Shirshov bases for universal linear algebras".
- Pierre-Louis Curien, CNRS and University of Paris VII, France. "On some applications of operadic ideas and techniques in the field of the semantics of programming languages".
- Vladimir Dotsenko, Dublin Institute for Advanced Studies and Trinity College, Ireland. "Operadic homological algebra via shuffle operads".
- Alessandra Frabetti, Lyon University, France. "Combinatorial Hopf Algebras from renormalization".
- Yael Frégier, University of Luxembourg, Luxembourg. "Layer cake and homotopy representations : formal geometry approach".
- Yves Guiraud, INRIA, France. "Higher-dimensional rewriting strategies and acyclic polygraphs".
- Joseph Hirsh, City University of New York, USA. "Homotopy-coherent morphisms of algebraic structures".
- Ralf Holtkamp, University of Hamburg, Germany. "On sub-operads of ComMag".
- Yi-Zhi Huang, Rutgers University at New Brunswick, USA. "Operads in two-dimensional conformal field theory applications of these formulations and studies".
- Liang Kong, Tsinghua University, China. "On the classification of rational open-closed conformal field theories".
- Haisheng Li, Rutgers University at Camden, USA. "Formal groups and vertex algebras".
- Dong Liu, Huzhu Teachers College, China. "Alternative dialgebras and Leibniz algebras".
- Muriel Livernet, University of Paris 13, France. "A Tor interpretation of E_n-homology".

- Philippe Malbos, Lyon University, France. "Higher-dimensional rewriting strategies and acyclic polygraphs".
- Abdenacer Makhlouf, Mulhouse University, France. "An overview of Hom-algebras, cohomologies and deformations".
- Miguel Mendez, IVIC and University of Central Venezuela, Venezuela. "A formula for the antipode of the natural Hopf algebra associated to a set-operad".
- Joan Millès, University of Nice, France. "Curved Koszul duality theory and homotopy unital associative algebras".
- Xiang Ni, Nankai University, China. "Nonabelian generalized Lax pairs, O-operators and various successors of some binary algebraic operads".
- Maria Ronco, University of Valparaiso, Chile. "Algebra and coalgebra structures on the generalized associahedron".
- C. Selvaraj, Periyar University, India. "Topological conditions of 2-primal rings".
- Bruno Vallette, University of Nice, France. "Homotopy Batalin-Vilkovisky algebras".
- Christine Vespa, University of Strasbourg, France. "Quadratic functors on pointed categories".
- Shoufeng Wang, Southwest University, China. "A survey of a topic on universal algebra: Characterizing and generalizing regular languages by semifilter-congruences".
- Changchang Xi, Beijing Normal University, China. "Auslander-Yoneda algebras and derived equivalences".
- Wen-Li Yang, Northwest University, China. "Differential operator realizations of (super)algebras and free-field representations of corresponding current algebras".
- Yong Zhang, Zhejiang University, China. "Variations of the dendriform quadri-algebra".

Participants and Photos

Chengming Bai	Nankai University, China
Olivia Bellier	University of Nice, France
Zhen Bian	Beijing Normal University, China
Alexandre Bouayad	Ecole Polytechnique, France
Emily Burgunder	Toulouse University, France
Wenjing Chang	Capital Normal University, China
Kun Chen	Capital Normal University, China
Sheng Chen	Harbin Inst. of Technology, China
Xiaojun Chen	University of Michigan, USA
Xiaomei Chen	Harbin Inst. of Technology, China
Xiuli Chen	Zhejiang University, China
Yin Chen	Northeast Normal University, China
Yongshan Chen	South China Normal University, China
Yuqun Chen	South China Normal University, China
Zhiqi Chen	Nankai University, China
Chenghao Chu	Johns Hopkins University, USA
Pierre-Louis Curien	CNRS, University of Paris 7, France
Vladimir Dotsenko	Dublin Institute, Ireland
Alessandra Frabetti	Lyon University, France
Yal Frégier	University of Luxembourg, Luxembourg
Mo-Lin Ge	Nankai University, China
Yves Guiraud	INRIA, Lyon, France
Li Guo	Rutgers University, Newark, USA
Kun Hao	Northwest University, China
Manfred Hartl	University of Valenciennes, France
Hai-Long Her	Nanjing Normal University, China
Joseph Hirsh	City University of New York, USA
Ralf Holtkamp	University of Hamburg, Germany
Wei Hong	Peking University, China
Yanyong Hong	Zhejiang University, China
Dongping Hou	Nankai University, China
Jiapeng Huang	South China Normal University, China

Yi-Zhi Huang	Rutgers University, New Brunswick, USA
Donghua Huo	Harbin Inst. of Technology, China
Xiaoyu Jia	Capital Normal University, China
Adam Katz	University of California, Riverside, USA
Liang Kong	Tsinghua University, China
Yuan Kong	Nankai University, China
Roald Koudenburg	University of Sheffield, U.K
Seunghun Lee	Konkuk University, Korea
Peng Lei	Lanzhou University, China
Haisheng Li	Rutgers University, Camden, USA
Jing Li	South China Normal University, China
Yu Li	South China Normal University, China
Muriel Livernet	University of Paris 13, France
Dong Liu	Huzhou Teachers Colleage, China
Fang Liu	Harbin Institute of Technology, China
Minjie Liu	Wuzhou College, China
Yun Liu	Yuxi Normal University, China
Jean-Louis Loday	CNRS, Univ/of Strasbourg, France
Abdenacer Makhlouf	Mulhouse University, France
Philippe Malbos	Lyon University, France
Miguel Mendez	IVIC, Venezuela
Joan Milles	University of Nice, France
Qiuhui Mo	South China Normal University, China
Xiang Ni	Nankai University, China
Yufeng Pei	Shanghai Normal University
Fei Qi	Nankai University, China
Maria Ronco	University of Valparaiso, Chile
C. Selvaraj	Periyar University, India
Longgang Sun	Zhejiang University, China
Shanzhong Sun	Capital Normal University, China
Bruno Vallette	University of Nice, France
Christine Vespa	University of Strasbourg, France
Bin Wang	South China Normal University, China
Guojun Wang	Zhejiang University, China
Liyun Wang	Nankai University, China
Lizhong Wang	Peking University China
Shoufeng Wang	Southwest University, China
Xiuling Wang	Nankai University, China

Ke Wu	Capital Normal University, China
Changchang Xi	Beijing Normal University, China
Bingshu Xia	Nankai University, China
Chao Xia	Harbin Institute of Technology, China
Xinli Xiao	Nankai University, China
Jinwei Yang	Rutgers University, USA
Wen-Li Yang	Northwest University, China
Zifeng Yang	Capital Normal University, China
Pan Zhan	Capital Normal University, China
Luyi Zhang	Nankai University, China
Runxuan Zhang	Nankai University, China
Yong Zhang	Zhejiang University, China
Chenyan Zhou	Wuhan Technological College of Communications, China

Fig. 1. Group Photo of the Preparatory School on "Operads and Universal Algebra"

Fig. 2. Group Photo of the Summer School on "Operads and Universal Algebra"

Fig. 3. Group Photo of the International Conference on "Operads and Universal Algebra"

Contents

Preface v

Organizing Committees vii

Speakers and Lectures viii

Participants and Photos xi

Gröbner-Shirshov Bases for Categories 1
L. A. Bokut, Yuqun Chen and Yu Li

Operads, Clones, and Distributive Laws 25
Pierre-Louis Curien

Leibniz Superalgebras Graded by Finite Root Systems 51
Naihong Hu, Dong Liu and Linsheng Zhu

Tridendriform Algebras Spanned by Partitions 69
Daniel Jimènez and Marìa Ronco

Generalized Disjunctive Languages and Universal Algebra 89
Yun Liu

Koszul Duality of the Category of Trees and Bar Constructions for Operads 107
Muriel Livernet

Some Problems in Operad Theory 139
Jean-Louis Loday

Hom-dendriform Algebras and Rota-Baxter Hom-algebras 147
Abdenacer Makhlouf

Free Field Realizations of the Current Algebras Associated with (Super) Lie Algebras 173
Wen-Li Yang

Free TD Algebras and Rooted Trees 199
Chenyan Zhou

Encyclopedia of Types of Algebras 2010 217
G. W. Zinbiel

Author Index 299

Gröbner-Shirshov Bases for Categories*

L. A. Bokut†
School of Mathematical Sciences, South China Normal University,
Guangzhou 510631, P. R. China
Sobolev Institute of Mathematics, Russian Academy of Sciences
Siberian Branch, Novosibirsk 630090, Russia
E-mail: bokut@math.nsc.ru

Yuqun Chen‡ and Yu Li
School of Mathematical Sciences, South China Normal University,
Guangzhou 510631, P. R. China
E-mail: yqchen@scnu.edu.cn
liyu820615@126.com

In this paper, we give a short survey of Gröbner-Shirshov bases in algebra. We establish Composition-Diamond lemma for categories that is closely related to non-commutative Gröbner bases for quotients of path algebras. Gröbner-Shirshov bases for simplicial category and cyclic category are obtained.

Keywords: Gröbner-Shirshov basis; simplicial category; cyclic category.

1. Introduction

This paper devotes to Gröbner-Shirshov bases for small categories (all categories below are supposed to be small) presented by a graph (=quiver) and defining relations (see, Maclane[58]). As important examples, we use the simpicial and the cyclic categories (see, for example, Maclane,[59] Gelfand, Manin[42]). In an above presentation, a category is viewed as a "monoid with several objects". A free category $C(X)$, generated by a graph X, is just "free partial monoid of partial words" $u = x_{i_1} \dots x_{i_n},\ n \geq 0,\ x_i \in X$ and all product defined in $C(X)$. A relation is an expression $u = v,\ u, v \in C(X)$,

*Supported by the NNSF of China (Nos.10771077, 10911120389).
†Supported by RFBR (project 09–01–00157), Integration Project SB RAS No.94, and the Federal Target Grant "Scientific and educational staff of innovation Russia" for 2009–2013 (contracts 02.740.11.5191, 02.740.11.0429, 14.740.11.0346).
‡Corresponding author.

where sources and targets of u, v are coincident respectively. The same as for semigroups, we may use two equivalent languages: Gröbner-Shirshov bases language or rewriting systems language. Since we are using the former, we need a Composition-Diamond lemma (CD-lemma for short) for a free associative partial algebra $kC(X)$ over a field k, where $kC(X)$ is just a linear combination of uniform (with the same sources and targets) partial words. Then it is a routing matter to establish CD-lemma for $kC(X)$. It is a free category ("semigroup") partial algebra of a free category. Remark that in the literature one is usually used a language of rewriting system, see, for example, Malbos.[54] Let us stress that the partial associative algebras presented by directed multi-graphs and defining relations are closely related to the well-known quotients of the path algebras from representation theory of finitely dimensional algebras, see, for example, Assem, Simson, Skowroński.[1] In this aspect Gröbner-Shirshov bases for categories are closely related to non-commutative Gröbner bases for quotients of path algebras, see Farkas, Feustel, Green.[41] Rewriting system language for non-commutative Gröbner bases of quotients of path algebras was used by Kobayashi.[47] Main new results of this paper are Gröbner-Shirshov bases for the simplicial and cyclic categories.

All algebras assume to be over a field.

2. A short survey on Gröbner-Shirshov bases

What is now called Gröbner and Gröbner-Shirshov bases theory was initiated by A. I. Shirshov (1921-1981),[66,67] 1962 for non-associative and Lie algebras, by H. Hironaka,[44,45] 1964 for quotients of commutative infinite series algebras (both formal and convergent), and by B. Buchberger,[31,32] 1965, 1970 for commutative algebras.

English translation of selected works of A. I. Shirshov, including,[66,67] is recently pulished.[68]

Remark that Shirshov's approach was a most universal as we understand now since Lie algebra case becomes a model for many classes of non-commutative and non-associative algebras (with multiple operations), starting with associative algebras (see below). Hironaka's papers on resolution of singularities of algebraic varieties become famous very soon and Hironaka got Fields Medal due to them few years latter. B. Buchberger's thesis influenced very much many specialists in computer sciences, as well as in commutative algebras and algebraic geometry, for huge important applications of his bases, named him under his supervisor W. Gröbner (1898-1980).

Shirshov's approach for Lie algebras,[67] 1962, based on a notion of composition $[f,g]_w$ of two monic Lie polynomials f,g relative to associative word w, i.e., f,g are elements of a free Lie algebra $Lie(X)$ regarded as the subspace of Lie polynomials of the free associative algebra $k\langle X\rangle$ and $w \in X^*$, where X^* is the free monoid generated by X. The definition of Lie composition relies on a definition of associative composition $(f,g)_w$ as (monic) associative polynomials (after worked out into f,g by Lie brackets $[x,y] = xy - yx$) relative to degree-lexicographical ordering on X^*. Namely, $(f,g)_w = fb - ag$, where $w = acb$, $\bar{f} = ac$, $\bar{g} = cb$, $a,b,c \in X^*$, $c \neq 1$. Here $\bar{f}$ means the leading (maximal) associative word of f. Then $(f,g)_w$ belongs to the associative ideal $Id(f,g)$ of $k\langle X\rangle$ generated by f,g, and the leading word of $(f,g)_w$ is less than w. Now we need to put some Lie brackets $[fb]-[ag]$ on $fb-ag$ in such a way that the result would belong to the Lie ideal generated by f,g (so we can not trouble bracketing into f,g) and the leading associative word of $[fb]-[ag]$ must be less than w. To overcome these obstacles Shirshov used his previous paper,[64] 1958 with a new linear basis of the free Lie algebra $Lie(X)$. As it happened the same linear basis of $Lie(X)$ was discovered in the paper Chen, Fox, Lyndon,[33] 1958. Now this basis is called Lyndon-Shirshov basis, or, by a mistake, Lyndon basis. It consists of non-associative Lyndon-Shirshov words (NLSW) $[u]$ in X, that are in one-one correspondence with associative Lyndon-Shirshov words (ALSW) u in X. The latter is defined as by a property $u = vw > wv$ for any $v, w \neq 1$. Shirshov,[64] 1958 introduced and used the following properties of both associative and non-associative Lyndon-Shirshov words:

1) For an ALSW u, there is a unique bracketing $[u]$ such that $[u]$ is a NLSW.

There are two algorithms for bracketing an ALSW. He mostly used "down-to-up algorithm" to rewrite an ALSW u on a new alphabet $X_u = \{x_i(x_\beta)^j,\ i > \beta,\ j \geq 0,\ x_\beta \text{ is the minimal letter in } u\}$; the result u' is again ALSW on X_u with the lex-ordering $x_i > x_i x_\beta > ((x_i x_\beta)x_\beta) > \ldots$.

It is Shirshov's rewriting or elimination algorithm from his famous paper Shirshov,[63] 1953, on what is now called Shirshov-Witt theorem (any subalgebra of a free Lie algebra is free). This rewriting was rediscovered by Lazard,[49] 1960 and now called as Lazard elimination (it is better to call Lazard-Shirshov elimination).

There is "up-to-down algorithm" (see Shirshov,[65] 1958, Chen, Fox, Lyndon,[33] 1958): $[u] = [[v][w]]$, where w is the longest proper end of u that is ALSW and in this case v is also an ALSW.

2) Leading associative word of NLSW $[u]$ is just u (with the coefficient 1).

3) Leading associative word of any Lie polynomial is an associative Lyndon-Shirshov word.

4) A non-commutative polynomial f is a Lie polynomial if and only if

$$f = f_0 \to \cdots \to f_i \to f_{i+1} \to \cdots \to f_n = 0,$$

where $f_i \to f_{i+1} = f_i - \alpha_i[u_i]$, $\bar{f_i} = u_i$ is an ALSW, α_i is the leading coefficient of f_i, $i = 0, 1, \ldots$.

5) Any associative word $c \neq 1$ is the unique product of (not strictly) increasing sequence of associative Lyndon-Shirshov words: $c = c_1c_2\cdots c_n$, $c_1 \leq \cdots \leq c_n$, c_i are ALSW's.

6) If $u = avb$, where u, v are ALSW, $a, b \in X^*$, then there is a relative bracketing $[u]_v = [a[v]b]$ of u relative to v, such that the leading associative word of $[u]_v$ is just u. Namely, $[u] = [a[vc]d]$, $cd = b$, $[u]_v = [a[[v[c_1]] \ldots [c_n]]d]$, $c = c_1c_2\cdots c_n$ as above.

7) If ac and cb are ASLW's and $c \neq 1$ then acb is an ALSW as well. If a, b are ALSW's and $a > b$, then ab is an ALSW as well.

Property 5) was unknown to Chen-Fox-Lyndon,[33] 1958 (cf. Berstel-Perrin,[6] 2007).

Lyndon,[53] 1954 was actually the first for definition of associative "Lyndon-Shirshov" words. To the best of our knowledge no one use the term "Lyndon words" before Lothaire,[52] 1983. For example, Schützenberger,[62] 1965 and Viennot,[69] 1978 mentioned Chen-Fox-Lyndon,[33] 1958 and Shirshov,[64] 1958 as a sorce of ALSW. On the other hand, there were dozens of papers and some books on Lie algebras that mentioned both associative and non-associative "Lyndon-Shirshov" words as Shirshov's regular words, see, for example, P. M. Cohn,[37] 1965 and Bahturin,[2] 1978.

Now a Lie composition $[f, g]_w$ of monic Lie polynomials f, g relative to a word $w = \bar{f}b = a\bar{g} = acb$, $c \neq 1$ is defined by Shirshov,[67] 1962 as follows

$$[f, g]_w = [fb]_{\bar{f}} - [ag]_{\bar{g}},$$

where $[fb]_{\bar{f}}$ means the result of substitution f for $[\bar{f}]$ into the relative bracketing $[w]_{ac}$ of w with respect to $\bar{f} = ac$ and the same for $[ag]_{\bar{g}}$.

According to the definition and properties above, any Lie composition $[f, g]_w$ is an element of the Lie ideal generated by f, g, and the leading associative word of the composition is less than w.

The composition above is now called composition of intersection. Shirshov avoided what is now called composition of inclusion

$$[f, g]_w = f - [agb]_{\bar{g}}, \quad w = \bar{f} = a\bar{g}b,$$

assuming that any system S of Lie polynomials is reduced (irreducible) in a sense that leading associative word of any polynomial from S does not contain leading words of another polynomials from S. This assumption relies on his algorithm of elimination of leading words for Lie polynomials below.

For associative polynomials the elimination algorithm is just non-commutative version of the Euclidean elimination algorithm. For Lie polynomial case Shirshov,[67] 1962 defined the elimination of a leading word as follows:

If $w = avb$, where w, v are ALSW's, and v is the leading word of some monic Lie polynomial f, then the transformation $[w] \mapsto [w] - [afb]_v$ is called an elimination of leading word of f into w. The result of Lie elimination is a Lie polynomial with a leading associative word less than w.

Then Shirshov,[67] 1962 formulated an algorithm to add to an initial reduced system of Lie polynomials S a "non-trivial" composition $[f, g]_w$, where f, g belong to S. Non-triviality of a Lie polynomial h relative to S means that h is not going to zero using "elimination of leading words of S". Actually, he defines to add to S not just a composition but rather the result of elimination of leading words of S into the composition in order to have a reduced system as well.

Then Shirshov proved the following.

Composition Lemma. Let S be a reduced subset of $Lie(X)$. If f belongs to the Lie ideal generated by S, then the leading associative word $\bar{f}$ contains, as a subword, some leading associative word of a reduced multi-composition of elements of S.

He constantly got the following corollary.

Corollary. The set of all irreducible NLSW's $[u]$ such that u does not contain any leading associative word of a reduced multi-composition of elements of S is a linear basis of the quotient algebra $Lie(X)/Id(S)$.

Some later (see Bokut,[9] 1972) the Shirshov Composition lemma was reformulated in the following form: Let S be a set of monic Lie polynomials closed under compositions (it means that any composition of intersection and inclusion of elements of S is trivial). If $f \in Id(S)$, then $\bar{f} = a\bar{s}b$ for some $s \in S$. In other words, S-irreducible NLSW's is a linear basis of the quotient algebra $Lie(X)/Id(S)$.

The modern form of Shirshov's lemma is the following (see, for example, Bokut, Chen[13]).

Shirshov's Composition-Diamond lemma for Lie algebras. Let $Lie(X)$ be a free Lie algebra over a field, S monic subset of $Lie(X)$ relative to some monomial ordering on X^*. Then the following conditions are equivalent:

1) S is a Gröbner-Shirshov basis (i.e., any composition of intersection and inclusion of elements of S is trivial).

2) If $f \in Id(S)$, then $\bar{f} = a\bar{s}b$ for some $s \in S$, $a, b \in X^*$.

3) $Irr(S) = \{[u] | [u]$ is a NLSW and u does not contain any $\bar{s}, s \in S\}$ is a linear basis of the Lie algebra $Lie(X|S)$ with defining relations S.

The proof of the Shirshov's Composition-Diamond lemma for Lie algebras becomes a model for proofs of number of Composition-Diamond lemmas for many classes of algebras. An idea of his proof is to rewrite any element of Lie ideal generated by S in a form

$$\sum \alpha_i [a_i s_i b_i],$$

where each $s_i \in S$, $a_i, b_i \in X^*$, $\alpha_i \in k$ such that

i) leading words of each $[a_i s_i b_i]$ is equal to $a_i \overline{s_i} b_i$ (in this case an expression $[asb]$ is called normal Lie S-word in X) and

ii) $a_1 \overline{s_1} b_1 > a_2 \overline{s_2} b_2 > \ldots$.

Now let $F(X)$ be a free algebra of a variety (or category) of algebras. Following the idea of Shirshov's proof, one needs

1) to define appropriate linear basis (normal words) of F(X),

2) to define monomial ordering of normal words or some related words,

3) to define compositions of element of S (they may be compositions of intersection, inclusion and left (right) multiplication, or may be else),

4) prove two key lemmas:

Key Lemma 1. Let S be a Gröbner-Shirshov basis (any composition of polynomials from S is trivial). Then any S-word is a linear combination of normal S-words.

Key Lemma 2. Let S be a Gröbner-Shirshov basis, $[a_1 s_1 b_1]$ and $[a_2 s_2 b_2]$ normal S-words, $s_1, s_2 \in S$. If $w = a_1 \overline{s_1} b_1 = a_2 \overline{s_2} b_2$, then $[a_1 s_1 b_1] \equiv [a_2 s_2 b_2] \ mod(S, w)$, i.e., $[a_1 s_1 b_1] - [a_2 s_2 b_2]$ is a linear combination of normal S-words with leading terms less than w.

There are number of CD-lemmas that realized Shirshov's approach to them.

Shirshov,[67] 1962 assumed implicitly that his approach, based on the definition of composition of any (not necessary Lie) polynomials, is equally valid for associative algebras as well (the first author is a witness that Shirshov understood it very clearly and explicitly; only lack of non-trivial applications prevents him from publication this approach for associative algebras). Explicitly it was done by Bokut[10] and Bergman.[5]

CD-lemma for associative algebras is formulated and proved in the same way as for Lie algebras.

Composition-Diamond lemma for associative algebras. Let $k\langle X\rangle$ be a free associative algebra over a field k and a set X. Let us fix some monomial ordering on X^*. Then the following conditions are equivalent for any monic subset S of $k\langle X\rangle$:

1) S is a Gröbner-Shirshov basis (that means any composition of intersection and inclusion is trivial).

2) If $f \in Id(S)$, then $\bar{f} = a\bar{s}b$ for some $s \in S,\ a, b \in X^*$.

3) $Irr(S) = \{u \in X^* | u \neq a\bar{s}b, s \in S, a, b \in X^*\}$ is a linear basis of the factor algebra $k\langle X|S\rangle = k\langle X\rangle/Id(X)$.

There are a lot of applications of Shirshov's CD-lemmas for Lie and associative algebras. Let us mention some connected to the Malcev embedding problem for semigroup algebras (Bokut,[7,8] 1969, there is a semigroup S such that the multiplication semigroup of the semigroup algebra $k(S)$, where k is a field, is embeddable into a group, but $k(S)$ is not embeddable into any division algebra), the unsolvability of the word problem for Lie algebras (Bokut[9]), Gröbner-Shirshov bases for semisimple Lie algebras (Bokut, Klein[26,27]), Kac-Moody algebras,[60] finite Coxeter groups (Bokut, Shiao[29]), braid groups in different set of generators (Bokut, Chainikov, Shum,[24] Bokut,[11] Bokut[12]), quantum algebra of type A_n (Bokut, Malcolmson[28]), Chinese monoids (Chen, Qiu[34]).

There are applications of Shirshov's CD-lemma,[66] 1962 for free anti-commutative non-associative algebras: there are two anti-commutative Gröbner-Shirshov bases of a free Lie algebra, one gives the Hall basis (Bokut, Chen, Li[18]), another the Lyndon-Shirshov basis (Bokut, Chen, Li[19]).

Bokut, Chen, Mo[21] proved and reproved some embedding theorems for associative algebras, Lie algebras, groups, semigroups, differential algebras, using Shirshov's CD-lemmas for associative and Lie algebras.

Bahturin, Olshanskii[3] found embeddings without distortion of finitely generated associative and Lie algebras into 2-generated simple algebras. They also used Shirshov's CD-lemmas for associative and Lie algebras.

Mikhalev[55] used Shirshov's approach and CD-lemma for associative algebras to prove CD-lemma for colored Lie super-algebras.

Mikhalev, Zolotykh[57] proved CD-lemma for free associative algebra over a commutative algebra. A free object in this category is $k[Y] \otimes k\langle X\rangle$, tensor product of a polynomial algebra and a free associative algebra. Here one needs to use several compositions of intersection and inclusion.

Bokut, Fong, Ke[25] proved CD-lemma for free associative conformal (a sense of V. Kac[46]) algebra $C(X,(n),n=0,1,\ldots,D,N(a,b),a,b\in X)$ of a fixed locality $N(a,b)$. A linear basis of free associative conformal algebra was constructed by M. Roitman.[61] Any normal conformal word has a form $[u]=a_1(n_1)[a_2(n_2)[\ldots[a_k(n_k)D^ia_{k+1}]\ldots]]$, where $a_j\in X, n_j < N(a_j,a_{j+1})$, $i\geq 0$. The same word without brackets is called the leading associative word of $[u]$. One needs to use external multi-operator semigroup as a set of leading associative words of conformal polynomials (it is the same as for Lie algebras), several compositions of inclusion and intersection, and new compositions of left (right) multiplication (last compositions are absent into classical cases). Also in the CD-lemma for conformal algebras we have 1) $\Rightarrow$ 2) $\Leftrightarrow$ 3), but in general, 1) does not follow 2). Here conditions 2) and 3) are formulated in terms of associative leading words, the same as for Lie algebras. We see that CD-lemma for associative conformal algebras has a lot of in common with CD-lemma for Lie algebras, but there are also some differences. Though PBW-theorem is not valid for Lie conformal algebras (M. Roitman[61]), some generic intersection compositions for universal enveloping algebra $U(L)$ of any Lie conformal algebra L are trivial (it is called "1/2 PBW theorem").

Bokut, Chen, Zhang[23] proved CD-lemma for associative n-conformal algebras, where instead of one derivation D and polynomial algebra $k[D]$ one has n derivations $D_1,\ldots,D_n$ and polynomial algebra $k[D_1,\ldots,D_n]$. Here 2) follows 1) but not conversely. This case is treated in the same way as for $n=1$. A more general case, the associative H-conformal algebra (or H-pseudo-algebra in a sense of Bakalov, D'Andrea, Kac[4]), where H is any Hopf algebra, is still open.

Mikhalev, Vasilieva[56] proved CD-lemma for free supercommutative polynomial algebras. Here they use compositions of multiplication as well.

Bokut, Chen, Li[17] proved CD-lemma for free pre-Lie algebras (also known as Vinberg-Koszul-Gerstenhaber right-symmetric algebras).

Bokut, Chen, Liu[20] proved CD-lemma for free dialgebras in a sense of Loday.[50] Here 2) follows 1) but not conversely.

The cases of associative conformal algebras and dialgebras show that definition of Gröbner-Shirshov bases by condition 1) is in general preferable to one by condition 2).

Bokut, Shum[30] proved CD-lemma for free Γ-associative algebras, where Γ is a group. It has applications to the Malcev problem above and to Bruha normal forms for algebraic groups.

Eisenbud, Peeva, Sturmfels[40] found non-commutative Gröbner basis of any commutative algebra (extending any commutative Gröbner basis to non-commutative one).

Bokut, Chen, Chen[15] proved CD-lemma for Lie algebras over commutative algebras. Here one needs to establish Key Lemma 1 in a more strong form – any Lie S-word is a linear combination of S-words of the form $[asb]_{\bar{s}}$ in the sense of Shirshov's special Lie bracketing. As an application they proved Cohn's conjecture[36] for the case of characteristics 2, 3 and 5 (that some Cohn's examples of Lie algebras over commutative algebras are not embeddable into associative algebras over the same commutative algebras).

Bokut, Chen, Deng[16] proved CD-lemma for free associative Rota-Baxter algebras. As an application, Chen and Mo[35] proved that any dendriform algebra is embeddable into universal enveloping Rota-Baxter algebra. It was Li Guo's conjecture,[43]

Bokut, Chen, Chen[14] proved CD-lemma for tensor product of two free associative algebras. As an application they extended any Mikhalev-Zolotyh commutative-non-commutative Gröbner-Shirshov basis laying into tensor product $k[Y]\otimes k\langle X\rangle$ to non-commutative-non-commutative Gröbner-Shirshov basis laying into $k\langle Y\rangle\otimes k\langle X\rangle$ (a la Eisenbud-Peeva-Sturfels above). They also gave another proof of the Eisenbud-Peeva-Sturmfekls theorem above.

As we mentioned in introduction, Farkas, Feustel, Green[41] proved CD-lemma for path algebras.

Kobayashi[48] proved CD-lemma for algebras based on well-ordered semigroups.

Drensky, Holtkamp[39] proved CD-lemma for nonassociative algebras with multiple linear operators.

Bokut, Chen, Qiu[22] proved CD-lemma for associative algebras with multiple linear operators.

Dotsenko, Khoroshkin[38] proved CD-lemma for operads.

3. Composition-Diamond lemma for categories

3.1. *Free categories and category partial algebras*

Let$X = (V(X), E(X))$ be an oriented (multi) graph. Then the free category on X is $C(X) = (Ob(X), Arr(X))$, where $Ob(X) = V(X)$ and $Arr(X)$ is the set of all paths ("words") of X including the empty paths 1_v, $v \in V(X)$. It is easy to check $C(X)$ has the following universal property. Let $\mathscr{C}$ be a category and $\Gamma_{\mathscr{C}}$ the graph relative to $\mathscr{C}$, i.e., $V(\Gamma_{\mathscr{C}}) = Ob(\mathscr{C})$

and $E(\Gamma_{\mathscr{C}}) = mor(\mathscr{C})$. Let $e : X \to C(X)$ be a mono graph morphism of the graph X to the graph $\Gamma_{C(X)}$, where $e = (e_1, e_2)$, and e_1 is a mapping on $V(X)$, e_2 on $E(X)$, both e_1 and e_2 are mono. For any graph morphism b from X to $\Gamma_{\mathscr{C}}$, where $b = (b_1, b_2)$, and b_1 is mono, there exists a unique category morphism (a functor) $f : C(X) \to \mathscr{C}$, such that the corresponding diagram is commutative, i.e., $fe = b$. Therefore each category $\mathscr{C}$ is a homomorphic image of a free category $C(X)$ for some graph X and thus $\mathscr{C}$ is isomorphic to $C(X)/\rho(S)$ for some set S, where $S = \{(u,v)|u,v$ have the same sources and the same targets$\} \subseteq Arr(X) \times Arr(X)$ and $\rho(S)$ the congruence of $C(X)$ generated by S. If this is the case, X is called the generating set of $\mathscr{C}$ and S the relation set of $\mathscr{C}$ and we denote $\mathscr{C} = C(X|S)$.

Let $\mathscr{C}$ be a category, k a field, $k\mathscr{C}$ the set of elements $f = \sum_{i=1}^{n} \alpha_i \mu_i$, where $\alpha_i \in k$, $u_i \in mor(\mathscr{C})$, $n \geq 0$, u_i $(0 \leq i \leq n)$ have the same sources and the same targets.

Note that in $k\mathscr{C}$, for $f,\ g \in k\mathscr{C}$, $f+g$ is defined only if $f,\ g$ have the same sources and the same targets.

A multiplication $\cdot$ in $k\mathscr{C}$ is defined by linearly extending the usual compositions of morphisms of the category $\mathscr{C}$. Then $(k\mathscr{C}, \ \cdot)$ is called the category partial algebra over k relative to $\mathscr{C}$ and $kC(X)$ the free category partial algebra generated by the graph X.

3.2. *Composition-Diamond lemma for category partial algebras*

Let X be a oriented (multi) graph, $C(X)$ the free category generated by X and $kC(X)$ the free category partial algebra. Since we only consider the morphisms of the free category $C(X)$, we write $C(X)$ just for $Arr(X)$.

Note that for $f, g \in kC(X)$ if we write gf, then it means that the target of g is the source of f.

A well ordering $>$ on $C(X)$ is called monomial if it satisfies the following conditions: $u > v \Rightarrow uw > vw$ and $wu > wv$, for any $u, v, w \in C(X)$.

In fact, there are many monomial orderings on $C(X)$. For example, let $E(X)$ be a well-ordered set. We order $C(X)$ by the following way: for any words $u = x_1 \cdots x_m$, $v = y_1 \cdots y_n \in C(X)$, $u > v$ iff $m > n$ or ($m = n$ and $x_1 = y_1, x_2 = y_2, \ldots, x_t = y_t$, $x_{t+1} > y_{t+1}$ for some $0 \leq t < n$). It is easy to check that the above ordering, called degree lexicographical (deg-lex) ordering, is a monomial ordering on $C(X)$. In the next section, we will give other monomial orderings on $C(X)$.

Suppose that $>$ is a fixed monomial ordering on $C(X)$. Given a nonzero polynomial $f \in kC(X)$, it has a word $\bar{f} \in C(X)$ such that $f = \alpha\overline{f} + \sum \alpha_i u_i$, where $\overline{f} > u_i$, $\alpha, \alpha_i \in k$, $u_i \in C(X)$. We call $\overline{f}$ the leading term of f and f is monic if $\alpha = 1$.

Let $S \subset kC(X)$ be a set of monic polynomials, $s \in S$ and $u \in C(X)$. We define S-word u_s inductively:

(i) $u_s = s$ is an S-word of s-length 1.
(ii) Suppose that u_s is an S-word of s-length m and v is a word of length n, i.e., the number of edges in v is n (denoted by $|v| = n$). Then $u_s v$ and vu_s are S-words of s-length $m + n$.

Note that for any S-word $u_s = asb$, where $a, b \in C(X)$, we have $\overline{asb} = a\bar{s}b$.

Let f, g be monic polynomials in $kC(X)$. Suppose that there exist $w, a, b \in C(X)$ such that $w = \bar{f} = a\bar{g}b$. Then we define the composition of inclusion by

$$(f, g)_w = f - agb.$$

For the case that $w = \bar{f}b = a\bar{g}$, $w, a, b \in C(X)$ and $|w| < |\bar{f}| + |\bar{g}|$, the composition of intersection is defined by

$$(f, g)_w = fb - ag.$$

It is clear that

$$(f, g)_w \in Id(f, g) \quad and \quad \overline{(f, g)_w} < w,$$

where $Id(f, g)$ is the ideal of $kC(X)$ generated by f, g.

The composition $(f, g)_w$ is trivial modulo (S, w), if

$$(f, g)_w = \sum_i \alpha_i a_i s_i b_i$$

where each $\alpha_i \in k$, $a_i, b_i \in C(X)$, $s_i \in S$, $a_i s_i b_i$ an S-word and $a_i \bar{s}_i b_i < \bar{f}$. If this is the case, then we write $(f, g)_w \equiv 0 \ mod(S, w)$. In general, for $p, q \in kC(X)$, we write

$$p \equiv q \quad mod(S, w)$$

which means that $p - q = \sum \alpha_i a_i s_i b_i$, where each $\alpha_i \in k, a_i, b_i \in C(X)$, $s_i \in S$, $a_i s_i b_i$ an S-word and $a_i \bar{s}_i b_i < w$.

Definition 3.1. Let $S \subset kC(X)$ be a nonempty set of monic polynomials. Then S is called a Gröbner-Shirshov basis in $kC(X)$ if any composition $(f,g)_w$ with $f,g \in S$ is trivial modulo (S,w).

Lemma 3.1. *Let $a_1s_1b_1$, $a_2s_2b_2$ be S-words. If S is a Gröbner-Shirshov basis in $kC(X)$ and $w = a_1\overline{s_1}b_1 = a_2\overline{s_2}b_2$, then*

$$a_1s_1b_1 \equiv a_2s_2b_2 \ mod(S,w).$$

Proof. There are three cases to consider.

Case 1. Suppose that subwords $\bar{s}_1$ and $\bar{s}_2$ of w are disjoint, say, $|a_2| \geq |a_1| + |\bar{s}_1|$. Then, we can assume that $a_2 = a_1\bar{s}_1c$ *and* $b_1 = c\bar{s}_2b_2$ for some $c \in C(X)$, and so, $w = a_1\bar{s}_1c\bar{s}_2b_2$. Now,

$$\begin{aligned} a_1s_1b_1 - a_2s_2b_2 &= a_1s_1c\bar{s}_2b_2 - a_1\bar{s}_1cs_2b_2 \\ &= a_1s_1c(\bar{s}_2 - s_2)b_2 + a_1(s_1 - \bar{s}_1)cs_2b_2. \end{aligned}$$

Since $\overline{\overline{s_2} - s_2} < \bar{s}_2$ and $\overline{s_1 - \overline{s_1}} < \bar{s}_1$, we conclude that

$$a_1s_1b_1 - a_2s_2b_2 = \sum_i \alpha_iu_is_1v_i + \sum_j \beta_ju_js_2v_j$$

for some $\alpha_i, \beta_j \in k$, S-words $u_is_1v_i$ and $u_js_2v_j$ such that $u_i\bar{s}_1v_i, u_j\bar{s}_2v_j < w$.

This shows that $a_1s_1b_1 \equiv a_2s_2b_2 \quad mod(S,w)$.

Case 2. Suppose that the subword $\bar{s}_1$ of w contains $\bar{s}_2$ as a subword. We may assume that $\bar{s}_1 = a\bar{s}_2b$, $a_2 = a_1a$ and $b_2 = bb_1$, that is, $w = a_1a\bar{s}_2bb_1$ for some S-word as_2b. We have

$$\begin{aligned} a_1s_1b_1 - a_2s_2b_2 &= a_1s_1b_1 - a_1as_2bb_1 \\ &= a_1s_1 - as_2bb_1 \\ &= a_1(s_1,s_2)_{\overline{s_1}}b_1 \\ &\equiv 0 \quad mod(S,w) \end{aligned}$$

since S is a Gröbner-Shirshov basis.

Case 3. $\bar{s}_1$ and $\bar{s}_2$ have a nonempty intersection as a subword of w. We may assume that $a_2 = a_1a$, $b_1 = bb_2, w_1 = \bar{s}_1b = a\bar{s}_2$. Then, we have

$$\begin{aligned} a_1s_1b_1 - a_2s_2b_2 &= a_1s_1bb_2 - a_1as_2b_2 \\ &= a_1(s_1b - as_2)b_2 \\ &= a_1(s_1,s_2)_{w_1}b_2 \\ &\equiv 0 \quad mod(S,w) \end{aligned}$$

This completes the proof. □

Lemma 3.2. *Let $S \subset kC(X)$ be a subset of monic polynomials and $Irr(S) = \{u \in C(X) \mid u \neq a\bar{s}b,\ a,b \in C(X),\ s \in S\}$. Then for any $f \in kC(X)$,*

$$f = \sum_{u_i \leq \bar{f}} \alpha_i u_i + \sum_{a_j\overline{s_j}b_j \leq \bar{f}} \beta_j a_j s_j b_j$$

where each $\alpha_i, \beta_j \in k$, $u_i \in Irr(S)$ and $a_j s_j b_j$ an S-word.

Proof. Let $f = \sum_i \alpha_i u_i \in kC(X)$, where $0 \neq \alpha_i \in k$ and $u_1 > u_2 > \cdots$. If $u_1 \in Irr(S)$, then let $f_1 = f - \alpha_1 u_1$. If $u_1 \notin Irr(S)$, then there exist some $s \in S$ and $a_1, b_1 \in C(X)$, such that $\bar{f} = u_1 = a_1\bar{s_1}b_1$. Let $f_1 = f - \alpha_1 a_1 s_1 b_1$. In both cases, we have $\bar{f_1} < \bar{f}$. Then the result follows from the induction on $\bar{f}$. □

Theorem 3.1. *(Composition-Diamond lemma for categories) Let $S \subset kC(X)$ be a nonempty set of monic polynomials and $<$ a monomial ordering on $C(X)$. Let $Id(S)$ be the ideal of $kC(X)$ generated by S. Then the following statements are equivalent:*

(i) S is a Gröbner-Shirshov basis in $kC(X)$.
(ii) $f \in Id(S) \Rightarrow \bar{f} = a\bar{s}b$ for some $s \in S$ and $a, b \in C(X)$.
(ii)′ $f \in Id(S) \Rightarrow f = \alpha_1 a_1 s_1 b_1 + \alpha_2 a_2 s_2 b_2 + \ldots$, where each $\alpha_i \in k$, $a_i s_i b_i$ is an S-word and $a_1\bar{s}_1 b_1 > a_2\bar{s}_2 b_2 > \ldots$.
(iii) $Irr(S) = \{u \in C(X) \mid u \neq a\bar{s}b\ a,b \in C(X),\ s \in S\}$ is a linear basis of the partial algebra $kC(X)/Id(S) = kC(X|S)$.

Proof. $(i) \Rightarrow (ii)$. Let S be a Gröbner-Shirshov basis and $0 \neq f \in Id(S)$. Then, we have

$$f = \sum_{i=1}^{n} \alpha_i a_i s_i b_i$$

where each $\alpha_i \in k$, $a_i, b_i \in C(X)$, $s_i \in S$ and $a_i s_i b_i$ an S-word. Let

$$w_i = a_i\overline{s_i}b_i,\ w_1 = w_2 = \cdots = w_l > w_{l+1} \geq \cdots.$$

We will use the induction on l and w_1 to prove that $\overline{f} = a\overline{s}b$ for some $s \in S$ and $a, b \in C(X)$.

If $l = 1$, then $\overline{f} = \overline{a_1 s_1 b_1} = a_1\overline{s_1}b_1$ and hence the result holds. Assume that $l \geq 2$. Then, by Lemma 3.1, we have

$$a_1 s_1 b_1 \equiv a_2 s_2 b_2 \ \ mod(S, w_1).$$

Thus, if $\alpha_1+\alpha_2 \neq 0$ or $l > 2$, then the result holds. For the case $\alpha_1+\alpha_2 = 0$ and $l = 2$, we use the induction on w_1. Now, the result follows.

$(ii) \Rightarrow (ii)'$. Assume (ii) and $0 \neq f \in Id(S)$. Let $f = \alpha_1\overline{f} + \cdots$. Then, by (ii), $\overline{f} = a_1\overline{s_1}b_1$. Therefore,

$$f_1 = f - \alpha_1 a_1 s_1 b_1,\ \overline{f_1} < \overline{f},\ f_1 \in Id(S).$$

Now, by using induction on $\overline{f}$, we have $(ii)'$.

$(ii)' \Rightarrow (ii)$. This part is clear.

$(ii) \Rightarrow (iii)$. Suppose that $\sum_i \alpha_i u_i = 0$ in $kC(X|S)$, where $\alpha_i \in k$, $u_i \in Irr(S)$. It means that $\sum_i \alpha_i u_i \in Id(S)$ in $kC(X)$. Then all α_i must be equal to zero. Otherwise, $\overline{\sum_i \alpha_i u_i} = u_j \in Irr(S)$ for some j which contradicts (ii).

Now, by Lemma 3.2, (iii) follows.

$(iii) \Rightarrow (i)$. For any $f, g \in S$, by Lemma 3.2 and (iii), we have $(f,g)_w \equiv 0\ mod(S,w)$. Therefore, S is a Gröbner-Shirshov basis. □

Remark. If the category in Theorem 3.1 has only one object, then Theorem 3.1 is exactly Composition-Diamond lemma for free associative algebras.

4. Gröbner-Shirshov bases for the simplicial category and the cyclic category

In this section, we give Gröbner-Shirshov bases for the simplicial category and cyclic category respectively.

4.1. *Gröbner-Shirshov basis for the simplicial category*

For each non-negative integer p, let $[p]$ denote the set $\{0,1,2,\ldots,p\}$ of integers in their usual ordering. A (weakly) monotonic map $\mu : [q] \to [p]$ is a function on $[q]$ to $[p]$ such that $i \leq j$ implies $\mu(i) \leq \mu(j)$. The objects $[p]$ with morphisms all weakly monotonic maps μ constitute a category L called simplicial category. There is an equivalent definition for simplicial category given by Maclane.[59]

Let $X = (V(X), E(X))$ be an oriented (multi) graph, where $V(X) = \{[p] \mid p \in Z^+ \cup \{0\}\}$ and $E(X) = \{\varepsilon_p^i : [p-1] \to [p],\ \eta_q^j : [q+1] \to$

$[q] \mid p > 0, 0 \le i \le p, 0 \le j \le q\}$. Let $S \subseteq C(X) \times C(X)$ be the relation set consisting of the following:

$$f_{q+1,q}: \quad \varepsilon_{q+1}^{i}\varepsilon_{q}^{j-1} = \varepsilon_{q+1}^{j}\varepsilon_{q}^{i}, \quad j > i,$$

$$g_{q,q+1}: \quad \eta_{q}^{j}\eta_{q+1}^{i} = \eta_{q}^{i}\eta_{q+1}^{j+1}, \quad j \ge i,$$

$$h_{q-1,q}: \quad \eta_{q-1}^{j}\varepsilon_{q}^{i} = \begin{cases} \varepsilon_{q-1}^{i}\eta_{q-2}^{j-1}, & j > i, \\ 1_{q-1}, & i = j, \ i = j+1, \\ \varepsilon_{q-1}^{i-1}\eta_{q-2}^{j}, & i > j+1. \end{cases}$$

Then the simplicial category L is just the category $C(X|S)$ generated by X with defining relation set S, see Maclane,[59] Theorem VIII. 5.2. We will give another proof in what follows.

We order $C(X)$ by the following way.

Firstly, for any $\eta_{p}^{i}, \eta_{q}^{j} \in \{\eta_{p}^{i} | p \ge 0, 0 \le i \le p\}$, $\eta_{p}^{i} > \eta_{q}^{j}$ iff $p > q$ or ($p = q$ and $i < j$).

Secondly, for each $u = \eta_{p_1}^{i_1}\eta_{p_2}^{i_2}\cdots\eta_{p_n}^{i_n} \in \{\eta_{p}^{i} | p \ge 0, 0 \le i \le p\}^*$ (all possible words on $\{\eta_{p}^{i} | p \ge 0, 0 \le i \le p\}$, including the empty word 1_v, $v \in Ob(X)$), let $wt(u) = (n, \eta_{p_n}^{i_n}, \eta_{p_{n-1}}^{i_{n-1}}, \cdots, \eta_{p_1}^{i_1})$. Then for any $u, v \in \{\eta_{p}^{i} | p \ge 0, 0 \le i \le p\}^*$, $u > v$ iff $wt(u) > wt(v)$ lexicographically.

Thirdly, for any $\varepsilon_{p}^{i}, \varepsilon_{q}^{j} \in \{\varepsilon_{p}^{i}, | p \in Z^+, 0 \le i \le p\}$, $\varepsilon_{p}^{i} > \varepsilon_{q}^{j}$ iff $p > q$ or ($p = q$ and $i < j$).

Finally, for each $u = v_0\varepsilon_{p_1}^{i_1}v_1\varepsilon_{p_2}^{i_2}\cdots\varepsilon_{p_n}^{i_n}v_n \in C(X)$, $n \ge 0$, $v_j \in \{\eta_{p}^{i} | p \ge 0, 0 \le i \le p\}^*$, let $wt(u) = (n, v_0, v_1, \cdots, v_n, \varepsilon_{p_1}^{i_1}, \cdots, \varepsilon_{p_n}^{i_n})$. Then for any $u, v \in C(X)$,

$$u \succ_1 v \Leftrightarrow wt(u) > wt(v) \quad \text{lexicographically.}$$

It is easy to check that the $\succ_1$ is a monomial ordering on $C(X)$. Then we have the following theorem.

Theorem 4.1. *Let X, S be defined as the above. Then with the ordering $\succ_1$ on $C(X)$, S is a Gröbner-Shirshov basis for the category partial algebra $kC(X|S)$.*

Proof. According to the ordering $\succ_1$, $\bar{f}_{q+1,q} = \varepsilon_{q+1}^{i}\varepsilon_{q}^{j-1}$, $\bar{g}_{q,q+1} = \eta_{q}^{j}\eta_{q+1}^{i}$ and $\bar{h}_{q-1,q} = \eta_{q-1}^{j}\varepsilon_{q}^{i}$. So, all the possible compositions of S are the following:

(a) $(f_{q+2,q+1}, f_{q+1,q})_{\varepsilon_{q+2}^{k}\varepsilon_{q+1}^{i}\varepsilon_{q}^{j-1}}, \quad k \le i \le j-1;$

(b) $(g_{q-1,q}, g_{q,q+1})_{\eta_{q-1}^{k}\eta_{q}^{j}\eta_{q+1}^{i}}, \quad i \le j \le k;$

(c) $(h_{q,q+1}, f_{q+1,q})_{\eta_q^k \varepsilon_{q+1}^i \varepsilon_q^{j-1}}, \quad i \leq j-1;$

(d) $(g_{q-2,q-1}, h_{q-1,q})_{\eta_{q-2}^k \eta_{q-1}^j \varepsilon_q^i}, \quad j \leq k.$

We will prove that all possible compositions are trivial. Here, we only give the proof of the (b). For others cases, the proofs are similar.

Let us consider the following subcases of the case (b): (I) $i < j < k$; (II) $i < j, j = k$, or $j = k+1$; (III) $i < k, k+1 < j$; (IV) $j > k+1, i = k, k+1$; (V) $j > i > k+1$.

For (I),

$$\begin{aligned}(h_{q,q+1}, f_{q+1,q})_{\eta_q^k \varepsilon_{q+1}^i \varepsilon_q^{j-1}} &= \varepsilon_q^i \eta_{q-1}^{k-1} \varepsilon_q^{j-1} - \eta_q^k \varepsilon_{q+1}^j \varepsilon_q^i \\ &\equiv \varepsilon_q^i \varepsilon_{q-1}^{j-1} \eta_{q-2}^{k-2} - \varepsilon_q^j \eta_{q-1}^{k-1} \varepsilon_q^i \\ &\equiv 0 \quad mod(S, \eta_q^k \varepsilon_{q+1}^i \varepsilon_q^{j-1}).\end{aligned}$$

For (II),

$$\begin{aligned}(h_{q,q+1}, f_{q+1,q})_{\eta_q^k \varepsilon_{q+1}^i \varepsilon_q^{j-1}} &= \varepsilon_q^i \eta_{q-1}^{k-1} \varepsilon_q^{j-1} - \eta_q^k \varepsilon_{q+1}^j \varepsilon_q^i \\ &\equiv 0 \quad mod(S, \eta_q^k \varepsilon_{q+1}^i \varepsilon_q^{j-1}).\end{aligned}$$

For (III),

$$\begin{aligned}(h_{q,q+1}, f_{q+1,q})_{\eta_q^k \varepsilon_{q+1}^i \varepsilon_q^{j-1}} &= \varepsilon_q^i \eta_{q-1}^{k-1} \varepsilon_q^{j-1} - \eta_q^k \varepsilon_{q+1}^j \varepsilon_q^i \\ &\equiv \varepsilon_q^i \varepsilon_{q-1}^{j-2} \eta_{q-2}^{k-1} - \varepsilon_q^{j-1} \eta_{q-1}^k \varepsilon_q^i \\ &\equiv 0 \quad mod(S, \eta_q^k \varepsilon_{q+1}^i \varepsilon_q^{j-1}).\end{aligned}$$

For (IV),

$$\begin{aligned}(h_{q,q+1}, f_{q+1,q})_{\eta_q^k \varepsilon_{q+1}^i \varepsilon_q^{j-1}} &= \varepsilon_q^{j-1} - \eta_q^k \varepsilon_{q+1}^j \varepsilon_q^i \\ &\equiv \varepsilon_q^{j-1} - \varepsilon_q^{j-1} \eta_{q-1}^k \varepsilon_q^i \\ &\equiv 0 \quad mod(S, \eta_q^k \varepsilon_{q+1}^i \varepsilon_q^{j-1}).\end{aligned}$$

For (V),

$$\begin{aligned}(h_{q,q+1}, f_{q+1,q})_{\eta_q^k \varepsilon_{q+1}^i \varepsilon_q^{j-1}} &= \varepsilon_q^{i-1} \eta_{q-1}^k \varepsilon_q^{j-1} - \eta_q^k \varepsilon_{q+1}^j \varepsilon_q^i \\ &\equiv \varepsilon_q^{i-1} \varepsilon_{q-1}^{j-2} \eta_{q-2}^k - \varepsilon_q^{j-1} \eta_{q-1}^k \varepsilon_q^i \\ &\equiv 0 \quad mod(S, \eta_q^k \varepsilon_{q+1}^i \varepsilon_q^{j-1}).\end{aligned}$$

Therefore S is a Gröbner-Shirshov basis in $kC(X)$. □

By Theorem 3.1, $Irr(S) = \{\varepsilon_p^{i_1} \cdots \varepsilon_{p-m+1}^{i_m} \eta_{q-n}^{j_1} \cdots \eta_{q-1}^{j_n} | p \geq i_1 > \ldots > i_m \geq 0, \ \ 0 \leq j_1 < \ldots < j_n < q \text{ and } q - n + m = p\}$ is a linear basis of the category partial algebra $kC(X|S)$. Therefore, we have the following corollaries.

Corollary 4.1. *(Maclane,[59] Lemma VIII. 5.1) In the category $C(X|S)$, each morphism $\mu : [q] \to [p]$ can be uniquely represented as*

$$\varepsilon_p^{i_1} \cdots \varepsilon_{p-m+1}^{i_m} \eta_{q-n}^{j_1} \cdots \eta_{q-1}^{j_n},$$

where $p \geq i_1 > \ldots > i_m \geq 0$, $0 \leq j_1 < \ldots < j_n < q$ and $q - n + m = p$.

Corollary 4.2. *(Maclane,[59] Theorem VIII. 5.2) Let L be the simplicial category defined as before. Then $L = C(X|S)$.*

4.2. *Gröbner-Shirshov basis for the cyclic category*

The cyclic category is defined by generators and defining relations as follows, see.[42] Let $Y = (V(Y), E(Y))$ be an oriented (multi) graph, where $V(Y) = \{[p] \mid p \in Z^+ \cup \{0\}\}$ and $E(Y) = \{\varepsilon_p^i : [p-1] \to [p],\ \eta_q^j : [q+1] \to [q],\ t_q : [q] \to [q] \mid p > 0, 0 \leq i \leq p, 0 \leq j \leq q\}$. Let $S \subseteq C(Y) \times C(Y)$ be the set consisting of the following:

$$\begin{aligned}
f_{q+1,q} &: \quad \varepsilon_{q+1}^i \varepsilon_q^{j-1} = \varepsilon_{q+1}^j \varepsilon_q^i, \quad j > i, \\
g_{q,q+1} &: \quad \eta_q^j \eta_{q+1}^i = \eta_q^i \eta_{q+1}^{j+1}, \quad j \geq i, \\
h_{q-1,q} &: \quad \eta_{q-1}^j \varepsilon_q^i = \begin{cases} \varepsilon_{q-1}^i \eta_{q-2}^{j-1}, & j > i, \\ 1_{q-1}, & i = j, \ i = j+1, \\ \varepsilon_{q-1}^{i-1} \eta_{q-2}^j, & i > j+1, \end{cases} \\
\rho_1 &: \quad t_q \varepsilon_q^i = \varepsilon_q^{i-1} t_{q-1}, \quad i = 1, \ldots, q, \\
\rho_2 &: \quad t_q \eta_q^i = \eta_q^{i-1} t_{q+1}, \quad 1 = 1, \ldots, q, \\
\rho_3 &: \quad t_q^{q+1} = 1_q.
\end{aligned}$$

Then the category $C(Y|S)$ is called cyclic category, denoted by Λ. In the following, we give a Gröbner-Shirshov basis for the cyclic category.

Let us order $C(Y)$ by the following way.

Firstly, for any t_p^i, $t_q^j \in \{t_q | q \geq 0\}^*$, $(t_p)^i > (t_q)^j$ iff $i > j$ or ($i = j$ and $p > q$).

Secondly, for any $\eta_p^i, \eta_q^j \in \{\eta_p^i | p \geq 0, 0 \leq i \leq p\}$, $\eta_p^i > \eta_q^j$ iff $p > q$ or ($p = q$ and $i < j$).

Thirdly, for each $u = w_0 \eta_{p_1}^{i_1} w_1 \eta_{p_2}^{i_2} \cdots w_{n-1} \eta_{p_n}^{i_n} w_n \in \{t_q, \eta_p^i | q, p \geq 0, 0 \leq i \leq p\}^*$, where $w_i \in \{t_q | q \geq 0\}^*$, let $wt(u) = (n, w_0, w_1, \cdots, w_n, \eta_{p_n}^{i_n}, \eta_{p_{n-1}}^{i_{n-1}}, \cdots, \eta_{p_1}^{i_1})$. Then for any $u, v \in \{t_q, \eta_p^i | q, p \geq 0, 0 \leq i \leq p\}^*$, $u > v$ iff $wt(u) > wt(v)$ lexicographically.

Fourthly, for any $\varepsilon_p^i, \varepsilon_q^j \in \{\varepsilon_p^i, |p \in Z^+, 0 \leq i \leq p\}$, $\varepsilon_p^i > \varepsilon_q^j$ iff $p > q$ or $(p = q$ and $i < j)$.

Finally, for each $u = v_0\varepsilon_{p_1}^{i_1}v_1\varepsilon_{p_2}^{i_2}\cdots\varepsilon_{p_n}^{i_n}v_n \in C(Y)$, $n \geq 0$, $v_j \in \{t_q, \eta_p^i | q, p \geq 0, 0 \leq i \leq p\}^*$, let $wt(u) = (n, v_0, v_1, \cdots, v_n, \varepsilon_{p_1}^{i_1}, \cdots, \varepsilon_{p_n}^{i_n})$.

Then for any $u, v \in C(Y)$,

$$u \succ_2 v \Leftrightarrow wt(u) > wt(v) \quad \text{lexicographically.}$$

It is also easy to check the ordering $\succ_2$ is a monomial ordering on $C(Y)$, which is an extension of $\succ_1$. Then we have the following theorem.

Theorem 4.2. *Let Y, S be defined as the above. Let $S^c = S \cup \{\rho_4, \rho_5\}$, where*

$$\rho_4: \quad t_q\varepsilon_q^0 = \varepsilon_q^q,$$
$$\rho_5: \quad t_q\eta_q^0 = \eta_q^q t_{q+1}^2.$$

Then

(1) With the ordering $\succ_2$ on $C(Y)$, S^c is a Gröbner-Shirshov basis for the cyclic category partial algebra $kC(Y|S)$.

(2) For each morphism $\mu : [q] \to [p]$ in the cyclic category $\Lambda = C(Y|S)$, μ can be uniquely represented as

$$\varepsilon_p^{i_1}\cdots\varepsilon_{p-m+1}^{i_m}\eta_{q-n}^{j_1}\cdots\eta_{q-1}^{j_n}t_q^k,$$

where $p \geq i_1 > \ldots > i_m \geq 0$, $0 \leq j_1 < \ldots < j_n < q$, $0 \leq k \leq q$ and $q - n + m = p$.

Proof. It is easy to check that $\bar{f}_{q+1,q} = \varepsilon_{q+1}^i\varepsilon_q^{j-1}$, $\bar{g}_{q,q+1} = \eta_q^j\eta_{q+1}^i$, $\bar{h}_{q-1,q} = \eta_{q-1}^j\varepsilon_q^i$, $\bar{\rho_1} = t_q\varepsilon_q^i$, $\bar{\rho_2} = t_q\eta_q^i$, $\bar{\rho_3} = t_q^{q+1}$, $\bar{\rho_4} = t_q\varepsilon_q^0$ and $\bar{\rho_5} = t_q\eta_q^0$.

First of all, we prove $Id(S) = Id(S^c)$. It suffices to show $\rho_4, \rho_5 \in Id(S)$. Since $(\rho_3, \rho_1)_{t_q^{q+1}\varepsilon_q^q} = t_q^q\varepsilon_q^{q-1}t_{q-1} - \varepsilon_q^q \equiv t_q\varepsilon_q^0t_{q-1}^q - \varepsilon_q^q \equiv t_q\varepsilon_q^0 - \varepsilon_q^q = \rho_4$ and $(\rho_3, \rho_2)_{t_q^{q+1}\eta_q^q} = t_q^q\eta_q^{q-1}t_{q+1} - \eta_q^q \equiv t_q\eta_q^0t_{q+1}^q - \eta_q^q$, $\rho_4 \in Id(S)$ and $t_q\eta_q^0t_{q+1}^q - \eta_q^q \in Id(S)$. Clearly, the leading term of the polynomial $t_q\eta_q^0t_{q+1}^q - \eta_q^q$ is $t_q\eta_q^0t_{q+1}^q$. Therefore $(t_q\eta_q^0t_{q+1}^q - \eta_q^q, \rho_3)_{t_q\eta_q^0t_{q+1}^{q+2}} = -t_q\eta_q^0 + \eta_q^qt_{q+1}^2$ and thus $\rho_5 \in Id(S)$.

Secondly, we prove that all possible compositions of S^c are trivial which are the following:

(a) $(f_{q+2,q+1}, f_{q+1,q})_{\varepsilon_{q+2}^k\varepsilon_{q+1}^i\varepsilon_q^{j-1}}$, $\quad k \leq i \leq j-1$;

(b) $(g_{q-1,q}, g_{q,q+1})_{\eta_{q-1}^k\eta_q^j\eta_{q+1}^i}$, $\quad i \leq j \leq k$;

(c) $(h_{q,q+1}, f_{q+1,q})_{\eta_q^k \varepsilon_{q+1}^i \varepsilon_q^{j-1}}$, $i \leq j-1$;
(d) $(g_{q-2,q-1}, h_{q-1,q})_{\eta_{q-2}^k \eta_{q-1}^j \varepsilon_q^i}$, $j \leq k$;
(e) $(\rho_1, f_{q+1,q})_{t_{q+1}\varepsilon_{q+1}^i \varepsilon_q^{j-1}}$, $j > i$ and $i = 1, 2, \ldots, q$;
(f) $(\rho_3, \rho_1)_{t_q^{q+1}\varepsilon_q^i}$, $i = 1, 2, \ldots, q$;
(g) $(\rho_2, g_{q,q+1})_{t_q \eta_q^j \eta_{q+1}^i}$, $j \geq i$ and $j = 1, 2, \ldots, q$;
(h) $(\rho_2, h_{q,q+1})_{t_q \eta_q^j \varepsilon_{q+1}^i}$, $j \geq i$ and $j = 1, 2, \ldots, q$;
(i) $(\rho_3, \rho_2)_{t_q^{q+1}\eta_q^i}$, $i = 1, 2, \ldots, q$;
(j) $(\rho_3, \rho_4)_{t_q^{q+1}\varepsilon_q^0}$;
(k) $(\rho_3, \rho_5)_{t_q^{q+1}\eta_q^0}$;
(l) $(\rho_4, f_{q+1,q})_{t_{q+1}\varepsilon_{q+1}^0 \varepsilon_q^{j-1}}$, $j > 0$;
(m) $(\rho_5, g_{q,q+1})_{t_q \eta_q^0 \eta_{q+1}^0}$;
(n) $(\rho_5, h_{q,q+1})_{t_q \eta_q^0 \varepsilon_{q+1}^i}$, $i \geq 0$.

Here, we only give the proof of the case (n) $(\rho_5, h_{q,q+1})_{t_q \eta_q^0 \varepsilon_{q+1}^i}$. The others can be similarly proved. Let us consider the following subcases of the case (n): (I) $i = 0$; (II) $i = 1$; (III) $i > 1$.

For (I),

$$\begin{aligned}
(\rho_5, h_{q,q+1})_{t_q \eta_q^0 \varepsilon_{q+1}^0} &= \eta_q^q t_{q+1}^2 \varepsilon_{q+1}^0 - t_q \\
&\equiv \eta_q^q t_{q+1} \varepsilon_{q+1}^{q+1} - t_q \\
&\equiv \eta_q^q \varepsilon_{q+1}^q t_q - t_q \\
&\equiv 0 \quad mod(S, t_q \eta_q^0 \varepsilon_{q+1}^0).
\end{aligned}$$

For (II),

$$\begin{aligned}
(\rho_5, h_{q,q+1})_{t_q \eta_q^0 \varepsilon_{q+1}^1} &= \eta_q^q t_{q+1}^2 \varepsilon_{q+1}^1 - t_q \\
&\equiv \eta_q^q t_{q+1} \varepsilon_{q+1}^0 t_q - t_q \\
&\equiv \eta_q^q \varepsilon_{q+1}^{q+1} t_q - t_q \\
&\equiv 0 \quad mod(S, t_q \eta_q^0 \varepsilon_{q+1}^1).
\end{aligned}$$

For (III),

$$\begin{aligned}
(\rho_5, h_{q,q+1})_{t_q \eta_q^0 \varepsilon_{q+1}^i} &= \eta_q^q t_{q+1}^2 \varepsilon_{q+1}^i - t_q \varepsilon_q^{i-1} \eta_{q-1}^0 \\
&\equiv \eta_q^q \varepsilon_{q+1}^{i-2} t_q^2 - \varepsilon_q^{i-2} t_{q-1} \eta_{q-1}^0 \\
&\equiv 0 \quad mod(S, t_q \eta_q^0 \varepsilon_{q+1}^i).
\end{aligned}$$

Thus S^c is a Gröbner-Shirshov basis in $kC(Y)$.

Now, by Theorem 3.1, for each morphism $\mu : [q] \to [p]$ in $\Lambda = C(Y|S)$ can be uniquely represented as

$$\varepsilon_p^{i_1} \cdots \varepsilon_{p-m+1}^{i_m} \eta_{q-n}^{j_1} \cdots \eta_{q-1}^{j_n} t_q^k,$$

where $p \geq i_1 > \ldots > i_m \geq 0$, $0 \leq j_1 < \ldots < j_n < q$, $0 \leq k \leq q$ and $q - n + m = p$. $\square$

Remark. According to Loday,[51] the uniqueness property in Theorem 4.2 (2) was known.

Acknowledgement: The authors would like to thank the referee for comments.

References

1. I. Assem, D. Simson and A. Skowroński, Elements of the representation theory of associative algebras (London Mathematical Society Student Texts 65, 2006).
2. Y. A. Bahturin, Lectures on Lie Algebras (Akademie-Verlag, Berlin, 1978).
3. Y. A. Bahturin and A. Olshanskii, Filtrations and distortion in infinite-dimensional algebras, arXiv:1002.0015.
4. B. Bakalov, A. D'Andrea and V. G. Kac, Theory of finite pseudoalgebras, *Adv. Math.* **162**, 1 (2001), pp. 1–140.
5. G. M. Bergman, The diamond lemma for ring theory, *Adv. Math.* **29**, (1978), pp. 178–218.
6. Jean Berstel and Dominique Perrin, The origins of combinatorics on words, *Eur. J. Comb.* **28**, 3 (2007), pp. 996–1022.
7. L. A. Bokut, Groups of fractions of multiplication semigroups of certain rings I, II, III, *Sibir. Math. J.* **10**, (1969), pp. 246–286, 744–799, 800–819.
8. L. A. Bokut, On the Malcev problem, *Sibir. Math. J.* **10**, 5 (1969), pp. 965–1005.
9. L. A. Bokut, Unsolvability of the word problem, and subalgebras of finitely presented Lie algebras, *Izv. Akad. Nauk. SSSR Ser. Mat.* **36**, (1972), pp. 1173–1219.
10. L. A. Bokut, Imbeddings into simple associative algebras, *Algebra i Logika* **15**, (1976), pp. 117–142.
11. L. A. Bokut, Gröbner-Shirshov bases for braid groups in Artin-Garside generators, *J. Symbolic Computation* **43**, (2008), pp. 397–405.
12. L. A. Bokut, Gröbner-Shirshov bases for the braid group in the Birman-Ko-Lee generators, *J. Algebra* **321** (2009), pp 361–379.
13. L. A. Bokut and Y. Q Chen, Gröbner-Shirshov bases for Lie algebras: after A.I. Shirshov, *Southeast Asian Bull. Math.* **31**, (2007), pp. 1057–1076.
14. L. A. Bokut, Y. Q. Chen and Y. S. Chen, Composition-Diamond lemma for tensor product of free algebras, *J. Algebra* **323**, (2010), pp. 2520–2537.

15. L. A. Bokut, Y. Q. Chen and Y. S. Chen, Gröbner-Shirshov bases for Lie algebras over a commutative algebra, *J. Algebra* **337**, (2011), pp. 82–102.
16. L. A. Bokut, Y. Q. Chen and X. M. Deng, Gröbner-Shirshov bases for Rota-Baxter algebras, *Siberian Mathematical Journal* **51**, 6 (2010), pp. 978–988.
17. L. A. Bokut, Y. Q. Chen and Y. Li, Gröbner-Shirshov bases for Vinberg-Koszul-Gerstenhaber right-symmetric algebras, *Fundamental and Applied Mathematics* **14**, 8 (2008), pp. 55–67. (in Russian)
18. L. A. Bokut, Y. Q. Chen and Y. Li, Anti-commutative Gröbner-Shirshov basis of a free Lie algebra, *Sci. China* **52**, (2009), pp. 244–253.
19. L. A. Bokut, Y. Q. Chen and Y. Li, Anti-commutative Gröbner-Shirshov basis of a free Lie algebra relative to Lyndon-Shirshov words, preprint.
20. L. A. Bokut, Y. Q. Chen and C. H. Liu, Gröbner-Shirshov bases for dialgebras, *Internat. J. Algebra Comput.* **20**, 3 (2010), pp. 391–415.
21. L. A. Bokut, Y. Q. Chen and Q. H. Mo, Gröbner-Shirshov bases and embeddings of algebras, *Internat. J. Algebra Comput.* **20**, 7 (2010), pp. 875–900.
22. L. A. Bokut, Y. Q. Chen and J. J. Qiu, Gröbner-Shirshov bases for associative algebras with multiple operators and free Rota-Baxter algebras, *Journal of Pure and Applied Algebra* **214** (2010), pp. 89–100.
23. L. A. Bokut, Y. Q. Chen and G. L. Zhang, Composition-Diamond lemma for associative n-conformal algebras, arXiv:0903.0892
24. L. A. Bokut, V. V. Chainikov and K. P. Shum, Markov and Artin normal form theorem for braid groups, *Comm. Algebra* **35**, (2007), pp. 2105–2115.
25. L. A. Bokut, Y. Fong and W.-F. Ke, Composition Diamond Lemma for associative conformal algebras, *J. Algebra* **272**, (2004), pp. 739–774.
26. L. A. Bokut and A. A. Klein, Serre relations and Gröbner-Shirshov bases for simple Lie algebras I, *Internat. J. Algebra Comput.* **6**, 4 (1996), pp. 389–400.
27. L. A. Bokut and A. A. Klein, Serre relations and Gröbner-Shirshov bases for simple Lie algebras II, *Internat. J. Algebra Comput.* **6**, 4 (1996), pp. 401–412.
28. L. A. Bokut and P. Malcolmson, Gröbner-Shirshov bases for relations of a Lie algebra and its enveloping algebra, *Algebra and Combinatorics* (Hong Kong), (Springer, Singapore, 1999), pp. 47–54.
29. L. A. Bokut and L.-S. Shiao, Gröbner-Shirshov bases for Coxeter groups, *Comm. Algebra* **29**, (2001), pp. 4305–4319.
30. L. A. Bokut and K. P. Shum, Relative Gröbner-Shirshov bases for algebras and groups, *St Petersburg Math. J.* **19**, 6 (2008), pp. 867–881.
31. B. Buchberger, An algorithm for finding a basis for the residue class ring of a zero-dimensional polynomial ideal [in German], Ph.D. thesis, University of Innsbruck, (Austria, 1965).
32. B. Buchberger, An algorithmical criteria for the solvability of algebraic systems of equations [in German], *Aequationes Math.* **4**, (1970), pp. 374–383.
33. K. T. Chen, R. Fox and R. Lyndon, Free differential calculus IV: The quotient group of the lower central series, *Ann. Math.* **68**, (1958), pp. 81–95.
34. Y. Q. Chen and J. J Qiu, Gröbner-Shirshov basis for the Chinese monoid, *Journal of Algebra and its Applications* **7**, 5 (2008), pp. 623–628.
35. Y. Q. Chen and Q. H. Mo, Embedding dendriform algebra into its universal enveloping Rota-Baxter algebra, *Proc. Amer Math. Soc.*, to appear. arxiv.org/abs/1005.2717

36. P. M. Cohn, A remark on the Birkhoff-Witt theorem, *Journal London Math. Soc.* **38**, (1963), pp. 197–203
37. P. M. Cohn, Universal Algebra, Harper and Row, (1965).
38. V. Dotsenko and A. Khoroshkin, Gröbner bases for operads, *Duke Mathematical Journal* **153**, 2 (2010), pp. 363–396.
39. V. Drensky and R. Holtkamp, Planar trees, free nonassociative algebras, invariants, and elliptic integrals, *Algebra Discrete Math.* **2**, (2008), pp. 1–41.
40. D. Eisenbud, I. Peeva and B. Sturmfels, Non-commutative Gröbner bases for commutative algebras, *Proc. Amer. Math. Soc.* **126** 3 (1998), pp. 687–691.
41. D. R. Farkas, C. D. Feustel and E. L. Green, Synergy in the theories of Gröbner bases and path algebras, *Can. J. Math.* **45**, (1993), pp. 727–739.
42. S. I. Gelfand and Y. I. Manin, Homological Algebra, (Springer-Verlag, 1999).
43. L. Guo, private communication, (2009).
44. H. Hironaka, Resolution of singulatities of an algebraic variety over a field if characteristic zero, I, Ann. Math., 79(1964), 109-203.
45. H. Hironaka, Resolution of singulatities of an algebraic variety over a field if characteristic zero, II, *Ann. Math.* **79**, (1964), pp. 205–326.
46. V. Kac, Vertex algebras for beginners, *University Lecture Series* Vol. 10, AMS, (Providence, RI, 1996).
47. Y. Kobayashi, Gröbner bases on path algebras and the Hochschild cohomology algebras, *Sci. Math. Japonicae* **64**, (2006), pp. 411–437.
48. Y. Kobayashi, Gröbner bases on algebras based on well-ordered semigroups, *Math. Appl. Sci. Tech.* to appear.
49. M. Lazard, Groupes, anneaux de Lie et problème de Burnside. Istituto Matematico dell' Università di Roma.
50. J.-L. Loday, Dialgebras, in Dialgebras and Related Operads, *Lecture Notes in Mathematics* Vol. 1763 (Springer Verlag, Berlin, 2001), pp. 7–66.
51. J.-L. Loday, private communication, (2010).
52. M. Lothaire, (Perrin, D.; Reutenauer, C.; Berstel, J.; Pin, J. E.; Pirillo, G.; Foata, D.; Sakarovitch, J.; Simon, I.; Schuzenberger, M. P.; Choffrut, C.; Cori, R.; Lyndon, Roger; Rota, Gian-Carlo). Combinatorics on words. Foreword by Roger Lyndon. (English) [B] Encyclopedia of Mathematics and Its Applications, Vol. 17. Reading, Massachusetts, etc.: Addison-Wesley Publishing Company, Advanced Book Program/World Science Division. XIX, 238 p. (1983).
53. R. C. Lyndon, On Burnside's problem I, *Trans. Amer. Math. Soc.* **77**, (1954), pp. 202–215.
54. P. Malbos, Rewriting systems and hochschild-mitchell homology, *Electr. Notes Theor. Comput. Sci.* **81**, (2003), pp. 59–72.
55. A. A. Mikhalev, A composition lemma and the word problem for color Lie superalgebras, *Moscow Univ. Math. Bull.* **44**, 5 (1989), pp. 87–90.
56. A. A. Mikhalev and E. A. Vasilieva, Standard bases of ideals of free supercommutative polynomail algebra (ε-Grobner bases), *Proc. Second International Taiwan-Moscow Algebra Workshop* (Springer-Verlag, 2003).
57. A. A. Mikhalev and A. A. Zolotykh, Standard Gröbner-Shirshov bases of free algebras over rings, I. Free associative algebras, *Internat. J. Algebra Comput.* **8**, 6 (1998), pp. 689–726.

58. S. Maclane, Categories for the working mathematician, Second Edition.
59. S. Maclane, Homology, (Springer-Verlag, 1963).
60. E. Poroshenko, Gröbner-Shirshov bases for Kac-Moody algebras of the type $A_n^{(1)}$, *Comm. Algebra* **30**, 6 (2002), pp. 2617–2637.
61. M. Roitman, On free conformal and vertex algebras, *J. Algebra* **217**, 2 (1999), pp. 496–527.
62. M. P. Schützenberger, On a factorization of free monoids, *Proc. Amer. Math. Soc.* **16**, (1965), pp. 21–24.
63. A. I. Shirshov, On the representation of Lie rings in associative rings, *Uspekhi Mat. Nauk N.S.* **5**, 57 (1953), pp. 173–175.
64. A. I. Shirshov, On free Lie rings, *Mat. Sb.* **45**, 2 (1958), pp. 113–122 (in Russian).
65. A. I. Shirshov, Some Problems in the theory of rings that are nearly associative, *Uspekhi Mat. Nauk* **6**, 84 (1958), pp. 3–20.
66. A. I. Shirshov, Some algorithmic problem for ε-algebras, *Sibirsk. Mat. Z.* bf 3, (1962), pp. 132–137.
67. A. I. Shirshov, Some algorithmic problem for Lie algebras, *Sibirsk. Mat. Z.* **3**, 2 (1962), pp. 292–296 (in Russian); English translation in SIGSAM Bull., **33**, (2)(1999), pp. 3–6.
68. Selected works of A.I. Shirshov, Eds L. A. Bokut, V. Latyshev, I. Shestakov, E. Zelmanov, Trs M. Bremner, M. Kochetov, Birkhäuser, Basel, (Boston, Berlin, 2009).
69. G. Viennot, Algebras de Lie libres et monoid libres. Bases des Lie algebres et facrorizations des monoides libres. *Lecture Notes in Mathematics* 691., (Berlin-Geldelberg-New York, Springer-Verlag, 124 p. 1978).

Operads, Clones, and Distributive Laws

Pierre-Louis Curien*
PPS Laboratory, CNRS & University Paris Diderot,
Paris, France
**E-mail: curien@pps.jussieu.fr*

We show how non-symmetric operads (or multicategories), symmetric operads, and clones, arise from three suitable monads on **Cat**, each extending to a (pseudo-)monad on the bicategory of categories and profunctors. We also explain how other previous categorical analyses of operads (via Day's tensor products, or via analytical functors) fit with the profunctor approach.

1. Introduction

Operads, in their coloured and non-symmetric version, are also known as multicategories, since they are like categories, but with morphisms which have a single target as codomain but a multiple source (more precisely, a list of sources) as domain. These morphisms are often called operations (whence the term "operad" for the whole structure). Operads for short are the special case where the multicategory has just one object, and hence the arity of a morphism is just a number n (the length of the list), while the coarity is 1. There are many variations.

- non-symmetric operads versus symmetric operads versus clones (these variations concern the way in which operations are combined to form compound operations);
- operations versus cooperations (one input, several outputs), or "biop-erations" (several inputs, several outputs);
- operations whose shape, or arity, is more structured than a list (it could be a tree, etc...).

In this paper, we deal principally with the first variation, and touch on the second briefly. Our goal is to contribute to convey the idea that these variations can be smoothly and rather uniformly understood using some categorical abstractions.

We are aware of two approaches for a general categorical account of such variations. The first one, which is also the earliest one, is based on spans, while the second one is based on profunctors. In both approaches, one abstracts the details of the variant in a monad, and one then extends the monad to the category of spans, or to the category of profunctors. Here, we take the profunctor approach.

- Spans are pairs of morphisms with the same domain, in a suitable category **C**. Burroni has shown [8] that every cartesian* monad T on **C** extends to a monad on spans, and then, for example, multicategories arise as endomorphisms endowed with a monoid structure in the (co)Kleisli (bi)category of a category of spans.
- Profunctors are functors $\Phi : \mathbf{C} \times \mathbf{C}'^{op} \to \mathbf{Set}$. As pointed out by Cheng [11] , every monad in **Cat** (the 2-category of categories) satisfying a certain distributive law can be extended to the category of profunctors (whose objects are categories and whose morphisms are profunctors). We exhibit here how

  non-symmetric and symmetric operads, and clones, arise (again) as monoids in the (co)Kleisli (bi)category associated with the respective extended monad.

This is all rather "heavy" categorical vocabulary. We shall unroll this slowly in what follows. We offer the following intuitions for why monads, spans, and profunctors are relevant here.

- Monad. Consider the following composition of operations:

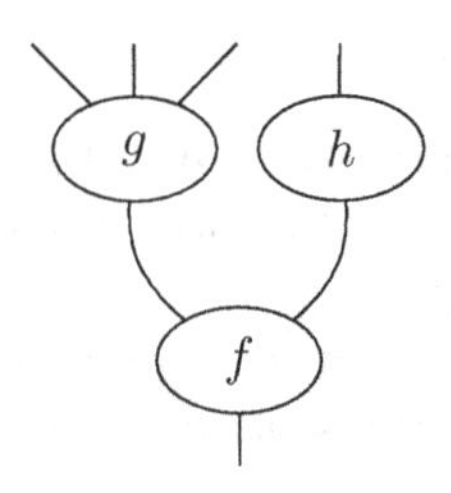

  By plugging the output of g and h on the inputs of f in parallel, we put together the three inputs of g and the input of h, yielding a compound

*A monad T is cartesian if it preserves pullbacks (hence **C** is required to have pullbacks) and if all naturality squares of the unit and multiplication of the monad are pullbacks.

operation with $3 + 1 = 4$ inputs. In other words, one may read two shapes on the upper part of the picture: $((\cdot,\cdot,\cdot),\cdot)$, and $(\cdot,\cdot,\cdot,\cdot)$. The first one remembers the construction, the second one flattens it. This flattening is a typical monad multiplication (monad of powersets, of lists, etc...).

- Span. Each bioperation has an arity and a coarity: the legs of the span are this arity and coarity mappings, respectively.
- Profunctor: A profunctor gives a family of sets of bioperations of fixed arity and coarity.

Arities (or coarities) are governed by the monad T. For example, an operad will be a span / profunctor from $T\mathbf{1}$ to $\mathbf{1}$, where T takes care of the multiplicity of inputs ($\mathbf{1}$ is the category with one object and one morphism). Similarly, a cooperad will go from 1 to $T\mathbf{1}$.

The span approach and the profunctor approach should be related, since one goes from profunctors to spans via the "element" or so-called Grothendieck construction. But under this correspondence the (bi)category of profunctors is (bi)equivalent to a subcategory of spans only, the discrete fibrations, while on the other hand the span approach leaves a lot of freedom on the choice of the underlying category $\mathbf{C}$. As a matter of fact, the two approaches have led to different types of successes. The current state of the art seems to be that:

(1) using spans, an impressive variety of shapes in the *non-symmetric case* have been covered. We refer to [23] for a book-length account;
(2) using profunctors, one may cover the two other kinds of variations mentioned in this introduction.

In the sequel, we use (and introduce) the profunctor road. In Section 2, we introduce several monads on **Cat**. After recalling the notion of Kan extension in Section 3, we present two classical categorical accounts of operads in Sections 4 and 5. We proceed then to profunctors in Section 6. A plan of the rest of the paper is given at the end of that section.

Notation We shall use juxtaposition to denote functor application. Moreover, if F is a functor from, say $\mathbf{C}$ to $\mathbf{Set}^{\mathbf{D}}$, we write FXD for $(FX)D$, etc... .

2. Three useful combinators

Consider the following three operations, or combinators. Think of them as morphisms in a category whose objects are (possibly empty) sequences $(C_1, \ldots, C_n)$, where the C_i's range over the objects of some category $\mathbf{C}$.

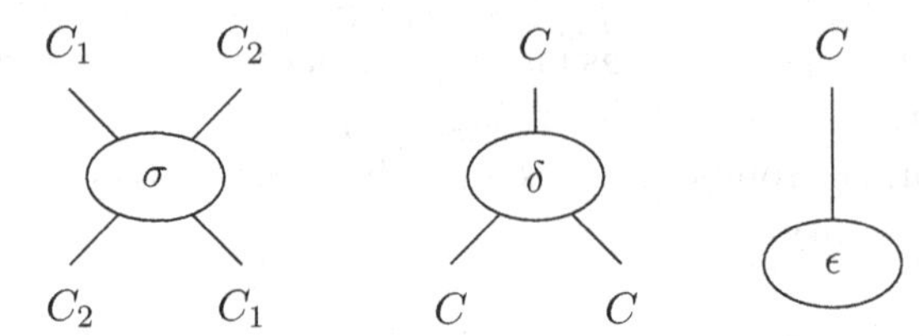

These combinators should satisfy equations:

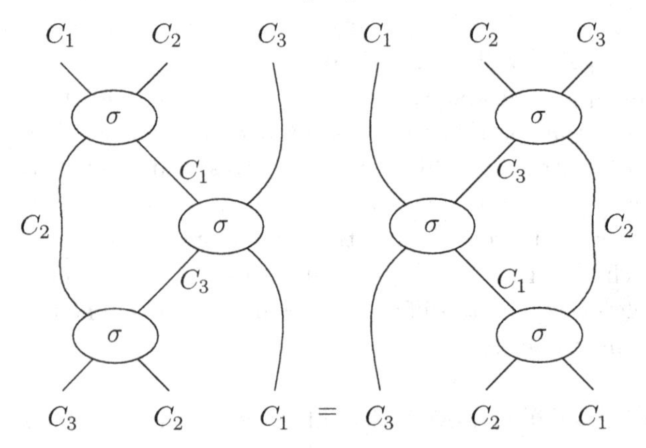

(compare with the familiar equations of transpositions, and notice that here we do not have to index them over natural numbers).

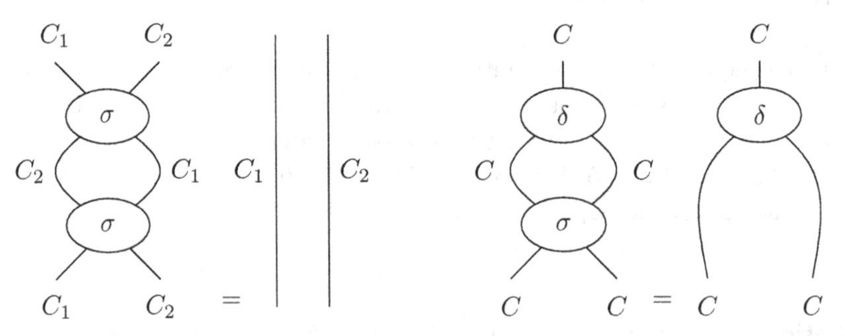

The other equations are the familiar ones for comonoids, and (mutatis mutandis) for distributive laws (see below). For a complete list, we refer the reader to [9,22].

Take any $X \subseteq \{\sigma, \delta, \epsilon\}$. We build a category $!_X\mathbf{C}$ as follows: objects are sequences $(C_1, \ldots, C_n)$ of objects of $\mathbf{C}$.[†] Morphisms are *string diagrams* built out of the combinators taken from X and out of the morphisms of $\mathbf{C}$, quotiented over all the equalities that concern the combinators of X, including their naturality, and the equalities

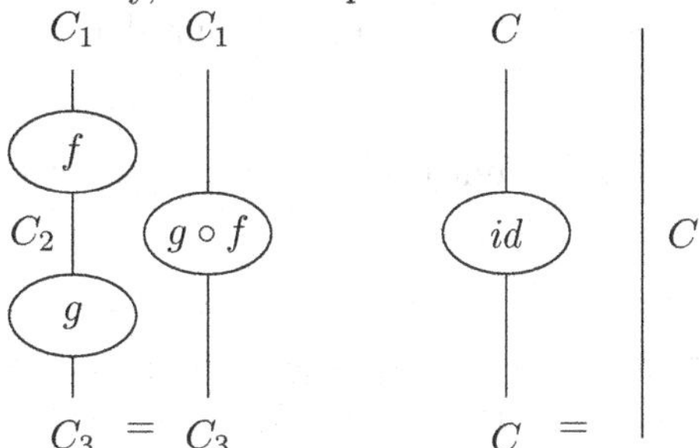

which allow to embed $\mathbf{C}$ functorially into $!_X\mathbf{C}$. String diagrams are combinations of vertical and horizontal combinations of the basic combinators $\sigma, \delta, \epsilon, f$, such as the ones drawn above.

For six of the eight choices of X, the functor $!_X$ is equipped with a monad structure: the unit takes C to (C) and $f : C \to D$ to $f : (C) \to (D)$ and the multiplication on objects is the usual flattening, which takes $((C_1^1, \ldots, C_1^{i_1}), (C_n^1, \ldots, C_n^{i_n}))$ to $(C_1^1, \ldots, C_n^{i_n})$. A little care is needed to define the "flattening" on morphisms. A morphism of $!!\mathbf{C}$ is an assembling of boxes connected by combinators typed in $!\mathbf{C}$. When we remove the boxes, we need to turn these combinators into (assemblings of) combinators typed in $\mathbf{C}$. We call this an expansion. A σ is expanded by means of σ's, an ϵ is expanded by putting ϵ's in parallel, but we need δ's *and* σ's to expand a δ. For example, at type (C_1, C_2), the expansion of $\delta : ((C_1, C_2)) \to ((C_1, C_2), (C_1, C_2))$ is the following morphism from (C_1, C_2) to (C_1, C_2, C_1, C_2):

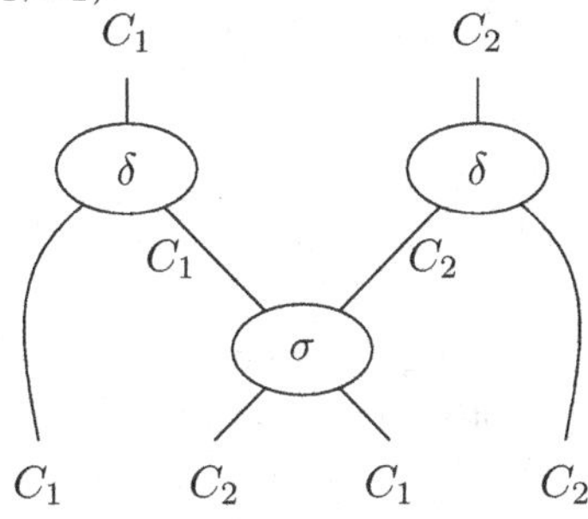

[†] This notation, that comes from linear logic [18] , stresses the idea of multiple input (see [14] for explicit links with linear logic).

We therefore exclude $!_{\{\delta\}}$ and $!_{\{\delta,\epsilon\}}$ from our treatment. But all the other ! constructions are fine.

Three of the six remaining combinations are of particular interest. If $X = \emptyset$ (resp. $X = \{\sigma\}$, $X = \{\sigma, \delta, \epsilon\}$), then $!\mathbf{C}$ is the free strict monoidal (resp. symmetric monoidal, cartesian) category over $\mathbf{C}$. (By cartesian category, we mean "category with specified finite products"). We write them $!_m, !_s, !_f$, respectively:

monoidal	symmetric monoidal	cartesian
$!_m$	$!_s$	$!_f$

When $X = \{\sigma, \delta, \epsilon\}$, the !-algebras are the cartesian categories, and when $X = \emptyset$ (resp. $X = \{\sigma\}$), the (pseudo-) !-algebras are the monoidal (resp. symmetric monoidal) categories. We refer to [23] for details.

We define $?\mathbf{C} = (!(\mathbf{C}^{op}))^{op}$. Graphically, this amounts to reversing the basic combinators σ, δ, ϵ (which could then, if we cared, be called σ, μ, η), while maintaining the orientation of the combinators f imported from $\mathbf{C}$ (since they become themselves again after two ^{op}'s). But it is more convenient to stick with σ, δ, ϵ, to reverse the direction of the f's, and to read the diagrams in the bottom-to-top direction. In particular, when $\mathbf{C} = \mathbf{1}$ (the terminal category), there is no f to reverse...

When $X = \emptyset$ or $X = \{\sigma\}$, then $?\mathbf{C}$ is (isomorphic to) $!\mathbf{C}$. When $X = \{\sigma, \delta, \epsilon\}$, $?\mathbf{C}$ is the free cocartesian category over $\mathbf{C}$. The objects of $?\mathbf{1}$ are written $0, 1, 2, \ldots$, standing for $(), (\cdot), (\cdot, \cdot, \cdot), \ldots$, where $\cdot$ is the unique object of $\mathbf{1}$.

Note that for any choice of ?, there is a faithful functor from $?\mathbf{1}$ to $\mathbf{Set}$. For any morphism from m to n, i.e., for any string diagram constructed out of the combinators in X that has n input wires and m output wires, one constructs a function from $\{0, \ldots, m-1\}$ to $\{0, \ldots, n-1\}$, by the following rules.

- σ: the transposition on $\{0, 1\}$;
- ϵ: the unique function from the empty set to $\{0\}$;
- δ: the unique function from $\{0, 1\}$ to $\{0\}$;
- vertical composition of string diagrams: function composition;
- horizontal composition of string diagrams: their categorical sum, i.e., for $f : \{0, \ldots, m-1\} \to \{0, \ldots, n-1\}$ and $g : \{0, \ldots, p-1\} \to \{0, \ldots, q-1\}$, $f + g : \{0, \ldots, m+p-1\} \to \{0, \ldots, n+q-1\}$ is defined by $(f+g)(i) = f(i)$ if $i < m$ and $f + g(i) = g(i) + n$ if $i \geq m$.

More synthetically, the function f associated with a string diagram is obtained by naming the output and input wires $0, 1, \ldots, m-1$ and $0, 1, \ldots, n-1$, respectively, and then computing $f(i)$ as the name of the input wire reached when starting from input wire i, going up in δ nodes, and going up left or right in σ nodes according to whether this node is reached from down right or down left, respectively. For example, the picture

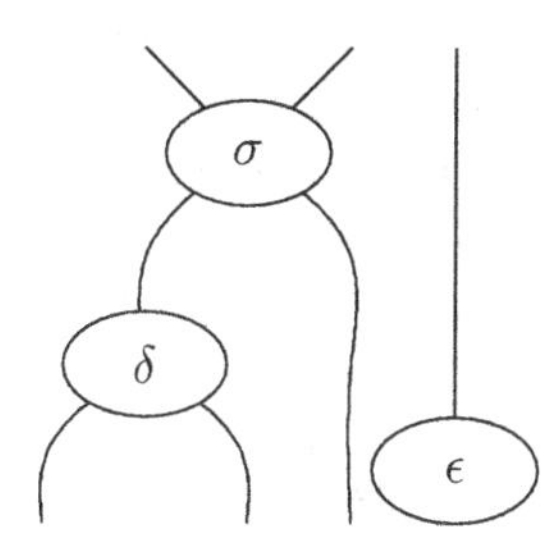

represents the function $f : \{0, 1, 2\} \to \{0, 1, 2\}$ defined by

$$f(0) = f(1) = 1 \quad f(2) = 0$$

Note on this example that ϵ witnesses the lack of surjectivity. More precisely, as first shown by Burroni [9] , the functor, when suitably corestricted, gives the following isomorphims and equivalences of categories:

$?_{\emptyset}$	identity functions
$?_{\{\sigma\}}$	bijections
$?_{\{\sigma,\delta,\epsilon\}}$	all functions
$?_{\{\delta,\epsilon\}}$	monotone functions
$?_{\{\sigma,\delta\}}$	surjective functions
$?_{\{\sigma,\epsilon\}}$	injective functions

where, say, the last line should be read as follows: the category $?_{\{\sigma,\epsilon\}}(\mathbf{1})$ is equivalent to the category of finite sets and injective functions, and isomorphic to its full subcategory whose objects are the sets $\{0, \ldots, i-1\}$ $(i \in \mathbb{N})$. For $?_{\{\delta,\epsilon\}}(\mathbf{1})$, one must take the category of all finite total orders.

3. Kan extensions

We recall that given a functor $K : \mathbf{M} \to \mathbf{C}$, a left Kan extension of a functor $T : \mathbf{M} \to \mathbf{A}$ along K is a pair of a functor $Lan_K(T) : \mathbf{C} \to \mathbf{A}$ and a natural transformation $\eta_T : T \to Lan_K(T)K$ such that $(Lan_K(T), \eta_T)$ is universal from T to $\mathbf{A}^K$. This means that for any other pair of a functor $S : \mathbf{C} \to \mathbf{A}$ and a natural transformation $\alpha : T \to SK$ there exists a unique natural transformation $\mu : Lan_K(T) \to S$ such that $\alpha = \mu K \circ \eta$. Here we shall only need to know that the solution of this universal problem is given by the following formula for $Lan_K(T)$:

$$Lan_K TC = \int^m \mathbf{C}[KM, C] \cdot TM$$

where we use Mac Lane's notation for coends [24] . Coends are sorts of colimits (or inductive limits), adapted to the case of diagrams which vary both covariantly and contravariantly over some parameter: here, M appears contravariantly in $\mathbf{C}[KM, C]$ and covariantly in TM. In the formula, $\mathbf{C}[KM, C] \cdot TM$ stands for the coproduct of as many copies of TM as there are morphisms from KM to C.

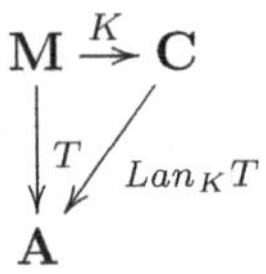

$$\begin{array}{ccc} \mathbf{M} & \xrightarrow{K} & \mathbf{C} \\ {\scriptstyle T}\downarrow & \swarrow {\scriptstyle Lan_K T} & \\ \mathbf{A} & & \end{array}$$

When K is full and faithful, then η is iso, and in particular the triangle commutes (up to isomorphism). When moreover K is the Yoneda embedding, we get

$$Lan_{\mathcal{Y}} TX = \int^M Xm \cdot TM$$

When moreover $\mathbf{A}$ is a presheaf category $\mathbf{Set}^{\mathbf{D}}$, we have (since limits and colimits of presheaves are pointwise):

$$Lan_{\mathcal{Y}} TXD = \int^M XM \times TMD$$

or, making the quotient involved in the coend explicit:

$$Lan_{\mathcal{Y}} TXD = (\sum_M XM \times TMD)/\approx$$

When $\approx$ is termwise, i.e. is the disjoint union of equivalences $\approx_M$ each on $XM \times TMD$, we get a formula which will look more familiar to algebraic operadists:

$$Lan_{\mathcal{Y}} TXD = \sum_M (XM \otimes_M TMD)$$

(where we have written $(XM \times TMD)/{\approx_M}$ as $XM \otimes_M TMD$). In our examples, $\approx$ will not always be termwise (it will be in the operad cases, but not in the clone case).

4. Kelly's account of operads

In 1972, Kelly gave the following description of operads [21] (see also [13]):

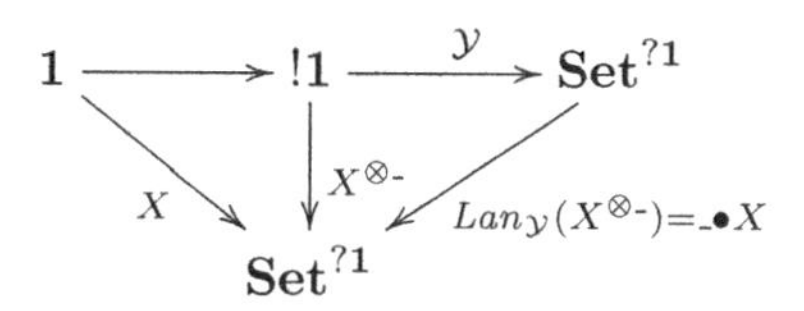

In this diagram, the functor $X^{\otimes}$- associates with n the nth iterated tensor of X, with respect to the following tensor product structure on $\mathbf{Set}^{?1}$ due (in a more general setting) to Day [12] :

$$(X \otimes Y)p = \int^{m,n} Xm \times Yn \times ?\mathbf{1}[m+n,p]$$

The n-ary tensor product is described by the following formula (which also gives the unit of the tensor product, for $m = 0$):

$$(X_1 \otimes X_2 \otimes \ldots \otimes X_m)p = \int^{n_1,\ldots,n_m} X_1 n_1 \times \ldots \times X_m n_m \times ?\mathbf{1}[n_1 + \ldots + n_m, p]$$

Hence the formula for $Lan_{\mathcal{Y}}(X^{\otimes}\text{-})Y$, which we write $Y \bullet X$, is

$$\begin{aligned}(Y \bullet X)p &= \textstyle\int^m Ym \times X^{\otimes m}p \\ &= \textstyle\int^{m,n_1,\ldots,n_m} Ym \times Xn_1 \times \ldots \times Xn_m \times ?\mathbf{1}[n_1 + \ldots + n_m, p]\end{aligned}$$

In the case of $!_f$ this boils down to

$$(Y \bullet X)p = \int^m Ym \times (Xp)^m$$

This is because in a cocartesian category, a morphism $f : n_1 + \ldots + n_m \to p$ amounts to morphisms $f_i \in ?\mathbf{1}[n_i, p]$ which allows us to define a map from $Xn_1 \times \ldots \times Xn_m \times ?\mathbf{1}[n_1 + \ldots + n_m, p]$ to $(Xp)^m$ as follows:

$$(a_1, \ldots, a_m, f) \mapsto (Xf_1a_1, \ldots, Xf_ma_m)$$

and then to "extend" cocones indexed over m to cocones indexed over $m, n_1, \ldots, n_m$.

One can check rather easily that the *substitutions operation* $\bullet$, together with the unit defined by

$$In = ?\mathbf{1}[1, n]$$

provide a (non-symmetric) monoidal structure on $\mathbf{Set}^{?1}$ (for $I \bullet X \approx X$, one uses the fact that $X^{\otimes}$- is functorial).

When ! is $!_m, !_s, !_f$, respectively, a monoid for this structure $(I, \bullet)$ is an operad, a symmetric operad, a clone, respectively.

5. Operads from analytic functors

Another way to arrive at the operation $\bullet$ just defined is via a different Kan extension:

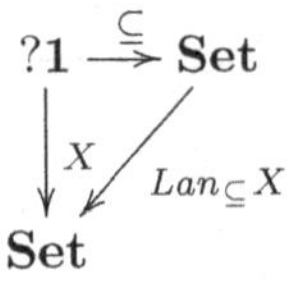

$$\begin{array}{ccc} ?\mathbf{1} & \stackrel{\subseteq}{\longrightarrow} & \mathbf{Set} \\ {\scriptstyle X}\downarrow & \swarrow {\scriptstyle Lan_{\subseteq}X} & \\ \mathbf{Set} & & \end{array}$$

where $\subseteq$ is the faithful (and non full) functor described at the end of Section 2. This is the approach taken by Joyal (for $!_s$) [20] . In Joyal's language, X is called a *species of structure*, and $Lan_{\subseteq}X$ is the associated *analytic functor*, whose explicit formula is (for any set z):

$$Lan_{\subseteq}X z = \int^{m} z^m \times Xm$$

It can be shown that $Lan_{\subseteq}$ is faithful. Then $Y \bullet X$ is characterised by the following equality:

$$Lan_{\subseteq}(Y \bullet X) = Lan_{\subseteq}Y \circ Lan_{\subseteq}X$$

which evidences the fact that $\bullet$ is a composition operation. Indeed, we have:

$$\begin{array}{l} Lan_{\subseteq}Y(Lan_{\subseteq}X z) \\ \quad = \int^m (\int^n z^n \times Xn)^m \times Ym \\ \quad = \int^{m,n_1,\ldots,n_m} z^{n_1+\ldots+n_m} \times Xn_1 \times \ldots \times Xn_m \times Ym \\ \quad = \int^p z^p \times (\int^{m,n_1,\ldots,n_m} Ym \times Xn_1 \times \ldots \times Xn_m \times ?\mathbf{1}[n_1+\ldots+n_m, p]) \\ \quad = Lan_{\subseteq}(Y \bullet X)z \end{array}$$

(the summation over p is superfluous since from $f \in ?\mathbf{1}[n_1 + \ldots + n_m, p]$ we get $z^{\subseteq(f)}$ from z^p to $z^{n_1+\ldots+n_m}$).

6. Profunctors

Recall that a profunctor (or distributor) [5,6] Φ from $\mathbf{C}$ to $\mathbf{C}'$ is a functor

$$\Phi : \mathbf{C} \times \mathbf{C}'^{op} \to \mathbf{Set}$$

We write $\Phi : \mathbf{C} \to\!\!\!\!\!+ \mathbf{C}'$. Composition of profunctors is given by the following formula:

$$(\Psi \circ \Phi)(C, C'') = \int^{C'} \Psi(C', C'') \times \Phi(C, C')$$

Therefore, profunctors compose only up to isomorphism. Categories, profunctors, and natural transformations form thus, not a 2-category, but a bicategory [4] .

The bicategory **Prof** of profunctors is self-dual, via the isomorphism ${}^{op} : \mathbf{Prof} \to \mathbf{Prof}^{op}$ which maps $\mathbf{C}$ to $\mathbf{C}^{op}$ and $\Phi : \mathbf{C_1} \multimap \mathbf{C_2}$ to

$$\Phi^{op} = ((C_2, C_1) \mapsto \Phi(C_1, C_2)) : \mathbf{C_2}^{op} \multimap \mathbf{C_1}^{op}$$

The composition of profunctors can be synthesised via Kan extensions: Given $\Phi : \mathbf{C} \multimap \mathbf{C}'$ and $\Psi : \mathbf{C}' \multimap \mathbf{C}''$, consider their "twisted curried"[‡] versions $\Phi' : \mathbf{C}'^{op} \to \mathbf{Set}^{\mathbf{C}}$ and $\Psi' : \mathbf{C}''^{op} \to \mathbf{Set}^{\mathbf{C}'}$, defined by $\Phi' C' C = \Phi(C, C')$ and $\Psi' C'' C' = \Psi(C', C'')$. Then it is immediate to check that $\Psi \circ \Phi$ is the uncurried version of $(Lan_{\mathcal{Y}} \Phi') \circ \Psi'$, as illustrated in the following diagram:

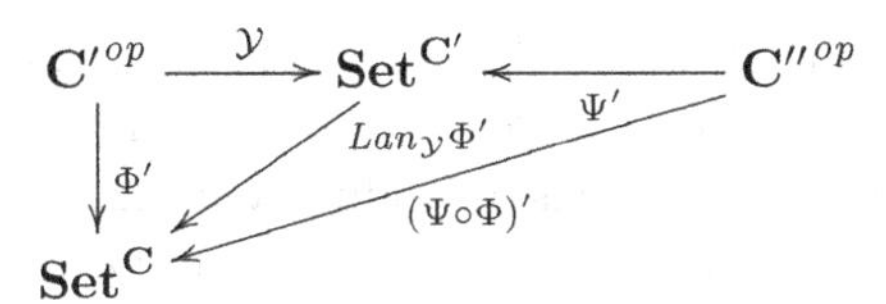

Now we take "profunctor glasses" to look at Kelly's and Joyal's diagrams.

- A presheaf $X :?\mathbf{1} \to \mathbf{Set}$ can be viewed as a profunctor

$$(n \mapsto (\cdot \mapsto Xn)) :?\mathbf{1} \multimap \mathbf{1}$$

- $X^{\otimes}$- can be viewed as a profunctor from $?\mathbf{1}$ to $?\mathbf{1}$ (since $(?\mathbf{1})^{op} = !(\mathbf{1}^{op}) = !\mathbf{1}$). Thus the transformation

$$\frac{X :?\mathbf{1} \multimap \mathbf{1}}{X^{\otimes}\text{-} :?\mathbf{1} \multimap ?\mathbf{1}}$$

  suggests to consider ? as a comonad over **Prof**.
- The diagram and the formula defining $Y \bullet X$ in Section 4 exhibit $Y \bullet X$ as the profunctor composition of $Y :?\mathbf{1} \multimap \mathbf{1}$ and $X^{\otimes}\text{-} :?\mathbf{1} \multimap ?\mathbf{1}$. But a better way to put it is that $Y \bullet X$ is the composition of Y and X in the coKleisli bicategory $\mathbf{Prof}_?$, and we shall see that it is indeed the case.
- Joyal's construction amounts to "taking points". We first observe that $?\emptyset = \mathbf{1}$, where $\emptyset$ is the empty category. (This holds for any of the eight choices for !.) It follows that $\emptyset$ is terminal in $\mathbf{Prof}_?$. In this category, the points of an object $\mathbf{C}$ are the morphisms from $\emptyset$ to $\mathbf{C}$, i.e., the

[‡]After the name of Curry, who defined a calculus of functions called combinatory logic, based on application and a few combinators, and where functions of several arguments are expressed through repeated applications.

profunctors from $\mathbf{1}$ to $\mathbf{C}$, i.e. the presheaves over $\mathbf{C}$. In particular, the points of $\mathbf{1}$ are just sets. We shall see that

$$\mathbf{Prof}_?[\emptyset, X] = Lan_{\subseteq} X$$

i.e., that the analytic functor associated with the presheaf X describes its pointwise behaviour in $\mathbf{Prof}_?$.

Our goal is thus to figure out ? as a comonad on **Prof**. We shall do this in three steps:

- In Section 7, we show that profunctors arise as a Kleisli category for the presheaf construction $\mathbf{C} \mapsto \mathbf{Set}^{\mathbf{C}^{op}}$.
- In Section 9, we show that all the monads ! of Section 2 distribute over *Psh*. Distributive laws are recalled in Section 8. In Section 10, we pause to explain how both Day's tensor product and this distributive law can be synthesised out of considerations of structure preservation.
- in Section 11, we show that this distributive law allows us to extend ! to a monad on **Prof**, and by self-duality we obtain the comonad ? that we are looking for.

Warning. In what follows, we ignore coherence issues for simplicity, and we partially address size issues.

- Coherence issues arise from the fact that we shall compose morphisms using coend formulas, which make sense only up to iso. In particular, *our "distributive law" will in fact be a "pseudo-distributive law"* [10,16,25].
- Size issues arise from the fact that the presheaf construction is dramatically size increasing. It is therefore in fact simply not rigorous to call *Psh* a monad (or even a pseudo-monad) on **Cat**. But the issue is fortunately not too severe. In a forthcoming paper, Fiore, Gambino, Hyland, and Winskel propose a general notion of Kleisli structure, which we shall sketch here (ignoring again coherence issues), and in which *Psh* fits.

7. Profunctors as a Kleisli category

If we write a profunctor $\Phi : \mathbf{C_1} \nrightarrow \mathbf{C_2}$ in (untwisted) curried form

$$C_1 \mapsto (C_2 \mapsto \Phi(C_1, C_2)) : \mathbf{C_1} \to \mathbf{Set}^{\mathbf{C_2}^{op}}$$

this suggests us to look at the operation *Psh* on the objects of **Cat** defined by $Psh(\mathbf{C}) = \mathbf{Set}^{\mathbf{C}^{op}}$. The idea is to exhibit *Psh* as a (pseudo-)monad, so that profunctors arise as a Kleisli category. We have a good candidate for the unit η, namely: $\mathcal{Y} : \mathbf{C} \to Psh(\mathbf{C})$.

But the Yoneda functor makes sense only for a locally small category (one in which all homsets are sets), while it is not clear at all whether *Psh* keeps us within the realm of locally small categories, i.e. if $\mathbf{C}$ is locally small, then $\mathbf{Set}^{\mathbf{C}^{op}}$ is not necessarily locally small. It is neverheless tempting to go on with a multiplication $\mu : \mathbf{Set}^{(\mathbf{Set}^{\mathbf{C}^{op}})^{op}} \to \mathbf{Set}^{\mathbf{C}^{op}}$ given by a left Kan extension

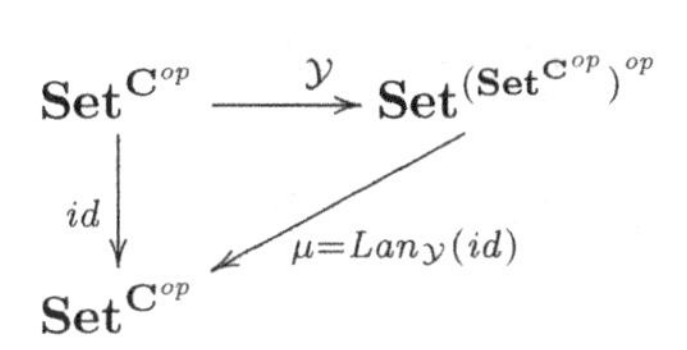

with explicit formula $\mu(H)C = \int^F HF \times FC$. But why should a coend on such a vertiginous indexing collection exist?

We pause here to recall that an equivalent presentation of a monad on a category $\mathbf{C}$ is by means of the following data:

- for each object C of $\mathbf{C}$, an object TC of $\mathbf{C}$, and a morphism $\eta_C \in \mathbf{C}[C, TC]$;
- for all objects C, D of $\mathbf{C}$ and each morphism $f \in \mathbf{C}[C, TD]$, a morphism $f^\# \in \mathbf{C}[TC, TD]$;

satisfying the equations $f = f^\# \circ \eta = f, \ , \eta^\# = id, \ , (g^\# \circ f)^\# = g^\# \circ f^\#$.

Let us also recall the definition of the associated Kleisli category $\mathbf{C}_T$: its objects are the objects of $\mathbf{C}$, and one sets $\mathbf{C}_T[C, D] = \mathbf{C}[C, TD]$, with composition easily defined using the composition in $\mathbf{C}$ and the $\_^\#$ operation.

It turns out that under this guise, the definition of monad can be generalised in a way that will fit our purposes. Ignoring coherence issues, a *Kleisli structure* (as proposed in [15]) is given by the following data:

- a collection $\mathcal{A}$ of objects of $\mathbf{C}$;
- for each object A in $\mathcal{A}$ an object TA of $\mathbf{C}$, and a morphism $\eta_A \in \mathbf{C}[A, TA]$;
- for all objects A, B in $\mathcal{A}$ and each morphism $f \in \mathbf{C}[A, TB]$, a morphism $f^\# \in \mathbf{C}[TA, TB]$;

satisfying the same equations. The associated Kleisli category $\mathbf{C}_T$ has $\mathcal{A}$ as collection of objects, and one sets $\mathbf{C}_T[A, B] = \mathbf{C}[A, TB]$. We recover the monads when every object of $\mathbf{C}$ is in $\mathcal{A}$. But in general there is no such thing as a multiplication, since it is not granted that TTC exists for all C. We also note that in a Kleisli structure, T is still a functor, not from $\mathbf{C}$ to $\mathbf{C}$, but from $\mathbf{C}\downarrow_{\mathcal{A}}$ (the full subcategory spanned by $\mathcal{A}$) to $\mathbf{C}$.

In our setting, we have $\mathbf{C} = \mathbf{Cat}$, and we can take $\mathcal{A}$ to consist of the small categories (in which the objects form a set, as well as all homsets). Then we have no worry about the coends that we shall write. We complete the definition of the Kleisli structure with the definition of the $\_^{\#}$ operation.

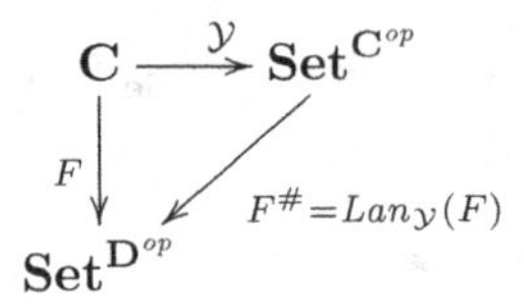

and the explicit formula is $F^{\#}XD = \int^{C} XC \times FCD$. It is then an easy exercise to check that the three equations of a Kleisli structure are satisfied, and that the composition in the associated Kleisli category coincides with the composition of profunctors as defined in the previous section.

We end the section with a description of the functorial action of Psh (on, say, functors between small categories):

$$Psh(F)XD = (\eta_{Psh} \circ F)^{\#}XD = \int^{C} XC \times \mathbf{D}[D, FC]$$

8. Distributive laws

Recall that a distributive law [3] (see also [2]) is a natural transformation $\boldsymbol{\lambda} : TS \to ST$, where (S, η_S, μ_S) and (T, η_T, μ_T) are two monads over the same category $\mathbf{C}$, satisfying the following laws, expressed in the language of string diagrams:

EQUATION $(\boldsymbol{\lambda} - \eta_T)$:

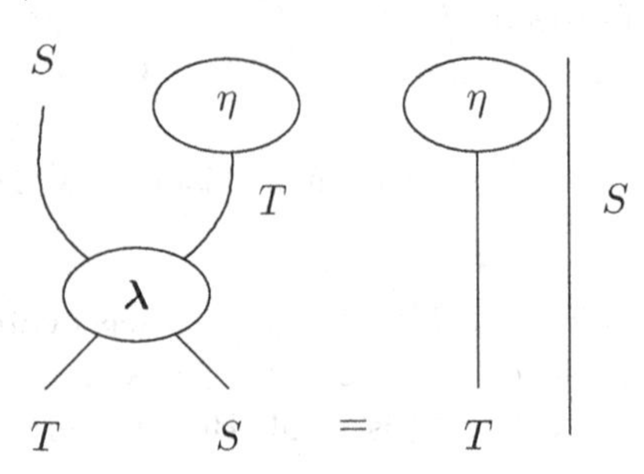

EQUATION ($\boldsymbol{\lambda} - \mu_T$):

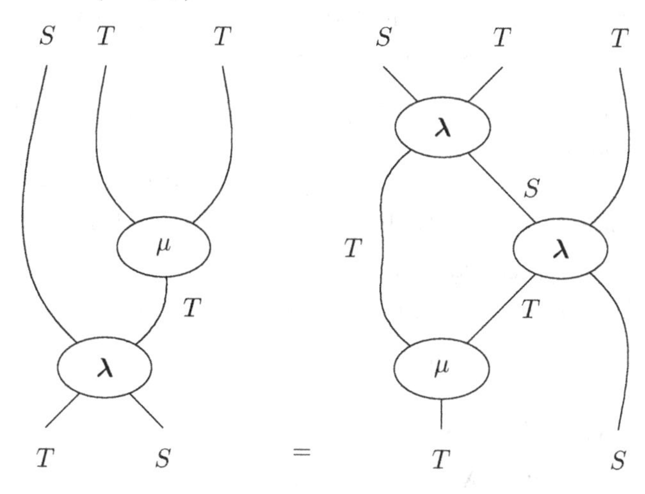

and two similar equations ($\boldsymbol{\lambda} - \mu_S$) and ($\boldsymbol{\lambda} - \eta_S$).

A distributive law allows us to extend T to $\mathbf{C}_S$, the Kleisli category of S (and conversely, such a lifting induces a distributive law such that the two constructions are inverse to each other).[§] The extended T acts on objects as the old T. On morphisms, its action is described as follows:

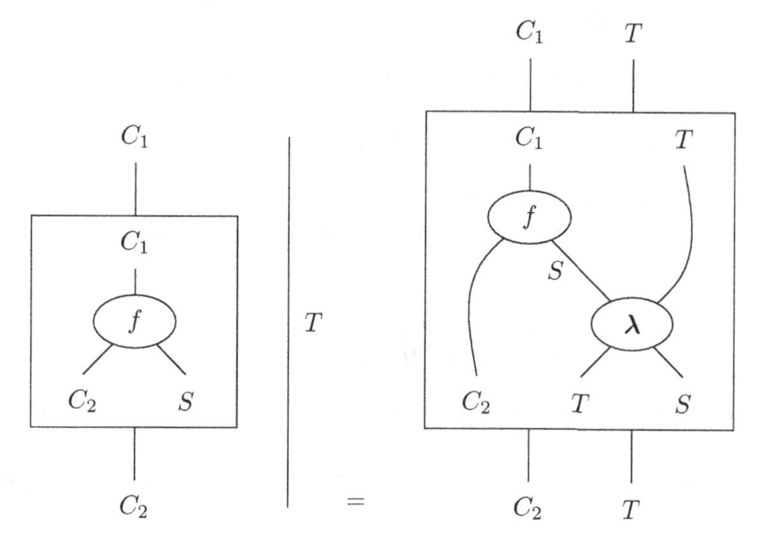

[§] From another standpoint, a distributive law $\boldsymbol{\lambda} : TS \to ST$ also amounts to lifting S to the category of T-algebras. See Section 10.

where the box separates the inside which lives in $\mathbf{C}$ from the outside which lives in $\mathbf{C}_S$. It is a nice exercise to check that the old and the new T satisfy

$$T \circ F_S = F_S \circ T$$

where $F_S : \mathbf{C} \to \mathbf{C}_S$ is defined by $F_S(C) = C$ and $F_S(f) = \eta_S \circ f$, and that the new T is a monad.

In the next section, we want to apply this to $S = Psh$ and $T = !$. But S is only a Kleisli structure. Only slight adjustments is needed:

(1) make sure that $\mathcal{A}$ is stable under T;
(2) replace the equation $(\boldsymbol{\lambda} - \mu_S)$ by the following one (for $f : A \to SB$):

$$\boldsymbol{\lambda}_B \circ T(g^{\#}) = (\boldsymbol{\lambda}_B \circ Tg)^{\#} \circ \boldsymbol{\lambda}_A$$

9. A $!/Psh$ distributive law

We define a transformation $\boldsymbol{\lambda} : ! \circ Psh \to Psh \circ !$, as follows:

- on objects:

$$\boxed{\begin{array}{l}(F_1, \ldots, F_n) \mapsto ((C_1, \ldots, C_m) \mapsto \\ \qquad \int^{D_1,\ldots,D_n} F_1 D_1 \times \ldots \times F_n D_n \times !\mathbf{C}[(C_1, \ldots, C_m), (D_1, \ldots, D_n)])\end{array}}$$

- on generating morphisms:
  - $\alpha : F \to G$. There is an obvious map from $\int^D FD \times !\mathbf{C}[C_1, \ldots, C_m, (D)]$ to $\int^D GD \times !\mathbf{C}[C_1, \ldots, C_m, (D)]$.
  - Combinators, say $\delta : (F) \to (F, F)$. We obtain a map

$$\begin{array}{l}\text{from } \int^D FD \times !\mathbf{C}[C_1, \ldots, C_m, (D)] \\ \text{to } \int^{D_1,D_2} FD_1 \times FD_2 \times !\mathbf{C}[(C_1, \ldots, C_m, (D_1, D_2)]\end{array}$$

  by going

$$\begin{array}{l}\text{from } FD \times !\mathbf{C}[C_1, \ldots, C_m, (D)] \\ \text{to } FD \times FD \times !\mathbf{C}[(C_1, \ldots, C_m, (D, D)]\end{array}$$

We verify three of the four equations (leaving the last one – as stated at the end of Section 8 – to the reader):

$(\boldsymbol{\lambda} - \eta_!)$. The left hand side is computed by taking the case $n = 1$ in the definition of $\boldsymbol{\lambda}$, thus we obtain

$$F \mapsto ((C_1, \ldots, C_m) \mapsto \int^D FD \times !\mathbf{C}[(C_1, \ldots, C_m), (D)])$$

which is the formula for $Psh(\eta_!)$.

$(\boldsymbol{\lambda} - \eta_{Psh})$. We have (at $(A_1, \ldots, A_n)$, $(C_1, \ldots C_m)$):

$$\begin{aligned}&\int^{D_1,\ldots,D_n} \mathbf{C}[D_1, A_1] \times \ldots \times \mathbf{C}[D_n, A_n] \times !\mathbf{C}[(C_1, \ldots, C_m), (D_1, \ldots, D_n)] \\ &\quad = !\mathbf{C}[(C_1, \ldots, C_m), (A_1, \ldots, A_n)] = \mathcal{Y}(A_1, \ldots, A_n)(C_1, \ldots C_m)\end{aligned}$$

$(\boldsymbol{\lambda} - \mu_!)$. The left hand side is (at $((F_1^1, \ldots, F_1^{i_1}), \ldots (F_n^1, \ldots, F_n^{i_n}))$, $(C_1, \ldots, C_m)$):

$$\int^{D_1^1,\ldots,D_n^{i_n}} F_1^1 D_1^1 \times \ldots \times F_n^{i_n} D_n^{i_n} \times !\mathbf{C}[\vec{C}, (D_1^1, \ldots, D_n^{i_n})]$$

We first compute the upper part $(\boldsymbol{\lambda}!) \circ (!\boldsymbol{\lambda})$ of the right hand side (at $(\vec{F_1}, \ldots, \vec{F_n})$, $((A_1^1, \ldots, A_1^{j_1}), \ldots, (A_p^1, \ldots, A_p^{j_p})))$:

$$\int^{\vec{B_1},\ldots\vec{B_n}} \int^{D_1^1,\ldots,D_n^{i_n}} F_1^1 D_1^1 \times \cdots \times F_n^{i_n} D_n^{i_n} \times !\mathbf{C}[\vec{B_1}, \vec{D_1}] \times \cdots \times !\mathbf{C}[\vec{C_n}, \vec{D_n}] \times !!\mathbf{C}[\vec{\vec{A}}, \vec{\vec{B}}]$$

$$= \int^{D_1^1,\ldots,D_n^{i_n}} F_1^1 D_1^1 \times \ldots \times F_n^{i_n} D_n^{i_n} \times !!\mathbf{C}[\vec{\vec{A}}, \vec{\vec{D}}]$$

Finally, applying $Psh(\mu)$, we get (at $\vec{\vec{F}}$, $\vec{C}$):

$$\int^{\vec{A_1},\ldots,\vec{A_p}} \int^{D_1^1,\ldots,D_n^{i_n}} F_1^1 D_1^1 \times \cdots \times F_n^{i_n} D_n^{i_n} \times !!\mathbf{C}[\vec{\vec{A}}, \vec{\vec{D}}] \times !\mathbf{C}[\vec{C}, (A_1^1, \ldots, A_p^{j_p})]$$

and we conclude using $\mu_!$.

Summarizing, we have the following result:

Proposition 9.1. *The transformation $\boldsymbol{\lambda}$ defines a (pseudo)-distributive law for any choice among the six $!_X$ monads over the presheaf Kleisli structure.*

A similar proposition is proved in [26] : there, the existence of $\boldsymbol{\lambda}$ is derived from the fact ! and *Psh* are both free constructions (the latter being the free cocompletion, i.e. the free category with all colimits) and commute in the sense that *Psh* lifts to the category of !-algebras (cf. Section 2). For completeness, we sketch this more conceptual approach in the following section.

On the other hand, the advantage of a "symbol-pushing" proof like the one presented above is to highlight plainly the "uniformity" in the choice of any combination X of Burroni's combinators.

10. Intermezzo

The reader may have noticed that the formula defining $\boldsymbol{\lambda}$ proposed in the last section "looks like" the formula for Day's tensor product (cf. Section 4). In this section, we actually synthesise the latter formula from considerations of cocontinuity, and then the former one from considerations of monad lifting. Since this section has merely an explanatory purpose, we limit ourselves here to the case $! = !_\emptyset$, and we disregard size issues.

Cocontinuity is at the heart of the presheaf construction. It is well-known that $\mathbf{Set}^{\mathbf{C}^{op}}$ is the free cocomplete category over $\mathbf{C}$, and that the unique cocontinuous (i.e., colimit preserving) extension of a functor $F : \mathbf{C} \to \mathbf{D}$ (where $\mathbf{D}$ is cocomplete, i.e., has all colimits) is its left Kan extension $Lan_{\mathcal{Y}}(F)$:

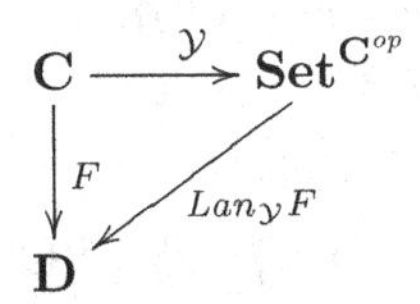

In particular, the fact that $Lan_{\mathcal{Y}}(F)$ preserves colimits follows from its being left adjoint to $(C \mapsto (D \mapsto \mathbf{D}[FC, D]))$ (easy check).

We could have defined the "monad" *Psh* more economically and more conceptually as the monad arising from the adjunction between **Cat** and the category of cocomplete categories. The "brute force" construction of last section is in any case a good exercise in coend computations.

Although we shall not need it here, it is worth pointing out that the adjunction is monadic, i.e., that the (pseudo-) *Psh*-algebras are actually the cocomplete categories (and hence that there can be at most one *Psh*-algebra structure on a given category). This can be proved using Beck's characterization of monadic adjunctions (Theorem 4.4.4 of [7]).

Consider a slightly more general version of Day's tensor product than that given in Section 4, with now some monoidal category $\mathbf{C}$ in place of $!\mathbf{1}$

$$(X \otimes Y)C = \int^{C_1, C_2} XC_1 \times XC_2 \times \mathbf{C}[C, C_1 \otimes C_2]$$

(thus Day's tensor depends on the tensor on $\mathbf{C}$). We show that this definition is entirely determined from the requirements that

- $\otimes$ extends the tensor product of $\mathbf{C}$ (via the Yoneda embedding), and
- $\otimes$ is cocontinuous in each argument.

And, of course, Day's product does satisfy these requirements (easy proof left to the reader). Recall that $\mathcal{Y}$ is dense, i.e., that every preseheaf is a colimit of representable presheaves (i.e. presheaves of the form YC), and, more precisely, that $Lan_{\mathcal{Y}}\mathcal{Y} = id$:

$$X = \int^{D} XD \cdot \mathcal{Y}D = colim_{(D,z \in XD)}\mathcal{Y}D$$

Then, we must have

$$\begin{aligned} X \otimes Y &= (colim_{(D_1,z\in XD_1)}\mathcal{Y}D_1) \otimes Y \\ &= colim_{(D_1,z\in XD_1)}(\mathcal{Y}D_1 \otimes Y) && \text{(by requirement 2)} \\ &= colim_{(D_1,z\in XD_1),(D_2,z\in XD_2)}(\mathcal{Y}D_1 \otimes \mathcal{Y}D_2) && \text{(by requirement 2)} \\ &= colim_{(D_1,z\in XD_1),(D_2,z\in XD_2)}\mathcal{Y}(D_1 \otimes D_2) && \text{(by requirement 1)} \\ &= \int^{D_1,D_2}(XD_1 \times XD_2) \cdot \mathcal{Y}(D_1 \otimes D_2) \end{aligned}$$

In other words, Day's tensor is determined, as announced.

It is shown in [19] that the (2-) adjunction between categories and cocomplete categories specialises to an adjunction between monoidal categories and cocomplete monoidal categories, and that $\mathbf{Set}^{\mathbf{C}^{op}}$, equipped with Day's tensor product, is the free such one. We content ourselves here with the key verification, namely that when $F : \mathbf{C} \to \mathbf{D}$ is monoidal, then so is its unique cocontinuous extension $\tilde{F} = Lan_{\mathcal{Y}}F$. We first compute $\tilde{F}X_1 \otimes \tilde{F}X_2$ and $\tilde{F}(X_1 \otimes X_2)$:

$$\tilde{F}X_1 \otimes \tilde{F}X_2 = \int^{C_1,C_2}(X_1C_1 \times X_2C_2) \cdot (FC_1 \otimes FC_2) \quad \text{(by cocontinuity)}$$

$$\begin{aligned} \tilde{F}(X_1 \otimes X_2) &= \int^{C} \int^{C_1,C_2} X_1C_1 \times X_2C_2 \times \mathbf{C}[C, C_1 \otimes C_2] \cdot FC \\ &= \int^{C_1,C_2} X_1C_1 \times X_2C_2 \cdot F(C_1 \otimes C_2) \end{aligned}$$

From this, it follows that the maps $FC_1 \otimes FC_2 \to F(C_1 \otimes C_2)$ induce a map $\tilde{F}X_1 \otimes \tilde{F}X_2 \to \tilde{F}(X_1 \otimes X_2)$.

As a consequence *Psh* lifts to (pseudo-) !-algebras, i.e. to monoidal categories. Given two monads T, S, a lifting of S to the category of T-algebras consists of a monad S' on the category of T-algebras such that S, S' commute with the forgetful functor, and such that the unit and the multiplication of S' are mapped by the forgetful functor to the unit and the multiplication of S. Here, we take $S = Psh$ and $T =!$. The lifting S' maps a monoidal structure $\otimes$ on $\mathbf{C}$ to the corresponding Day's monoidal structure on $Psh(\mathbf{C})$.

We turn back to the general situation. A lifting of S to T-algebras is equivalent to giving a distributive law $\boldsymbol{\lambda} : TS \to ST$ (cf. Footnote §). One defines $\boldsymbol{\lambda}$ from the data of the lifting as follows:

$$\boldsymbol{\lambda}_C = S'(\mu_C) \circ TS\eta_C \qquad (\mu, \eta \text{ relative to } T)$$

(recall that μ_C is a T-algebra – the free one). This allows us to derive the distributive law from Day's tensor product. First, for $S = Psh$ and $T =!$, $S'(\mu_{\mathbf{C}})$ is the Day's tensor product $(G_1, \ldots, G_n) \mapsto G_1 \otimes \ldots \otimes G_n$ associated with the (flattening) tensor product on $!\mathbf{C}$:

$$(G_1 \otimes \ldots \otimes G_n)\vec{C} = \int^{\vec{A_1},\ldots,\vec{A_n}} G_1\vec{A_1} \times \ldots \times G_n\vec{A_n} \times !\mathbf{C}[\vec{C}, (A_1^1, \ldots, A_n^{i_n})]$$

Then $\boldsymbol{\lambda}$ is obtained by replacing G_i by $Psh(\eta)(F_i)$:

$$\int^{\vec{A_1},\ldots,\vec{A_n}} z_1 \times \ldots \times z_n \times !\mathbf{C}[\vec{C}, (A_1^1, \ldots, A_n^{i_n})]$$

with $z_1 = \int^{D_1} !F_1 D_1 \times !\mathbf{C}[\vec{A_1}, (D_1)]$, $\ldots$, $z_n = \int^{D_n} F_n D_n \times !\mathbf{C}[\vec{A_n}, (D_n)]$. This simplifies to the formula given in Section 9.

11. The (bi)category $\mathbf{Prof}_?$

We next consider the coKleisli bicategory $\mathbf{Prof}_? = (\mathbf{Cat}_{Psh})_?$. From the previous sections we know that the monad ! on **Cat** extends to a (pseudo-) monad ! on **Prof**, which under the self-duality of **Prof** gives a (pseudo-) comonad ? on **Prof**. Just as a monad gives rise to a Kleisli category, a comonad gives rise to a coKleisli category. Here, $\mathbf{Prof}_?$ has categories as objects, while its (1-)morphisms are defined by

$$\mathbf{Prof}_?[\mathbf{C}, \mathbf{C}'] = \mathbf{Prof}[?\mathbf{C}, \mathbf{C}']$$

For $X = \{\sigma\}$, this is the category generalised species of structures of Fiore, Gambino, Hyland, and Winskel [14] . Also with $X = \{\sigma\}$, the endomorphisms in $\mathbf{Prof}[?\mathbf{C}, \mathbf{C}]$ are the C-profiles of Baez and Dolan [1] . ¶ We are

now in a position to round the circle and to show (cf. Theorem 4.1 of [14] for the case $X = \{\sigma\}$) that the composition in this category coincides with the substitution operation as defined in Section 4. We decompose all the steps of the construction.

¶ Baez and Dolan use a variation to synthesise the composition of **C**-signatures: they note that $\mathbf{Set}^{?\mathbf{C}}$ is a free symmetric monoidal cocomplete construction (where the tensor preserves the colimits in each argument, cf. Section 10), which allows them to identify C-profiles with the endofunctors on $\mathbf{Set}^{?\mathbf{C}}$ that preserve tensor and colimits, and hence to inherit composition from usual composition of functors.

(1) Let $\Phi : \mathbf{C} \multimap \mathbf{C}'$. Then $!\Phi : !\mathbf{C} \multimap !\mathbf{C}'$ is (using $\boldsymbol{\lambda}$):

$$!\Phi((C_1,\ldots,C_n),\vec{C'}) = \int^{D'_1,\ldots,D'_n} \Phi C_1 D'_1 \times \ldots \times \Phi C_n D'_n \times !\mathbf{C}'[\vec{C'},\vec{D'}]$$

$(1)^{op}$ We can then define $?\Phi = (!\Phi^{op})^{op}$:

$$?\Phi(\vec{C},(C'_1,\ldots,C'_p)) = \int^{D_1,\ldots,D_p} \Phi D_1 C'_1 \times \ldots \times \Phi D_p C'_p \times ?\mathbf{C}[\vec{D},\vec{C}]$$

(2) The multiplication of ! viewed as a monad on **Prof** at **C** is

$$(((C_1^1,\ldots,C_1^{i_1}),\ldots,(C_n^1,\ldots,C_n^{i_n})),\vec{D}) \mapsto !\mathbf{C}[\vec{D},(C_1^1,\ldots,C_n^{i_n})] : !!\mathbf{C} \multimap !\mathbf{C})$$

$(2)^{op}$ Hence the comultiplication $\delta : ?\mathbf{C} \multimap ??\mathbf{C}$ of $? = (\mathbf{C} \mapsto (!(\mathbf{C}^{op})^{op})$ is

$$((\vec{C},((D_1^1,\ldots,D_1^{i_1}),\ldots,(D_n^1,\ldots,D_n^{i_n}))) \mapsto ?\mathbf{C}[(D_1^1,\ldots,D_n^{i_n}),\vec{C}])$$

(3) Let $\Phi : ?\mathbf{C} \multimap \mathbf{C}'$. Then $\Phi^\flat = (?\Phi) \circ \delta : ?\mathbf{C} \multimap ?\mathbf{C}'$ is given by the following explicit formula (on $\vec{C},(C'_1,\ldots,C'_p)$):

$$\int^{\vec{A_1},\ldots,\vec{A_m}} ?\mathbf{C}[(A_1^1,\ldots,A_m^{j_m}),\vec{C}] \times (\int^{\vec{D_1},\ldots,\vec{D_p}} \Phi\vec{D_1}C'_1 \times \cdots \times \Phi\vec{D_p}C'_p \times ??\mathbf{C}[\vec{\vec{D}},\vec{\vec{A}}])$$

$$= \int^{\vec{D_1},\ldots,\vec{D_p}} \Phi\vec{D_1}C'_1 \times \cdots \times \Phi\vec{D_p}C'_p \times ?\mathbf{C}[(D_1^1,\ldots,D_p^{i_p}),\vec{C}]$$

The simplification comes from the functoriality of the multiplication of the monad ! on **Cat** (we have $\mu_{\mathbf{C}^{op}} : !!\mathbf{C}^{op} \to !\mathbf{C}^{op}$, which we can view as a functor from $??\mathbf{C} = (!!\mathbf{C}^{op})^{op}$ to $?\mathbf{C} = (!\mathbf{C}^{op})^{op}$).

Finally, we can compose $\Phi : ?\mathbf{C} \multimap \mathbf{C}'$ and $\Psi : ?\mathbf{C}' \multimap \mathbf{C}''$: $\Psi \bullet \Phi = \Psi \circ \Phi^\flat$. Explicitly, $(\Psi \bullet \Phi)(\vec{C},C'')$ is given by the following formula:

$$\int^{C'_1,\ldots,C'_p} \Psi(\vec{C'},C'') \times \int^{\vec{D_1},\ldots,\vec{D_p}} \Phi\vec{D_1}C'_1 \times \cdots \times \Phi\vec{D_p}C'_p \times ?\mathbf{C}[(D_1^1,\ldots,D_p^{i_p}),\vec{C}]$$

As promised, we recover Kelly's (Section 4) and Joyal's (Section 5) settings as instances:

(1) When $\mathbf{C} = \mathbf{C}' = \mathbf{C}'' = \mathbf{1}$, then Ψ and Φ are presheaves Y, X, and the definition boils down to $Y \bullet X$.
(2) When $\mathbf{C} = \mathbf{0}$ (the initial category) and $\mathbf{C}' = \mathbf{C}'' = \mathbf{1}$, then Ψ is a presheaf X, Φ is a set z, and the definition boils down to $Lan_{\subseteq} X z$.

It is also easily checked that the identity I as defined in Section 4 is indeed the identity of $\mathbf{Prof}_?$.

We end the section by giving a formula for $\Psi \bullet \Phi$ when ! is now an *arbitrary* monad $(!, \eta_!, \mu_!)$ over **Cat** given together with a distributive law: $\boldsymbol{\lambda} : ! \circ Psh \to Psh \circ !$. This could open the way for handling variations on the shapes of operations (the third kind of variation considered in the introduction) in a profunctor setting.

Recall (Section 6) that given $\Phi : \mathbf{C} \nrightarrow \mathbf{C}'$ we write Φ' for its presentation as a functor from $\mathbf{C}'^{op}$ to $\mathbf{Set}^{\mathbf{C}}$. Then we can give abstract versions of the steps $(1)^{op}$, $(2)^{op}$ and (3) above, as follows: (we still write $?\mathbf{C} = (!\mathbf{C}^{op})^{op}$):

$(1)^{op}$ Given $\Phi : \mathbf{C} \nrightarrow \mathbf{C}'$, we can define $?\Phi : ?\mathbf{C} \nrightarrow ?\mathbf{C}'$ by the equation:

$$(?\Phi)' = \boldsymbol{\lambda} \circ !(\Phi')$$

where on the right hand side we apply ! as a functor on **Cat**.

$(2)^{op}$ The comultiplication $\delta : ?\mathbf{C} \nrightarrow ??\mathbf{C}$ is defined by the formula

$$\delta' = \eta_{Psh} \circ (\mu_!)_{\mathbf{C}^{op}}$$

(3) The composition in **Prof** of $\Phi' : \mathbf{C}'^{op} \to \mathbf{Set}^{\mathbf{C}}$ and $\Psi' : \mathbf{C}''^{op} \to \mathbf{Set}^{\mathbf{C}'}$ is defined as $\Phi'^{\#} \circ \Psi'$. Hence, given $\Phi : ?\mathbf{C} \nrightarrow \mathbf{C}'$, we obtain the following formula for $\Phi^\flat : ?\mathbf{C} \nrightarrow ?\mathbf{C}'$:

$$\Phi^{\flat'} = (\eta_{Psh} \circ (\mu_!)_{\mathbf{C}^{op}})^{\#} \circ \boldsymbol{\lambda} \circ (!\Phi') = Psh((\mu_!)_{\mathbf{C}^{op}}) \circ \boldsymbol{\lambda} \circ (!\Phi')$$

Finally, expanding the definition of Psh, we get the following formula for the composition $\Psi \bullet \Phi = \Psi \circ \Phi^\flat$ in $\mathbf{Prof}_?$:

$$(\Psi \bullet \Phi)(\gamma, C'') = \int^{\gamma'} \Psi(\gamma', C'') \times \int^{\delta} \boldsymbol{\lambda}((!\Phi')\gamma')\delta \times ?\mathbf{C}[\mu_!\delta, \gamma]$$

Following the style of [26] , one may also hide the distributive law and write

$$(\Psi \bullet \Phi)(\gamma, C'') = \int^{\gamma'} \Psi(\gamma', C'') \times \overline{\gamma'}(\Phi')\gamma$$

where $\overline{\gamma'}(\Phi')\gamma = \hat{\Phi}(\gamma, \gamma')$ can be described abstractly as follows. For each (pseudo-)!-algebra $(\mathbf{D}, \alpha : !\mathbf{D} \to \mathbf{D})$, for each category $\mathbf{C}$, and each object C of $!\mathbf{C}$, C induces a functor $\overline{C} = \mathbf{D}^{\mathbf{C}} \to \mathbf{D}$ defined as $(F \mapsto \alpha(!FC))$.

The above formula is obtained by instantiating $\mathbf{C}$, C, $\mathbf{D}$, and α as $\mathbf{C}'^{op}$, γ', $\mathbf{Set}^{!(\mathbf{C}^{op})}$, and $Psh(\mu_!) \circ \boldsymbol{\lambda}$, respectively. That this is a (pseudo)-!-algebra structure is established using the equalities satisfied by $\boldsymbol{\lambda}$.

This leads us to a generalised definition of operad.

Definition 11.1. Given a monad ! on **Cat** and a distributive law $\boldsymbol{\lambda} : \,! \circ Psh \to Psh \circ !$, a **C**-coloured !-operad is a monoid in $\mathbf{Prof}_?[\mathbf{C}, \mathbf{C}]$, that is, a functor $X : ?\mathbf{C} \times \mathbf{C}^{op} \to \mathbf{Set}$ given together with two natural transformations $e : id \to X$ and $m : X \bullet X \to X$ satisfying the monoid laws. A **1**-coloured !-operad is called a !-operad.

Hence the non-symmetric operads (resp. symmetric operads, clones) are the **1**-coloured $!_\emptyset$-operads (resp. $!_{\{\sigma\}}$-operads, $!_{\{\sigma,\delta,\epsilon\}}$-operads), and the non-symmetric coloured operads (resp. symmetric coloured operads) are the coloured $!_\emptyset$-operads (resp. coloured $!_{\{\sigma\}}$-operads). Indeed, coloured operads were defined in this way by Dolan and Baez [1] (though they do not explicate the underlying distributive law or lifting, cf. Footnote ¶).

12. Cooperads and properads

So far, we have addressed the variation on the "first axis" that goes from non-symmetric operads to clones. In this section, we address the variation on the second axis (cooperations, bioperations).

Cooperations are dual to operations. Sets of cooperations are organised into cooperads, the notion dual to that of operad. One works now in the Kleisli category $\mathbf{Prof}_!$, i.e., a cooperad is a profunctor $X : \mathbf{1} \not\rightarrow !\mathbf{1}$ with a monoid structure in $\mathbf{Prof}_!$,, and we can vary on the first axis as we did for operads.

For bioperations, the natural idea is to consider profunctors from $?\mathbf{1}$ to $!\mathbf{1}$. Such profunctors can compose provided there is a distribitutive law (another one!) $\boldsymbol{\lambda} : ?_X!_X \to !_X?_X$: given $\Phi, \Psi : ?\mathbf{1} \not\rightarrow !\mathbf{1}$, we define $\Psi \bullet \Phi$ by composing the comultiplication of ?, $?\Phi$, $\boldsymbol{\lambda}$, $!\Psi$, and the multiplication of !. The identity is the composition ot the counit of ? and of the unit of !.

This idea has been carried out in detail by Garner, in the case of $X = \{\sigma\}$ [17] . We set

$$\boldsymbol{\lambda}((m_1, \ldots, m_p), (n_1, \ldots, n_q)) \neq \emptyset \quad \text{iff} \quad m_1 + \ldots + m_p = n_1 + \ldots + n_q$$

and when the equality holds, $\boldsymbol{\lambda}((m_1, \ldots, m_p), (n_1, \ldots, n_q))$ is the set of permutations s from $m_1 + \ldots + m_p$ to $n_1 + \ldots + n_q$ such that the graph

- whose set of vertices is the disjoint union of

  $\{1,\ldots,p\}, \{1,\ldots m_1+\ldots+m_p\}, \{1,\ldots,n_1+\ldots+n_q\}$, and $\{1,\ldots,q\}$

- and whose set of edges is the union
  - of the edges between 1 and $1,\ldots m_1$, between 2 and $m_1+1,\ldots,m_2$, $\ldots$, and between p and $m_{p-1}+1,\ldots,m_p$,
  - of the edges between i and $s(i)$ $(i \leq m_1+\ldots+m_p)$, and
  - of the edges between $1,\ldots n_1$ and 1, $\ldots$, and between $n_{q-1}+1,\ldots,n_q$ and q.

is connected. This formalises the idea that when composing two sets of bi-operations, we are interested primarily in the compositions that preserve connectedness. More precisely, composition of properations has both a vertical aspect and a horizontal aspect. The distributive law takes care of the vertical aspect. Disjoint connected compositions can be placed in parallel and composed horizontally. Thus, a full categorical account requires a double category setting, that takes care of these two aspects. We refer to [17] for details.

Monoids for this composition operation provide an analogue of operads for bioperations, and for $X = \{\sigma\}$ and $\boldsymbol{\lambda}$ as described above, what we obtain (replacing **Set** by **Vect**) is exactly the notion of *properad* introduced by Vallette in his Thèse de Doctorat [27,28] .

Acknowledgements. I collected the material presented here for an invited talk at the conference Operads 2006. When preparing this talk, I benefited a lot from discussions with Marcelo Fiore and Martin Hyland. The string diagrams have been drawn with `strid`, a tool due to Samuel Mimram and Nicolas Tabareau (`http://strid.sourceforge.net/`). I also wish to thank the anonymous referee for his helpful remarks. He in particular pointed out an alternative way to address the size issue, namely to restrict attention to presheaves that are limits of small diagrams of representables.

References

1. J. Baez and J. Dolan, Higher-dimensional algebra III: n-categories and the algebra of opetopes, Advances in Mathematics 135, 145206 (1998).
2. M. Barr and C. Wells, Toposes, triples and theories, Springer-Verlag (1985).
3. J. Beck, Distributive laws, Lecture Notes in Mathematics 80, 119-140 (1969).
4. J. Bénabou, Introduction to bicategories, Lecture Notes in Mathematics 40, 1-77 (1967).
5. J. Bénabou, Les distributeurs, Université Catholique de Louvain, Institut de Mathématique Pure et Appliquée, Rapport 33 (1973).

6. F. Borceux, Handbook of categorical algebra, vol. I: basic category theory, Cambridge University Press (1994).
7. F. Borceux, Handbook of categorical algebra, vol. II: categories and structures, Cambridge University Press (1994).
8. A. Burroni, T-catégories (catégories dans un triple), Cahiers de Topologie et Géométrie Différentielle XII(3), 215-321 (1971).
9. A. Burroni, Higher Dimensional Word Problem, Theoretical Computer Science 115, 43-62 (1993).
10. E. Cheng, M. Hyland, J. Power, Pseudo-distributive laws, Electronic Notes in Theoretical Computer Science 83 (2004).
11. E. Cheng, Weak n-categories: opetopic and multitopic foundations, Journal of Pure and Applied Algebra 186, 109137 (2004).
12. B.J. Day, On closed categories of functors, Reports of the Midwest Category Seminar IV, Lecture Notes in Mathematics 137, 1-38 (1970).
13. M. Fiore, Mathematical models of computational and combinatorial structures. Invited address for Foundations of Software Science and Computation Structures (FOSSACS 2005), Lecture Notes in Computer Science 3441, 25-46 (2005).
14. M. Fiore, N.Gambino, M.Hyland, and G.Winskel, The cartesian closed bicategory of generalised species of structures, Journal of the London Mathematical Society 77 (2), 203-220 (2008).
15. M. Fiore, N. Gambino, M. Hyland, and G. Winskel, Kleisli Bicategories, in preparation.
16. N. Gambino, On the coherence conditions for pseudo-distributive laws, preprint, ArXiv:0907:1359 (2009).
17. R. Garner, Polycategories, PhD Thesis, Cambridge University (2005).
18. J.-Y. Girard, Linear logic, Theoretical Computer Science 50, 1-102 (1987).
19. G.B. Im and G.M. Kelly, A universal property of the convolution monoidal structure, Journal of Pure and Applied Algebra 43, 75-88 (1986).
20. A. Joyal, Foncteurs analytiques et espèces de structures, in Combinatoire énumérative, Lecture Notes in Mathematics 1234, 126-159 (1986).
21. M. Kelly, On the operads of J.P. May, Reprints in Theory and Applications of Categories 13, 1-13 (2005).
22. Y. Lafont, Towards an algebraic theory of Boolean circuits, Journal of Pure and Applied Algebra 184 (2-3), 257-310, Elsevier (2003).
23. T. Leinster, Higher operads, higher categories, London Mathematical Society Lecture Note Series 298, Cambridge University Press (2004).
24. S. Mac Lane, Categories for the working mathematician, Springer-Verlag (1971).
25. F. Marmolejo, Distributive laws for pseudomonads, Theory and Applications of Categories 5, 91-147 (1999).
26. M. Tanaka and J. Power, Pseudo-distributive laws and axiomatics for variable binding, Higher-Order and Symbolic Computation 19(2-3), 305 - 337 (2006).
27. B. Vallette, Dualité de Koszul des PROPS, Thèse de Doctorat, Université de Strasbourg (2003).
28. B. Vallette, A Koszul duality for props, Trans. Amer. Math. Soc. 359, 4865-4943 (2007).

Leibniz Superalgebras Graded by Finite Root Systems

Naihong Hu
Department of Mathematics, East China Normal University, Shanghai, 200062, China
E-mail: nhhu@gmath.ecnu.edu.cn

Dong Liu
Department of Mathematics, Huzhou Teachers College, Zhejiang Huzhou, 313000, China
E-mail: liudong@hutc.zj.cn, corresponding author

Linsheng Zhu
Department of Mathematics, Changshu Institute of Technology, Jiangsu Changshu, 215500, China
E-mail: lszhu@cslg.edu.cn

The structure of Lie algebras, Lie superalgebras and Leibniz algebras graded by finite root systems has been studied by several researches since 1992. In this paper, we study the structure of Leibniz superalgebras graded by finite root systems, which gives an approach to study various classes of Leibniz superalgebras.

Keywords: Δ-graded; dialgebras; Steinberg Leibniz superalgebras.

1. Introduction

In Ref. 16, J.-L. Loday introduces a non-antisymmetric version of Lie algebras, whose bracket satisfies the Leibniz relation and therefore is called *Leibniz algebra.* The Leibniz relation, combined with antisymmetry, is a variation of the Jacobi identity, hence Lie algebras are anti-symmetric Leibniz algebras. In Ref. 17, Loday also introduces an *associative* version of Leibniz algebras, called *associative dialgebras*, equipped with two binary operations, $\vdash$ and $\dashv$, which satisfy the five relations (see the axiom (Ass) in Section 2). These identities are all variations of the associative law, so associative algebras are dialgebras for which the two products coincide. The peculiar point is that the bracket $[a, b] =: a \vdash b - b \dashv a$ defines a (left) Leibniz algebra which is not antisymmetric, unless the left and right products coincide. Hence dialgebras yield a commutative diagram of categories and functors

$$\begin{array}{ccc} \textbf{Dias} & \xrightarrow{-} & \textbf{Leib} \\ \downarrow & & \downarrow \\ \textbf{As} & \xrightarrow{-} & \textbf{Lie} \end{array}$$

Recently associative super dialgebras and Leibniz superalgebras were studied in Refs. 1,10,14, etc.. The structure of Lie algebras, Lie superalgebras and Leibniz algebras graded by finite root systems was studied in several papers (Refs. 2–9,13, etc.). In this paper we determine the structure of Leibniz superalgebras graded by the root systems of the basic classical simple Lie superalgebras.

The paper is organized as follows. In Section 2, we recall some notions of Leibniz superalgebras and superdialgebras. In Section 3, we give the definition and some properties of Leibniz superalgebras graded by finite root systems. In Section 4 and Section 5, we mainly determine the structure of Leibniz superalgebras graded by the root systems of types $A(m,n)$ and $C(n)$, $D(m,n)$, $D(2,1;\alpha)$, $F(4)$, $G(2)$. Throughout this paper, K denotes a field of characteristic 0.

2. Associative super dialgebras and leibniz superalgebras

We recall some notions of associative super dialgebras and Leibniz superalgebras and their (co)homology as defined in Refs. 10 and 14.

2.1. *Associative super dialgebras*

Definition 2.1.[17] An *associative dialgebra* D over K is a K-vector space with two operations $\dashv, \vdash$: $D \otimes D \to D$, called left and right products, satisfying the following five axioms:

$$(Ass) \qquad \begin{cases} a \dashv (b \dashv c) = (a \dashv b) \dashv c = a \dashv (b \vdash c), \\ (a \vdash b) \dashv c = a \vdash (b \dashv c), \\ (a \vdash b) \vdash c = a \vdash (b \vdash c) = (a \dashv b) \vdash c. \end{cases}$$

Obviously an associative dialgebra is an associative algebra if and only if $a \dashv b = a \vdash b = ab$.

An *associative super dialgebra* over K is a $\mathbb{Z}_2$-graded K-vector space D with two operations $\dashv, \vdash$: $D \otimes D \to D$, satisfying the axiom (Ass) and

$$D_\sigma \dashv D_{\sigma'}, D_\sigma \vdash D_{\sigma'} \subset D_{\sigma+\sigma'}, \quad \forall \sigma, \sigma' \in \mathbb{Z}_2.$$

An associative super dialgebra is called supercommutative if $a \vdash b = (-1)^{|a||b|} b \dashv a$ for all $a, b \in D$. A associative super dialgebra is called unital if it is given a specified bar-unit: an element $1 \in D$ which is a unit for the left and right products only on the bar-side, that is $1 \vdash a = a = a \dashv 1$, for any $a \in D$. Denote by **SDias, SAs** the categories of associative super dialgebras or associative superalgebras over K respectively. Then the category **SAs** is a full subcategory of **SDias**.

Examples. 1. Obviously, an associative super dialgebra is an associative superalgebra if $a \dashv b = a \vdash b = ab$. An associative dialgebra is a trivial associative super dialgebra.

2. *Super differential dialgebra.* Let $(A = A_{\bar{0}} \oplus A_{\bar{1}}, d)$ be a differential associative super algebra($|d| = \bar{0}$). So by hypothesis, $d(ab) = (da)b + adb$ and $d^2 = 0$. Define left and right products on A by the formulas $x \dashv y = xdy$ and $x \vdash y = (dx)y$. Then A equipped with these two products is an associative super dialgebra.

3. *Tensor product.* If D and D' are two associative super dialgebras, then the tensor product $D \otimes D'$ is also a super dialgebra with

$$(a \otimes a') \star (b \otimes b') = (-1)^{|a'||b|} (a \star b) \otimes (a' \star b') \tag{2.1}$$

for $\star = \dashv, \vdash$.

For instance, $M_{m+n}(D) := M_{m+n}(K) \otimes D$ is an associative super dialgebra if D is an associative super dialgebra and $M_{m+n}(K)$ is the superalgebra of all $(m+n) \times (m+n)$-matrices over K.

4. The *free associative super dialgebra* (see Ref. 14 in details) on a $\mathbb{Z}_2$-graded vector space V is the dialgebra $Dias(V) = T(V) \otimes V \otimes T(V)$ equipped with the induced $\mathbb{Z}_2$-gradation.

2.2. *Leibniz superalgebra*

Definition 2.2.[10] A Leibniz superalgebra is a $\mathbb{Z}_2$-graded vector space $L = L_{\bar{0}} \oplus L_{\bar{1}}$ over a field K equipped with a K-bilinear map $[-,-] : L \times L \to L$ satisfying

$$[L_\sigma, L_{\sigma'}] \subset L_{\sigma+\sigma'}, \quad \forall \sigma, \sigma' \in \mathbb{Z}_2$$

and the Leibniz superidentity

$$[[a,b],c] = [a,[b,c]] - (-1)^{|a||b|}[b,[a,c]], \quad \forall\, a,\, b,\, c \in L. \tag{2.2}$$

Obviously, $L_{\bar{0}}$ is a Leibniz algebra. Moreover any Lie superalgebra is a Leibniz superalgebra and any Leibniz algebra is a trivial Leibniz superalgebra. A Leibniz superalgebra is a Lie superalgebra if and only if

$$[a,b] + (-1)^{|a||b|}[b,a] = 0, \quad \forall\, a,b \in L. \tag{2.3}$$

Examples. 1. Let $\mathfrak{g}$ be a Lie superalgebra, D be a unital supercommutative associative super dialgebra, then $\mathfrak{g} \otimes D$ with Leibniz bracket $[x \otimes a, y \otimes b] = (-1)^{|a||y|}[x,y] \otimes (a \vdash b)$ is a Leibniz superalgebra. Let $\mathfrak{g}$ be a basic classical simple Lie superalgebra which is not of type $A(n,n)$, $n \geq 1$, then $\tilde{\mathfrak{g}} = \mathfrak{g} \otimes D \oplus \Omega_D^1$ with the bracket

$$[x\otimes a, y\otimes b] = (-1)^{|a||y|}\big([x,y]\otimes(a \vdash b) + (x,y)b \dashv da\big), \quad \forall a,b \in D, x,y \in \mathfrak{g}, \tag{2.4}$$

$$[\Omega_D^1,\ \tilde{\mathfrak{g}}] = 0 \tag{2.5}$$

is also a Leibniz superalgebra, where $(-,-)$ is an even invariant bilinear form of $\mathfrak{g}$, Ω_D^1 is defined in Ref. 12 (also see Ref. 14). In fact it is the universal central extension of $\mathfrak{g} \otimes D$ (see Ref. 14 in details).

2. Tensor product. Let $\mathfrak{g}$ be a Lie superalgebra, then the bracket

$$[x \otimes y, a \otimes b] = [[x,y],a] \otimes b + (-1)^{|a||b|} a \otimes [[x,y],b] \tag{2.6}$$

defines a Leibniz superalgebra structure on the vector space $\mathfrak{g} \otimes \mathfrak{g}$ (see Ref. 11 for that in Leibniz algebras case).

3. The *general linear Leibniz superalgebra* $\mathfrak{gl}(m,n,D)$ is generated by all $n \times n$ matrices with coefficients from a dialgebra D, and $m,n \geq 0, n+m \geq 2$ with the bracket

$$[E_{ij}(a), E_{kl}(b)] = \delta_{jk}E_{il}(a \vdash b) - (-1)^{\tau_{ij}\tau_{kl}}\delta_{il}E_{kj}(b \dashv a), \tag{2.7}$$

for all $a,b \in D$.

Clearly, $\mathfrak{gl}(m,n,D)$ is a Leibniz superalgebra. If D is an associative superalgebra, then $\mathfrak{gl}(m,n,D)$ becomes a Lie superalgebra.

By definition, the *special linear Leibniz superalgebra* with coefficients in D is

$$\mathfrak{sl}\,(m,n,D) := [\,\mathfrak{gl}(m,n,D), \mathfrak{gl}(m,n,D)\,].$$

Notice that if $n \neq m$ the Leibniz superalgebra $\mathfrak{sl}\,(m,n,D)$ is simple.

The special linear Leibniz superalgebra $\mathfrak{sl}\,(m,n,D)$ has generators $E_{ij}(a), 1 \leq i \neq j \leq m+n, a \in D$, which satisfy the following relations:

$$[E_{ij}(a), E_{kl}(b)] = 0 \text{ if } i \neq l \text{ and } j \neq k;$$
$$[E_{ij}(a), E_{kl}(b)] = E_{il}(a \vdash b) \text{ if } i \neq l \text{ and } j = k;$$
$$[E_{ij}(a), E_{kl}(b)] = -(-1)^{\tau_{ij}\tau_{kl}} E_{kj}(b \dashv a) \text{ if } i = l \text{ and } j \neq k,$$

4. The *Steinberg Leibniz superalgebra* $\mathfrak{stl}(m, n, D)$ (Ref. 12) is a Leibniz superalgebra generated by symbols $u_{ij}(a)$, $1 \leq i \neq j \leq n$, $a \in D$, subject to the relations

$$\begin{aligned}
&v_{ij}(k_1 a + k_2 b) = k_1 v_{ij}(a) + k_2 v_{ij}(b), \quad \text{for} \quad a, b \in D, \ k_1, k_2 \in K; \\
&[v_{ij}(a), v_{kl}(b)] = 0, \quad \text{if} \quad i \neq l \quad \text{and} \quad j \neq k; \\
&[v_{ij}(a), v_{kl}(b)] = v_{il}(a \vdash b) \quad \text{if} \quad i \neq l \text{ and } \quad j = k; \\
&[v_{ij}(a), v_{kl}(b)] = -(-1)^{\tau_{ij}\tau_{kl}} v_{kj}(b \dashv a) \quad \text{if} \quad i = l \text{ and } j \neq k,
\end{aligned}$$

where $1 \leq i \neq j \leq m + n$, $a \in D$. It is clear that the last two relations make sense only if $m + n \geq 3$. See Refs. 12 and 15 for more details about the Steinberg Leibniz superalgebra.

We also denote by **SLeib, SLie** the categories of Leibniz superalgebras and Lie superalgebras over K respectively.

For any associative super dialgebra D, if we define

$$[x, y] = x \vdash y - (-1)^{|x||y|} y \dashv x, \tag{2.8}$$

then D equipped with this bracket is a Leibniz superalgebra. We denoted it by D_L. The canonical morphism $D \to D_L$ induces a functor $(-)$: **SDias→SLeib**.

Remark. For an associative super dialgebra D, if we define

$$[x, y] = x \dashv y - (-1)^{|x||y|} y \vdash x, \tag{2.9}$$

then $(D, [,])$ is a right Leibniz superalgebra in the sense of Ref. 18.

For a Leibniz superalgebra L, let L_{LS} be the quotient of L by the ideal generated by elements $[x, y] + (-1)^{|x||y|}[y, x]$, for all $x, y \in L$. It is clear that L_{LS} is a Lie superalgebra. The canonical epimorphism : $L \to L_{LS}$ is universal among the maps from L to Lie superalgebras. In other words the functor $(-)_{LS}$: **SLeib→SLie** is left adjoint to *inc* : **SLie→SLeib**.

Moreover we have the following commutative diagram of categories and functors

$$\begin{array}{ccc}
\textbf{SDias} & \xrightarrow{-} & \textbf{SLeib} \\
\downarrow & & \downarrow \\
\textbf{SAs} & \xrightarrow{-} & \textbf{SLie}
\end{array}$$

As in the Leibniz algebra case, the *universal enveloping super associative dialgebra* (Ref. 14) of a Leibniz superalgebra L is

$$Ud(L) := (T(L) \otimes L \otimes T(L))/\{[x, y] - x \vdash y + (-1)^{|x||y|} y \dashv x | x, y \in L\}.$$

Proposition 2.1.[14] *The functor* Ud : **SLeib** $\to$ **SDias** *is left adjoint to the functor* $-$: **SDias** $\to$ **SLeib**. ∎

Let L be a Leibniz superalgebra. We call a $\mathbb{Z}_2$-graded space $M = M_{\bar{0}} \oplus M_{\bar{1}}$ a left L-module if there is a bilinear map:

$$[-,-] : L \times M \to M$$

satisfying the following axiom

$$[[x,y],m] = [x,[y,m]] - (-1)^{|x||y|}[y,[x,m]],$$

for any $m \in M$ and $x, y \in L$. In this case we also say $\varphi : L \to \mathrm{End}_K M$, $\varphi(x)(m) = [x,m]$ is a left representation of L.

Clearly, if L is a Lie superalgebra, then M is just the left L-module in the Lie superalgebra case.

Suppose that L is a Leibniz superalgebra over K. For any $z \in L$, we define $\mathrm{ad}\, z \in \mathrm{End}_k L$ by

$$\mathrm{ad}\, z(x) = [z,x], \quad \forall x \in L. \tag{2.10}$$

It follows from (2.2) that

$$\mathrm{ad}\, z([x,y]) = [\mathrm{ad}\, z(x), y] + (-1)^{|z||x|}[x, \mathrm{ad}\, z(y)] \tag{2.11}$$

for all $x, y \in L$. This says that $\mathrm{ad}\, z$ is a super derivation of degree $|z|$ of L. We also call it the inner derivation of L.

For a Lie superalgebra L, $HL^1(L,M) = H^1(L,M)$, where $H^1(L,M)$ (resp. $HL^1(L,M)$) denotes the first cohomology groups of L in the category of Lie (resp. Leibniz) superalgebras (see Ref. 16).

2.3. *Leibniz algebras graded by finite root systems*

Now we recall some notions of Leibniz algebras graded by finite root systems defined as in Ref. 13.

Definition 2.3.[13] A Leibniz algebra L over a field K of characteristic 0 is graded by the (reduced) root system Δ or is Δ-graded if

(1) L contains as a subalgebra a finite-dimensional simple Lie algebra $\dot{\mathfrak{g}} = H \oplus \bigoplus_{\alpha\in\Delta} \dot{\mathfrak{g}}_\alpha$ whose root system is Δ relative to a split Cartan subalgebra $H = \dot{\mathfrak{g}}_0$;

(2) $L = \bigoplus_{\alpha\in\Delta\cup\{0\}} L_\alpha$, where $L_\alpha = \{x \in L \mid \mathrm{ad}\, h(x) = [h,x] = \alpha(h)x, \forall h \in H\}$ for $\alpha \in \Delta \cup \{0\}$; and

(3) $L_0 = \sum_{\alpha\in\Delta}[L_\alpha, L_{-\alpha}]$.

Remarks.

1. The conditions for being a Δ-graded Leibniz algebra imply that L is a direct sum of finite-dimensional irreducible Leibniz representations of $\dot{\mathfrak{g}}$ whose highest weights are roots, hence are either the highest long root or high short root or are 0.

2. If L is Δ-graded then L is perfect (i.e. $[L, L] = L$). Indeed the result follows from $L_\alpha = [L_\alpha, H]$ for all $\alpha \in \Delta$ and (3) above.

In Ref. 13, the structure of Leibniz algebras graded by the root systems of types A, D, E was determined by using the methods in Ref. 9. In fact we have

Theorem 2.1.[13] *Let L be a Leibniz algebra over K graded by the root system Δ of type $X_l\,(l \geq 2)\,(X_l = A_l, D_l, E_l)$.*

(1) If $X_l = A_l, l \geq 3$, then there exists a unital associative K-dialgebra R such that L is centrally isogenous with $\mathfrak{sl}\,(l+1, R)$, .

(2) If $X_l = A_l, l = 2$, then there exists a unital alternative K-dialgebra R such that L is centrally isogenous with the Steinberg Leibniz algebra $\mathfrak{stl}\,(l+1, R)$, where $\mathfrak{stl}\,(l+1, R)$ defined in Ref. 12 (also see the example 4 in Section 2.2).

(3) If $X_l = D_l\,(l \geq 4), E_l\,(l = 6, 7, 8)$, then there exists a unital associative commutative K-dialgebra A such that L is centrally isogenous with the Leibniz algebra $\dot{\mathfrak{g}} \otimes R$.

Remark. Two perfect Lie algebras L_1 and L_2 are called *centrally isogenous* if they have the same universal central extension (up to isomorphism).

3. Leibniz superalgebras graded by finite root systems

In this section, we shall study the structure of Leibniz superalgebras graded by finite root systems.

Motivated the definitions of Lie superalgebras and Leibniz algebras graded by finite root systems defined in Refs. 4–6 and 13, we give the following definition.

Definition 3.1. A Leibniz superalgebra L over a field K of characteristic 0 is graded by the (reduced) root system Δ or is Δ-graded if

(1) L contains as a subsuperalgebra a finite-dimensional split simple basic Lie superalgebra $\mathfrak{g} = H \oplus \bigoplus_{\alpha \in \Delta} \mathfrak{g}_\alpha$ whose root system Δ is relative to a split Cartan subalgebra $H = \mathfrak{g}_0$;

(2) $L = \bigoplus_{\alpha \in \Delta \cup \{0\}} L_\alpha$, where $L_\alpha = \{x \in L \mid \operatorname{ad} h(x) = [h, x] = \alpha(h)x, \forall h \in H\}$ for $\alpha \in \Delta \cup \{0\}$; and

(3) $L_0 = \sum_{\alpha \in \Delta} [L_\alpha, L_{-\alpha}]$.

Remarks.

1. If L is Δ-graded then L is perfect. Indeed the result follows from $L_\alpha = [H, L_\alpha]$ for all $\alpha \in \Delta$ and (3) above.

2. The Steinberg Leibniz superalgebra $\mathfrak{stl}(m, n, D)\,(m \neq n)$ is graded by the root system of type $A_{m-1,n-1}$.

We would like to view L as a left $\mathfrak{g}$-module in order to determine the structure of L. The following result plays a key role in examining Δ-graded Leibniz superalgebras. It follows from the Lemma 2.2 in Ref. 6.

Lemma 3.1. *Let L be a Δ-graded Leibniz superalgebra, and let $\mathfrak{g}$ be its associated split simple basic Lie superalgebra. Then L is locally finite as a module for $\mathfrak{g}$.* ∎

As a consequence, each element of a Δ-graded Leibniz superalgebra L, in particular each weight vector of L relative to the Cartan subalgebra H of $\mathfrak{g}$, generates a finite-dimensional $\mathfrak{g}$-module. Such a finite-dimensional module has a $\mathfrak{g}$-composition series whose irreducible factors have weight which are roots of $\mathfrak{g}$ or 0. Next we determine which finite-dimensional irreducible $\mathfrak{g}$-modules have nonzero weights that are roots of $\mathfrak{g}$.

Throughout this paper we will identify the split simple Lie superalgebras $\mathfrak{g}$ of type $A(m, n)$, $m > n \geq 0$, with the special linear Lie superalgebra $\mathfrak{sl}(m+1, n+1)$. For simplicity of notation, set $p = m+1, q = n+1$, so $\mathfrak{g} = \mathfrak{sl}(p, q)$, $p > q \geq 1$.

Proposition 3.1.[4,5] *Let $\mathfrak{g}$ be a split simple Lie superalgebra of type $A(m, n)$, with $m \geq n \geq 0, m+n \geq 1$, or $C(n)$, $D(m, n)$, $D(2, 1; \alpha)$ $(\alpha \notin \{0, -1\})$, $F(4)$, $G(2)$. The only finite-dimensional irreducible left $\mathfrak{g}$-modules whose weights relative the Cartan subalgebra of diagonal matrices (modulo the center if necessary) are either roots or 0 are exactly the adjoint and the trivial modules (possibly with the parity changed).*

Proposition 3.2.[4,5] *Let $\mathfrak{g}$ be a split simple Lie superalgebra of type $A(m, n)$ $(m > n \geq 0)$ or $C(n)$, $D(m, n)$, $D(2, 1; \alpha)$ $(\alpha \notin \{0, -1\})$, $F(4)$, $G(2)$, with split Cartan subalgebra H. Assume V is a locally finite left $\mathfrak{g}$-module satisfying*

(i) H acts semisimply on V;

(ii) any composition factor of any finite-dimensional right submodule of V is isomorphic to the adjoint representation $\mathfrak{g}$ or to a trivial representation.

Then V is completely reducible. ∎

Proposition 3.3.[4] *Let $\mathfrak{g}$ be a split simple Lie superalgebra of type $A(m,n)$, with $m > n \geq 0$. Then $Hom_{\mathfrak{g}}(\mathfrak{g} \otimes \mathfrak{g}, \mathfrak{g})$ is two-dimensional and spanned by the Lie (super) bracket and by the map given by $(x, y) \mapsto x * y = xy + (-1)^{|x||y|}yx - \frac{2}{m-n} str(xy)I$, for any $x, y \in \mathfrak{g} = \mathfrak{sl}(m+1, n+1)$.* ∎

4. The structure of the $A(m,n)$-graded Leibniz superalgebras $(m > n)$

The results of Section 2 show that any $A(m,n)$-graded Leibniz superalgebra L with $m > n \geq 0$ is the direct sum of adjoint and trivial modules (possibly with a change of parity) for the grading subalgebra $\mathfrak{g}$. After collecting isomorphic summands, we may suppose that there are superspaces $A = A_{\bar{0}} \oplus A_{\bar{1}}$ and $D = D_{\bar{0}} \oplus D_{\bar{1}}$ so that $L = (\mathfrak{g} \otimes A) \oplus D$, and a distinguished element $1 \in A_{\bar{0}}$ which allows us to identify the grading subalgebras $\mathfrak{g}$ with $\mathfrak{g} \otimes 1$. Observe first that D is a subsuperalgebra of L since it is the (super) centralizer of $\mathfrak{g}$.

To determine the multiplication on L, we may apply the same type of arguments as in Ref. 5. Indeed, fix homogeneous basis elements $\{a_i\}_{i \in I}$ of A and choose a_i, a_j, a_k with $i, j, k \in I$, we see that the projection of the product $[\mathfrak{g} \otimes a_i, \mathfrak{g} \otimes a_j]$ onto $\mathfrak{g} \otimes a_k$ determines an element of $\mathrm{Hom}_{\mathfrak{g}}(\mathfrak{g} \otimes \mathfrak{g}, \mathfrak{g})$, which is spanned by the Leibniz supercommutator. Then there exist scalars $\xi_{i,j}^k$ and $\theta_{i,j}^k$ so that

$$[x \otimes a_i, y \otimes a_j]|_{\mathfrak{g} \otimes A} = (-1)^{|a_i||y|} \left([x, y] \otimes (\sum_{k \in I} \xi_{i,j}^k a_k) + x * y \otimes (\sum_{k \in I} \theta_{i,j}^k a_k) \right)$$

Define $\circ : A \times A \to A$ by $a_i \circ a_j = 2 \sum_{k \in I} \xi_{i,j}^k a_k$, and $[,] : A \times A \to A$ by $[a_i, a_j] = 2 \sum_{k \in I} \theta_{i,j}^k a_k$ and extending them bilinearly, we obtain two products "$\circ$" and "$[\ ,\ ]$" on A.

Taking into account that $\mathrm{Hom}_{\mathfrak{g}}(\mathfrak{g} \otimes \mathfrak{g}, K)$ is spanned by the supertrace, we see that there exists a bilinear form $\langle , \rangle : A \times A \to D$ and an even bilinear map $D \times A \to A : (d, a) \to da$ with $d1 = 0$ such that the multiplication in L is given by

$$[f \otimes a, g \otimes b] = (-1)^{|a||g|} \left([f, g] \otimes \frac{1}{2} a \circ b + f * g \otimes \frac{1}{2} [a, b] + \mathrm{str}(fg) \langle a, b \rangle \right), \tag{4.1}$$

$$[d, f \otimes a] = (-1)^{|d||f|} f \otimes da, \tag{4.2}$$

for homogeneous elements $f, g \in \mathfrak{g},\ a, b \in A, d \in D$. Additionally, $1 \circ a = 2a$ and $[1, a] = 0$ for all $a \in A$.

There are two unital multiplications $\dashv$ and $\vdash$ on A such that

$$a \circ b = a \vdash b + (-1)^{|a||b|} b \dashv a, \tag{4.3}$$

$$[a, b] = a \vdash b - (-1)^{|a||b|} b \dashv a, \tag{4.4}$$

for any homogeneous elements $a, b \in A$. Moreover by setting $a = 1$, we have $1 \vdash b = b \dashv 1 = b$, so the dialgebra A is unital.

Now the Jacobi superidentity $[[z_1, z_2], z_3] = [z_1, [z_2, z_3]] - (-1)^{|z_1||z_2|} [z_2, [z_1, z_3]]$, when specialized with homogeneous elements $d_1, d_2 \in D$ and $f \otimes a \in \mathfrak{g} \otimes A$, shows that $\phi : D \to \mathrm{End}_K(A) : \phi(d)(a) = da$, is a left representation of the Leibniz superalgebra D. When it is specialized with homogeneous elements $d \in D$ and $f \otimes a, g \otimes b \in \mathfrak{g} \otimes A$, we obtain

$$[d, [f \otimes a, g \otimes b]] = [[d, f \otimes a], g \otimes b] + (-1)^{|d|(|f|+|a|)} [f \otimes a, [d, g \otimes b]],$$

and using (4.1) and (4.2), we see that this is same as:

$$\begin{aligned}
&(-1)^{|d|(|f|+|g|)}(-1)^{|a||g|} \left([f, g] \otimes \frac{1}{2} d(a \circ b) + f * g \otimes \frac{1}{2} d([a, b]) + \mathrm{str}(fg)[d, \langle a, b \rangle]\right) \\
&= (-1)^{|d||f|}(-1)^{|d|+|a||g|} \left([f, g] \otimes \frac{1}{2} da \circ b) + f * g \otimes \frac{1}{2} [da, b] + \mathrm{str}(fg)\langle da, b \rangle\right) \\
&+ (-1)^{|d|(|f|+|g|+|a|)}(-1)^{|a||g|} \left([f, g] \otimes \frac{1}{2} a \circ db + f * g \otimes \frac{1}{2} [a, db] + \mathrm{str}(fg)\langle a, db \rangle\right).
\end{aligned}$$

When $f = E_{1,2}$ and $g = E_{2,1}$, the elements $[f, g]$ and $f * g$ are linearly independent and $\mathrm{str}(fg) = 1$. Hence we have

(i) $d(a \circ b) = (da) \circ b + (-1)^{|d||a|} a \circ (db)$,
(ii) $d([a, b]) = [da, b] + (-1)^{|d||a|} [a, db]$,
(iii) $[d, \langle a, b \rangle] = \langle da, b \rangle + (-1)^{|d||a|} \langle a, db \rangle$,
for any homogeneous $d \in D$ and $a, b \in A$.

Items (i) and (ii) can be combined to give that

$$\phi \text{ is a left representation as superderivations} : \phi : D \to \mathrm{Der}_K(A). \tag{4.5}$$

While (iii) says that

$$\langle\ ,\ \rangle \text{ is invariant under the action of } D. \tag{4.6}$$

For $f \otimes a, g \otimes b, h \otimes c \in \mathfrak{g} \otimes A$, the Jacobi superidentity is equivalent to the following two relations ($\mathfrak{g} \otimes A$ and D components):

$$\begin{aligned}
(\star) \quad &\left(\mathrm{str}([f, g]h)\langle \frac{1}{2} a \circ b, c \rangle + \mathrm{str}((f * g)h)\langle \frac{1}{2}[a, b], c \rangle\right) \\
&- \left(\mathrm{str}(f[g, h])\langle a, \frac{1}{2} b \circ c \rangle + \mathrm{str}(f(g * h))\langle a, \frac{1}{2}[b, c] \rangle\right)
\end{aligned}$$

$$+ (-1)^{(|f||g|+|a||b|)}\Big(\mathrm{str}(g[f,h])\langle b, \frac{1}{2}a\circ c\rangle + \mathrm{str}(g(f*h))\langle b, \frac{1}{2}[a,c]\rangle\Big)$$
$$= 0.$$

and

$$(\star\star)\quad \Big([[f,g],h]\otimes\frac{1}{4}(a\circ b)\circ c + [f,g]*h\otimes\frac{1}{4}[a\circ b,c]$$
$$+[f*g,h]\otimes\frac{1}{4}[a,b]\circ c + (f*g)*h\otimes\frac{1}{4}[[a,b],c] + \mathrm{str}(fg)h\otimes\langle a,b\rangle c\Big)$$
$$-\Big([f,[g,h]]\otimes\frac{1}{4}a\circ(b\circ c) + f*[g,h]\otimes\frac{1}{4}[a,b\circ c]$$
$$+[f,g*h]\otimes\frac{1}{4}a\circ[b,c] + f*(g*h)\otimes\frac{1}{4}[a,[b,c]]$$
$$+(-1)^{|a|(|g|+|h|)}\mathrm{str}(gh)[f\otimes a,\langle b,c\rangle]\Big)$$
$$+ (-1)^{|f||g|+|a||b|}\Big([g,[f,h]]\otimes\frac{1}{4}b\circ(a\circ c) + g*[f,h]\otimes\frac{1}{4}[b,a\circ c]$$
$$+[g,f*h]\otimes\frac{1}{4}b\circ[a,c] + g*(f*h)\otimes\frac{1}{4}[b,[a,c]]$$
$$+(-1)^{|b|(|f|+|h|)}\mathrm{str}(fh)[g\otimes b,\langle a,c\rangle]\Big)$$
$$= 0.$$

The formula $(\star)$ can be written as

$$\mathrm{str}(fgh)\big(\langle a\vdash b,c\rangle - \langle a,b\vdash c\rangle - (-1)^{|a|(|b|+|c|)}\langle b,c\dashv a\rangle\big)$$
$$- (-1)^{|g||h|+|a||b|}\mathrm{str}(fhg)\big(\langle b\dashv a,c\rangle - \langle b,a\vdash c\rangle - (-1)^{|b|(|a|+|c|)}\langle a,c\dashv b\rangle\big)$$
$$= 0.$$

Then we have

$$\langle a\vdash b,c\rangle = \langle a,b\vdash c\rangle + (-1)^{|a|(|b|+|c|)}\langle b,c\dashv a\rangle, \tag{4.7}$$

$$\langle a\vdash b,c\rangle = \langle a\dashv b,c\rangle. \tag{4.8}$$

The formula $(\star\star)$ can be written as:

$$(\star\star\star)\quad fgh\otimes\big((a\vdash b)\vdash c - a\vdash(b\vdash c)\big)$$
$$- (-1)^{|f||g|+|a||b|}gfh\otimes\big((b\dashv a)\vdash c - b\vdash(a\vdash c)\big)$$
$$+ (-1)^{(|f|+|g|)|h|+(|a|+|b|)|c|}hfg\otimes\big((c\dashv a)\dashv b - c\dashv(a\vdash b)\big)$$
$$+ (-1)^{(|h|+|g|)|f|+(|c|+|b|)|a|}ghf\otimes\big((b\vdash c)\dashv a - b\vdash(c\dashv a)\big)$$
$$+ (-1)^{|h||g|+|c||b|}fhg$$
$$\otimes\big((a\vdash c)\vdash b - a\vdash(c\dashv b)\big)$$

$$
\begin{aligned}
&- (-1)^{(|h||g|+|f||g|+|f||h|+|c||b|+|a||b|+|a||c|}hgf \\
&\quad \otimes\big((c \dashv b) \dashv a - c \dashv (b \dashv a)\big) \\
&- \mathrm{str}(fg)\left(\frac{1}{m-n}h \otimes [[a,b],c] - h \otimes \langle a,b\rangle c\right) \\
&+ \mathrm{str}(gh)\left(\frac{1}{m-n}f \otimes [a,[b,c]] - (-1)^{|a|(|g|+|h|)}[f \otimes a, \langle b,c\rangle]\right) \\
&- (-1)^{|f||g|+|a||b|}\mathrm{str}(fh) \\
&\quad \left(\frac{1}{m-n}g \otimes [b,[a,c]] - (-1)^{|b|(|f|+|h|)}[g \otimes b, \langle a,c\rangle]\right) \\
&- \frac{1}{m-n}\mathrm{str}(fgh)I \otimes \Big([a \vdash b, c] - [a, b \vdash c] - (-1)^{|a|(|b|+|c|)}[b, c \dashv a]\Big) \\
&+ (-1)^{|f||g|+|a||b|}\frac{1}{m-n}\mathrm{str}(gfh)I \\
&\quad \otimes\Big([b \dashv a, c] - (-1)^{|b|(|a|+|c|)}[a, c \dashv b] - [b, a \vdash c]\Big) \\
&= 0.
\end{aligned}
$$

Set $f = E_{12}, g = E_{23}, h = E_{31}$ in $(\star\star\star)$, if $m \geq 2$ then by $|f| = |g| = |h| = \bar{0}$ and the independent of $E_{11}, E_{22}, E_{33}, I$, we have:

$$(a \vdash b) \vdash c = a \vdash (b \vdash c), \tag{4.9}$$

$$(b \vdash c) \dashv a = b \vdash (c \dashv a), \tag{4.10}$$

$$(c \dashv a) \dashv b = c \dashv (a \vdash b). \tag{4.11}$$

Similarly by setting $f = E_{31}, g = E_{23}, h = E_{12}$, we can obtain

$$(c \dashv b) \dashv a = c \dashv (b \dashv a), \tag{4.12}$$

$$(b \dashv a) \vdash c = b \vdash (a \vdash c), \tag{4.13}$$

so A is an associative super dialgebra.

If $m = 1$, then $|f| = \bar{0}$ and $|g| = |h| = \bar{0}$. The expression in $(\star\star\star)$ is a linear combination of E_{11}, E_{22}, E_{33} with coefficients in A. By direct calculation we also obtain that A is associative.

Then $(\star\star\star)$ becomes

$$
\begin{aligned}
&- \mathrm{str}(fg)\left(\frac{1}{m-n}h \otimes [[a,b],c] - h \otimes \langle a,b\rangle c\right) \\
&+ \mathrm{str}(gh)\left(\frac{1}{m-n}f \otimes [a,[b,c]] - (-1)^{|a|(|g|+|h|)}[f \otimes a, \langle b,c\rangle]\right)
\end{aligned}
$$

$$
\begin{aligned}
&- (-1)^{|f||g|+|a||b|}\mathrm{str}(fh) \\
&\quad \left(\frac{1}{m-n} g \otimes [b, [a, c]] - (-1)^{|b|(|f|+|h|)}[g \otimes b, \langle a, c\rangle]\right) \\
&= 0.
\end{aligned}
$$

Then

$$\langle a, b\rangle c = \frac{1}{m-n}[[a, b], c] \tag{4.14}$$

and

$$[f \otimes a, \langle b, c\rangle] = \frac{1}{m-n} f \otimes [a, [b, c]] \tag{4.15}$$

since $|g| + |h| = \bar{0}$ if $\mathrm{str}(gh) \neq 0$.

In this way, we have arrived at our main Theorem. The last sentence in it is a consequence of condition (3) in Definition 3.1.

Theorem 4.1. *Assume* $L = \mathfrak{g} \otimes A \oplus D$ *is a superalgebra over a field* K *of characteristic* 0 *where* $\mathfrak{g} = \mathfrak{sl}(m+1, n+1), m > n \geq 0$, A *is a unital associative super dialgebra, and* D *is a Leibniz superalgebra, and with multiplication as in (4.1) and (4.2). Then* L *is a Leibniz superalgebra if and only if*

(1) A *is an associative super dialgebra,*

(2) D *is a Leibniz subsuperalgebra of* L *and* $\phi : D \to \mathrm{Der}_K(A)(\phi(d)a = da)$ *is a left representation of* D *as superderivations on the dialgebra* A,

(3) $[d, \langle a, b\rangle] = \langle da, b\rangle + (-1)^{|d||a|}\langle a, db\rangle$,

(4) (4.7), (4.8) and (4.14), (4.15) hold,

for any homogeneous elements $d \in D$ *and* $a, b, c \in A$.

Moreover, the $A(m, n)$*-graded Leibniz superalgebras (for* $m > n \geq 0$*) are exactly these superalgebras with the added constraint that*

$$D = \langle A, A\rangle.$$

Remark. Let A be any unital associative super dialgebra. Then $\mathrm{ad}_{[A,A]}$ is a subsuperalgebra of $\mathrm{Der}_K(A)$ (it is a Lie superalgebra). Consider the Leibniz superalgebra

$$\mathfrak{L}(A) := (\mathfrak{g} \otimes A) \oplus \mathrm{ad}_{[A,A]},$$

with $\mathfrak{g} = \mathfrak{sl}(m+1, n+1)(m > n \geq 0)$, with multiplication given by (4.1) and (4.2) in place of D and with $\langle a, b\rangle = \frac{1}{m-n}\mathrm{ad}_{[a,b]}$ for any $a, b \in A$. Then Theorem 4.1 shows that $\mathfrak{L}(A)$ is an $A(m, n)$-graded Leibniz superalgebra. Moreover for any $A(m, n)$-graded Leibniz superalgebra L with coordinate associative super dialgebra A, Theorem 4.1 implies that $L/Z(L) \cong \mathfrak{L}(A)$. Thus L is a cover of $\mathfrak{L}(A)$.

Corollary 4.1. *The $A(m,n)$-graded Leibniz superalgebras (for $m > n \geq 0$) are precisely the Leibniz superalgebras which are centrally isogeneous to the Leibniz superalgebra $\mathfrak{sl}(m+1, n+1, A)$ for a unital associative super dialgebra A.*

Remarks. 1. The situation when $m = n$ in $A(m,n)$ is much more involved than the case of $m \neq n$, due to the fact that the complete reducibility in Proposition 3.4 no longer valid in this case. However, using the similar consideration as above and that in Ref. 7, we can also obtain the following result.

Theorem 4.2. *Let L be a Leibniz superalgebra graded by $\mathfrak{g}$, which is a split simple classical Lie superalgebra of type $A(n,n)(n > 1)$. Then there exists a unital associative super dialgebra D such that it is centrally isogenous to*

$$\mathfrak{sl}(n+1, D) = [\mathfrak{gl}(n+1) \otimes D, \mathfrak{gl}(n+1) \otimes D].$$

2. For a unital associative dialgebra A, the universal central extension of the Leibniz superalgebra $\mathfrak{sl}(m, n, A)$ with $m \neq n$ and $m + n \geq 3$ has been shown to be the Steinberg Leibniz supergebra $\mathfrak{stl}(m, n, A)$ in Ref. 12.

5. The structure of Δ-graded Leibniz superalgebras of other types

It follows from Proposition 3.4 that every Leibniz superalgebra graded by the root system $C(n)$, $D(m,n)$, $D(2,1;\alpha)$ $(\alpha \notin \{0,-1\})$, $F(4)$, or $G(2)$ decomposes as a $\mathfrak{g}$-module into a direct sum of adjoint modules and trivial modules. The next general result describes the structure of Leibniz superalgebras L having such decompositions, which is essentially same as the Lemma 4.1 in Ref. 5.

Lemma 5.1. *Let L be a Leibniz superalgebra over K with a subsuperalgebra $\mathfrak{g}$, and assume that under the adjoint action of $\mathfrak{g}$, L is a direct sum of*

(1) *copies of the adjoint module $\mathfrak{g}$,*

(2) *copies of the trivial module K.*

Assume that

(3) *$\dim Hom_{\mathfrak{g}}(\mathfrak{g} \otimes \mathfrak{g}, \mathfrak{g}) = 1$ so that $Hom_{\mathfrak{g}}(\mathfrak{g} \otimes \mathfrak{g}, \mathfrak{g})$ is spanned by $x \otimes y \to [x,y]$.*

(4) *$Hom_{\mathfrak{g}}(\mathfrak{g} \otimes \mathfrak{g}, K) = K\kappa$, where κ is even, non-degenerate and supersymmetric, and the following conditions hold:*

(5) *There exist* $f, g \in \mathfrak{g}_{\bar{0}}$ *such that* $[f, g] \neq 0$ *and* $\kappa(f, g) \neq 0$,

(6) *There exist* $f, g, h \in \mathfrak{g}_{\bar{0}}$ *such that* $[f, h] = [g, h] = 0$ *and* $\kappa(f, h) = \kappa(g, h) = 0 \neq \kappa(f, g)$,

(7) *There exist* $f, g, h \in \mathfrak{g}_{\bar{0}}$ *such that* $[[f, g], h] = 0 \neq [[g, h], f]$.

Then there exist superspaces A *and* D *such that* $L \cong (\mathfrak{g} \otimes A) \oplus D$ *and*

(a) A *is a unital supercommutative associative super dialgebra;*

(b) D *is a trivial* $\mathfrak{g}$*-module and is a Leibniz superalgebra;*

(c) *Multiplication in* L *is given by*

$$[f \otimes a, g \otimes b] = (-1)^{|a||g|}\Big([f, g] \otimes (a \vdash b) + \kappa(f, g)\langle a, b\rangle\Big), \tag{5.1}$$

$$[d, f \otimes a] = (-1)^{|d||f|} f \otimes da, \tag{5.2}$$

for homogeneous elements $f, g \in \mathfrak{g}, a, b \in A, d \in D$, *where*

(i) D *is a Leibniz subsuperalgebra of* L *and* $\phi : D \to Der_K(A)(\phi(d)a = da)$ *is a left representation of* D *as superderivations on the dialgebra* A *with* $\langle A, A\rangle \subset Ker\phi$,

(ii) $[d, \langle a, b\rangle] = \langle da, b\rangle + (-1)^{|d||a|}\langle a, db\rangle$, *in particular,* $\langle A, A\rangle$ *is an ideal of* D,

(iii) $\langle a \vdash b, c\rangle = \langle a, b \vdash c\rangle + (-1)^{|a|(|b|+|c|)}\langle b, c \dashv a\rangle$ *and* $\langle a \vdash b, c\rangle = \langle a \dashv b, c\rangle$.

Conversely, the conditions above are sufficient to guarantee that a superspace $L = \mathfrak{g} \otimes A \oplus D$ *satisfied (a)–(c) is a Leibniz superalgebra.*

Proof. When a Leibniz superalgebra L is a direct sum of copies of adjoint modules and trivial module for $\mathfrak{g}$, then after collecting isomorphic summands, we may assume that there are superspaces $A = A_{\bar{0}} \oplus A_{\bar{1}}$ and $D = D_{\bar{0}} \oplus D_{\bar{1}}$ so that $L = \mathfrak{g} \otimes A \oplus D$. Suppose that such a superalgebra L satisfies conditions (1)—(4). Note that D is a super subalgebra of L.

Fixing homogeneous basis $\{a_i\}_{i\in I}$ of A and choose a_i, a_j, a_k with $i, j, k \in I$, we see that the projection of the product $[\mathfrak{g} \otimes a_i, \mathfrak{g} \otimes a_j]$ onto $\mathfrak{g} \otimes a_k$ determines an elements of $\mathrm{Hom}_{\mathfrak{g}}(\mathfrak{g} \otimes \mathfrak{g}, \mathfrak{g})$, which is spanned by the Leibniz supercommutator. Then there exist scalars $\xi_{i,j}^k$ and $\theta_{i,j}^k$ so that

$$[x \otimes a_i, y \otimes a_j]|_{\mathfrak{g}\otimes A} = (-1)^{|a_i||y|}[x, y] \otimes (\sum_{k\in I} \xi_{i,j}^k a_k).$$

Defining $\dashv, \vdash : A \times A \to A$ by $a_i \vdash a_j = \sum_{k\in I} \xi_{i,j}^k a_k = (-1)^{|a_i||a_j|a_j \vdash a_i}$ and extending it bilinearly, we have a supercommutative dialgebra structure on A.

By similar arguments, there exist bilinear maps $A \times A \to D, (a, b) \mapsto \langle a, b\rangle \in D$, and $D \times A \to A, (d, a) \mapsto da \in A$, such that the multiplication in L is as in (c).

First the Jacobi superidentity, when specialized with homogeneous elements $d_1, d_2 \in D$ and $f \otimes a \in \mathfrak{g} \otimes A$, and then with $d \in D$ and $f \otimes a, g \otimes b \in \mathfrak{g} \otimes A$ will show that $\phi(d)a = da$ is a representation of D as superderivations of A. We assume next that f, g are taken to satisfy (5). Then for homogeneous elements $d \in D$, $a, b \in A$, the Jacobi superidentity gives the condition $[d, \langle a, b\rangle] = \langle da, b\rangle + (-1)^{|d||a|}\langle a, db\rangle$.

The Jacobi superidentity, when specialized with homogeneous elements $f \otimes a, g \otimes b, h \otimes c \in \mathfrak{g} \otimes A$ and f, g, h as condition (7), gives that $\langle A, A\rangle$ is contained in the kernel of ϕ.

Finally the Jacobi superidentity, when specialized with homogeneous elements $f \otimes a, g \otimes b, h \otimes c \in \mathfrak{g} \otimes A$ and f, g, h as condition (8), gives the (iii) and associativity of dialgebra A (see Section 4 or Ref. 4).

The converse is a simple computation. ∎

From Ref. 4, we see that $\mathfrak{g}$ satisfies the above condition if $\mathfrak{g}$ is a split simple classical Lie superalgebra of type $C(n)(\geq 3)$, $D(m, n)(m \geq 2, n \geq 1)$, $D(2, 1; \alpha), \alpha \in K, \alpha \neq 0, 1$, $F(4)$ or $G(2)$. Then we have the following structure theorem for Leibniz superalgebras graded by the root system of type $\mathfrak{g}$ (compare Theorem 5.2 in Ref. 4).

Theorem 5.1. *Let $\mathfrak{g}$ be a split simple classical Lie superalgebra of type $C(n)\,(\geq 3)$, $D(m, n)\,(m \geq 2, n \geq 1)$, $D(2, 1; \alpha)$, $\alpha \in K$, $\alpha \neq 0, 1$, $F(4)$ or $G(2)$, then by Ref. 4 we have $\dim Hom_{\mathfrak{g}}(\mathfrak{g}\otimes\mathfrak{g}, \mathfrak{g}) = \dim Hom_{\mathfrak{g}}(\mathfrak{g}\otimes\mathfrak{g}, K) = 1$. Let L be a Δ-graded Leibniz superalgebra of type $\mathfrak{g}$, then there exists a unital supercommutative associative super dialgebra A and a K-superspace D such that $L \cong (\mathfrak{g} \otimes A) \oplus D$. Multiplication in L is given by*

$$[f \otimes a, g \otimes b] = (-1)^{|a||g|}\Big([f, g] \otimes (a \vdash b) + \kappa(f, g)\langle a, b\rangle\Big), \tag{5.3}$$

$$[d, L] = 0, \tag{5.4}$$

for homogeneous elements $f, g \in \mathfrak{g}$, $a, b \in A$, $d \in D$, where $\kappa(,)$ is a fix even nondegenerate supersymmetric invariant bilinear form on $\mathfrak{g}$ and $\langle,\rangle : A \otimes A \to D$ is a K-bilinear and satisfies the following conditions:

(1) $[d, \langle a, b\rangle] = \langle da, b\rangle + (-1)^{|d||a|}\langle a, db\rangle$,

(2) $\langle a \vdash b, c\rangle = \langle a, b \vdash c\rangle + (-1)^{|a|(|b|+|c|)}\langle b, c \dashv a\rangle$, $\quad \langle a \vdash b, c\rangle = \langle a \dashv b, c\rangle$,

and D is a Leibniz subsuperalgebra of L and $\phi : D \to Der_K(A)(\phi(d)a = da)$ is a left representation of D as superderivations on the dialgebra A.

Proof. The only point left is the proof of the centrality of D. Condition (3) of Definition 3.1 forces $D = \langle A, A\rangle$, which by Lemma 5.1 is contained in $\operatorname{Ker}\phi$. Therefore $D = \langle A, A\rangle$ is abelian and centralizes $\mathfrak{g} \otimes A$, hence it is central. ∎

As the similar arguments in Ref. 4, we also have

Corollary 5.1. *Let $\mathfrak{g}$ be a split simple classical Lie superalgebra of type $C(n)(\geq 3)$, $D(m,n)\,(m \geq 2, n \geq 1)$, $D(2,1;\alpha)$, $\alpha \in K$, $\alpha \neq 0,1$, $F(4)$ or $G(2)$ and L be a Δ-graded Leibniz superalgebra of type $\mathfrak{g}$, then there exists a unital supercommutative associative super dialgebra A such that L is a covering of the Leibniz superalgebra $\mathfrak{g} \otimes A$.* ∎

Let $\mathfrak{g}$ be a split simple classical Lie superalgebra of type $C(n)(\geq 3)$, $D(m,n)\,(m \geq 2,\, n \geq 1)$, $D(2,1;\alpha)$, $\alpha \in K$, $\alpha \neq 0,1$, $F(4)$ or $G(2)$, A a unital associative commutative dialgebra, the universal central extension of the Leibniz superalgebra $\mathfrak{g} \otimes A$ has been studied in Ref. 14.

Remark. Such researches for the cases of $B(m,n)$ and $A(1,1)$ are more complicated (see Refs. 6,7), which depends on the results of (non)associative dialgebras.

ACKNOWLEDGMENTS

N. Hu is supported in part by by the NNSF (Grant 10431040, 10271047), the TRAPOYT and the FUDP from the MOE of China, the SRSTP from the STCSM, the Shanghai Priority Academic Discipline from the SMEC. D. Liu is supported by the NNSF (No. 11071068, 10701019), the ZJNSF(No. D7080080,Y6100148), Qianjiang Excellence Project (No. 2007R10031), the "New Century 151 Talent Project" (2008) and the "Innovation Team Foundation of the Department of Education" (No. T200924) of Zhejiang Province. L. Zhu is supported by the NNSF (No. 11071068).

References

1. Albeverio, S.; Ayupov, Sh. A.; Omirov, B. A., *On nilpotent and simple Leibniz algebras*, Communications in Algebra, **33**(2005), 159172.
2. Allison, B. N; Benkart, G; Gao, Y., *Central extensions of Lie algebras graded by finite root systems*, Math. Ann. **316** (2000), no. 3, 499–527.
3. Allison, B. N; Benkart, G; Gao, Y., *Lie algebras graded by root systems $BC_r, r \geq 2$*, Mem. Amer. Math. Soc. 158 (2002), no.751,

4. Benkart, G.; Elduque, A., *Lie superalgebras graded by the root systems $C(n)$, $D(m,n)$, $D(2,1;\alpha)$, $F(4)$, $G(2)$*, Canad. Math, Bull. Vol. 45 (4), (2002), 509–524.
5. Benkart, G.; Elduque, A., *Lie superalgebras graded by the root systems $A(m,n)$*, J. Lie Theory 13(2) (2003), 387–400.
6. Benkart, G.; Elduque, A., *Lie superalgebras graded by the root systems $B(m,n)$*, Selecta Math. 9(2003), no. 3, 313–360.
7. Benkart, G.; Elduque, A.; Martnez, C., *$A(n,n)$-graded Lie superalgebras*, J. Reine Angew. Math. 573 (2004), 139–156.
8. Benkart, G.; Zelmanov, E., *Lie algebras graded by finite root systems and the intersection matrix algebras*, Invent. Math. **126** (1996), 1-45.
9. Berman, S.; Moody, R.V., *Lie algebras graded by finite root systems and the intersection matrix algebras of Slodowy*, Invent. Math. **108** (1992), 323-347.
10. Dzhumadil'daev, A.S., *Cohomologies of colour Leibniz superalgebras: pre-simplicial approach*, Lie Theory and Its Applications in Physics III, Proceeding of the Third International Workshop (1999), 124—135
11. Kurdiani, R.; Pirashvili, T., *A Leibniz algebra structure on the second tensor power*, Journal of Lie Theorey **12**(2) (2002), 583-596.
12. Liu, D., *Steinberg Leibniz algebras and superalgebras*, J. Alg., Vol. **283**(1) (2005), 199-221.
13. Liu, D.; Hu, N., *Leibniz algebras graded by the finite root systems*, Alg. Colloq. **17**(3) (2010) 431-446.
14. Liu, D.; Hu, N., *Leibniz superalgebras and central extensions*, Journal of Algebra and its Applications, Vol. **5**(6)(2006), 765-780.
15. Liu, D.; Hu, N., *Universal central extensions of the matrix Leibniz superalgebras $\mathfrak{sl}(m,n,A)$*, Comm. Alg., Vol.**35**(6)(2007), 1814 - 1823.
16. Loday, J.-L., *Une version non commutative des algèbres de Lie: Les algèbres de Leibniz*, Enseign. Math. **39** (1993) 269–294.
17. Loday, J.-L., *Dialgebras*, Springer, Lecture Notes in Mathematics, Vol. **1763** (2001) 7–66.
18. Loday, J.-L.; Pirashvili, T., *Universal enveloping algebras of Leibniz algebras and (co)-homology*, Math. Ann. **296** (1993) 138–158.

Tridendriform Algebras Spanned by Partitions

Daniel Jiménez

Depto. de Matemáticas
Fac. de Ciencias, Univ. de Valparaíso
Gran Bretaña 1091, Playa Ancha
Valparaíso, CHILE
E-mail: daniel.jimenez@uv.cl

María Ronco

Instituto de Matemáticas,
Universidad de Talca,
Campus Norte, Camino Lircay s/n,
Talca, CHILE
E-mail: mariaronco@inst-mat.utalca.cl

We define a tridendriform bialgebra structure on the vector space spanned by non-empty ordered partitions of finite sets. This construction, in term of partitions instead of maps between finite sets, permits us to define a Rota-Baxter structure on the vector space spanned by all ordered partitions of finite sets, whose associated tridendriform structure, restricted to non-empty partitions, coincides with the first one. The tridendriform structure on all ordered partitions induces a tridendriform structure on the space spanned by all maps $f : \{1, \ldots, n\} \to \mathbb{N}$, for all $n \geq 1$.

Keywords: Bialgebras; dendriform; Hopf algebra; partitions; Rota-Baxter algebras.

Introduction

Let $\mathbb{K}$ be a field. For any $\lambda \in \mathbb{K}$, a λ-Rota-Baxter algebra (also called a Baxter algebra, see[3] or[16]) is an associative algebra over $\mathbb{K}$, equipped with a linear map $R : A \longrightarrow A$ satisying the following equation:

$$P(a) \cdot P(b) = P(P(a) \cdot b + a \cdot P(b)) + \lambda\, P(a \cdot b),$$

for any pair of elements a and b of A.

Several works deal with the construction of free Rota-Baxter algebras. Free objects for commutative Rota-Baxter algebras were described in[16] and.[5] In,[8] L. Guo and W. Keigher constructed the free commutative Rota-Baxter algebra over a commutative unital algebra A, in terms of shuffles.

On the other hand, in,[10] J.-L. Loday introduced dendriform algebras, which are a particular case of associative algebras whose product is the sum of two binary operations. For this type of algebra, free objects are described in terms of planar binary trees and some decomposition of the shuffle product. This notion was generalized in,[11] where dendriform algebras are considered as a particular case of tridendriform algebras. This last notion can be easily extended to get λ-tridendriform algebras, for any λ in the base field, as is shown in.[4] The free tridendriform algebra on one generator is described in,[11] and the result is easily extended to obtain the free tridendriform algebra over any vector space. Moreover, any free tridendriform algebra has a natural structure of bialgebra.

M. Aguiar (see[1]) found out that any Rota-Baxter algebra of weight 0 gives rise to a dendriform algebra. His result was at the origin of many works relating λ-Rota-Baxter algebras and λ-tridendriform algebras, as,[76] and.[2]

Aguiar's construction extends naturally to a functor from the category of λ-Rota-Baxter algebras to the category of λ-tridendriform algebras, but it is not clear how to construct its left adjoint functor.

In,[1213] and,[15] many examples of tridendriform bialgebras are studied. These bialgebras are defined on vector spaces spanned by certain types of maps between finite sets, such as the bialgebra of surjective maps and the bialgebra of parking functions, and their tridendriform structures are not obtained from a Rota-Baxter algebra.

The main idea for our work comes from the description in terms of ordered partitions of the Hopf algebra of $\mathcal{M}$ permutations introduced in.[9] In[4] we proved that this bialgebra is in fact a 1-tridendriform bialgebra.

Based on this result, we describe in the second Section of the present work a λ-tridendriform bialgebra structure on the space spanned by the set of all non-empty ordered partitions of finite sets of type $\{1,\ldots,n\}$. In the next section, we define a λ-Rota-Baxter bialgebra structure on the space spanned by all ordered partitions, that is when we admit empty blocks in the partition. This Rota-Baxter structure induces the tridendriform bialgebra structure on the space of non-empty ordered partitions define in Section 2, and it also permits to define a tridendriform structure on the space spanned by all maps $f:\{1,\ldots,n\}\longrightarrow\mathbb{N}$ as pointed out in the last section.

In,[14] J.-C. Novelli, J.-Y. Thibon y L. Williams describe deformations of combinatorial Hopf algebras on spaces spanned by words on an ordered alphabet. Even if the underlying bialgebra of the tridendriform bialgebra **Part**(1) is isomorphic to the Hopf algebra of packed words **WQSym**, the result is not true for $\lambda \neq 1$.

Acknowledgements: Our joint work was supported by the Dipuv-Reg 22/2008 of the Universidad de Valparaíso and by the project MathAmSud 10-math-01 OPECSHA. The attendance of the second author to the International Conference Operads and Universal Algebra was funded by the Centre National de la Recherche Scientifique (C.N.R.S., France) and the National Science Foundation (N.S.F., China).

1. Preliminaries

All the vector spaces considered in the present work are over $\mathbb{K}$, where $\mathbb{K}$ is a field. For any set X, we denote by $\mathbb{K}[X]$ the vector space spanned by X.

Finite sets and partitions

Let n be a positive integer, we denote by $[n]$ the set $\{1, \dots, n\}$. If $J = \{j_1, \dots, j_k\} \subseteq [n]$ and $r \geq 1$, we denote by $J + r$ the set $\{j_1 + r, \dots, j_k + r\}$.

Let $\mathcal{F}_n$ be the set of maps from $[n]$ to the set $\mathbb{N}$ of natural numbers. We identify a function $f : [n] \longrightarrow \mathbb{N}$, with its image $(f(1), \dots, f(n))$.

Given a map $f : [n] \longrightarrow \mathbb{N}$ and a subset $J = \{i_1 < \cdots < i_k\} \subseteq [n]$, the restriction of f to J is the map $f|_J := (f(i_1), \dots, f(i_k))$. Similarly, for a subset K of $\mathbb{N}$, the co-restriction of f to K is the map $f|^K := (f(j_1), \dots, f(j_l))$, where $\{j_1 < \cdots < j_l\} := \{i \in [n] / f(i) \in K\}$.

Definition 1.1. For any set J, a *non-empty ordered partition* of J is a family of non-empty subsets $B = \{B_1, \dots, B_r\}$ of J such that $B_i \cap B_j = \emptyset$, for $i \neq j$, and $J = \bigcup_{j=1}^{r} B_j$. The subsets B_i are called the blocks of the partition B.

Notation 1.1. We denote by Part_J the set of all non-empty ordered partitions of J.

The set of ordered partitions of J is partially ordered by the relation: $B \leq D$ if D may be obtained from B by joining consecutive blocks.

For instance, the partition $B = \{(1,3), (4,5), (2), (6,7)\}$ of $[7]$, is smaller than $D = \{(1,3,4,5), (2,6,7)\}$.

Note that to any ordered partition $B = \{B_1, \ldots, B_r\}$ of J we may associate a surjective map $f_B : J \longrightarrow [r]$ given by $f_B(j) = i$ if, and only if, $j \in B_i$. The map $B \mapsto f_B$ is bijective from the set of ordered non-empty partitions of J to the set of surjective maps from J to some interval of integers $[r]$, for $1 \leq r \leq |J|$.

Let $B = \{B_1, \ldots, B_r\}$ be an ordered partition of J and $K \subseteq J$, then $B|^K$ is the partition of K obtained by making the intersections $\{B_1 \cap K, \ldots, B_r \cap K\}$ and taking off the empty sets.

To any ordered partition $B = \{B_1, \ldots, B_r\}$ of a finite set $J \subseteq \mathbb{N}$ it is possible to associate a partition $\mathrm{std}(B) = \{\overline{B}_1, \ldots, \overline{B}_r\}$ of $[n]$, where $|J| = n$, in the following way:

$$i \in \overline{B}_s, \text{ if, and only if } j_i \in B_s,$$

where $J = \{j_1 < \cdots < j_l\}$. For instance, if $B = \{(1,4),(2,7),(9,11,16)\}$, then $\mathrm{std}(B) = \{(1,3),(2,4),(5,6,7)\}$.

Notation 1.2. Given a finite subset $J \subseteq \mathbb{N}$ and any positive integer n, for any non-empty ordered partition $B = (B_1, \ldots, B_k)$ of J we denote by $B + n$ the partition of $J + n$ obtained by replacing the block B_i by $B_i + n$, for all $1 \leq i \leq k$.

Shuffles

Recall that the symmetric group of permutations of n elements is denoted by S_n. Given a composition $\underline{n} = (n_1, \ldots, n_r)$ of n, a $\underline{n}$*-shuffle* is a permutation $\sigma \in S_n$ such that $\sigma^{-1}(n_1 + \cdots + n_i + 1) < \cdots < \sigma^{-1}(n_1 + \cdots + n_{i+1})$, for $0 \leq i \leq r - 1$. We denote by $Sh(n_1, \ldots, n_r)$ the set of all $\underline{n}$-shuffles.

The following results about shuffles are well-known, for their proof see for instance.[17]

Given compositions $\underline{n} = (n_1, \ldots, n_r)$ of n and $\underline{m} = (m_1, \ldots, m_s)$ of m, the following formula is known as the *associativity of the shuffle*:

$$(Sh(n_1, \ldots, n_r) \times Sh(m_1, \ldots, m_s)) \cdot \mathrm{Sh}(n, m) = Sh(n_1, \ldots, n_r, m_1, m_s), \quad (*)$$

where $\cdot$ denotes the product in the group algebra $\mathbb{K}[S_{n+m}]$ and $\times : S_n \times S_n \hookrightarrow S_{n+m}$ is the concatenation of permutations, given by:

$$\sigma \times \tau := (\sigma(1), \ldots, \sigma(n), \tau(1) + n, \ldots, \tau(m) + n).$$

Moreover, for any permutation $\sigma \in S_n$ and any integer $1 \leq i \leq n$ there exist unique permutations $\sigma_{(1)} \in S_i$, $\sigma_{(2)} \in S_{n-i}$ and $\delta \in \mathrm{Sh}(i, n-i)$ such that $\sigma = \delta^{-1} \cdot (\sigma_{(1)} \times \sigma_{(2)})$.

2. Rota-Baxter algebras and tridendriform bialgebras

This section is devoted to recall basic results on Rota-Baxter algebras (see,[168] and[7]) and tridendriform algebras (see[11]). In the rest of the paper, λ is any element of the base field $\mathbb{K}$.

Definition 2.1. A λ-*Rota-Baxter* algebra is an associative algebra $(A, \cdot)$ equipped with a linear map $P : A \longrightarrow A$ satisfying that:

$$P(x) \cdot P(y) = P(P(x) \cdot y + x \cdot P(y)) + \lambda\, P(x \cdot y).$$

Definition 2.2. Let $\lambda \in \mathbb{K}$ be an element of the base field. A λ-*tridendriform algebra* is a vector space A over $\mathbb{K}$, equipped with three binary operations $\prec: A \otimes A \to A$, $\cdot : A \otimes A \to A$ and $\succ: A \otimes A \to A$, satisfying the following relations:

(1) $(a \prec b) \prec c = a \prec (b \prec c + b \succ c + \lambda\, b \cdot c)$,
(2) $(a \succ b) \prec c = a \succ (b \prec c)$,
(3) $(a \prec b + a \succ b + \lambda\, a \cdot b) \succ c = a \succ (b \succ c)$,
(4) $(a \cdot b) \cdot c = a \cdot (b \cdot c)$,
(5) $(a \succ b) \cdot c = a \succ (b \cdot c)$,
(6) $(a \prec b) \cdot c = a \cdot (b \succ c)$,
(7) $(a \cdot b) \prec c = a \cdot (b \prec c)$.

For the main examples of Rota-Baxter algebras we refer to[8] and,[7] while free tridendriform algebras are described in.[11]

Note that the operation $* := \prec + \lambda\ \cdot + \succ$ is associative. Moreover, given a λ-tridendriform algebra $(A, \prec, \cdot, \succ)$, the space A equipped with the binary operations $\prec$ and $\overline{\succ} := \lambda\ \cdot + \succ$ is a dendriform algebra, as defined by J.-L. Loday in.[10] Moreover, a dendriform algebra may be defined simply as a λ-tridendriform algebra such that the associative product $\cdot$ is trivial.

On the other hand, any λ-Rota-Baxter algebra $(A, \cdot, P)$ gives rise to a λ-tridendriform structure on A by keeping the associative product $\cdot$ and setting:

(1) $a \succ b := P(a) \cdot b$,
(2) $a \prec b := a \cdot P(b)$,

for $a, b \in A$.

Note that, if $(A, \circ)$ is an associative algebra and $(B, \prec, \cdot, \succ)$ is a λ-tridendriform algebra, then the space $A \otimes B$ is equipped with a natural structure of λ-tridendriform algebra, given by:

$$(a \otimes b) \prec_{A\otimes B} (a' \otimes b') := (a \circ a') \otimes (b \prec b'),$$
$$(a \otimes b) \cdot_{A\otimes B} (a' \otimes b') := (a \circ a') \otimes (b \cdot b'),$$
$$(a \otimes b) \succ_{A\otimes B} (a' \otimes b') := (a \circ a') \otimes (b \succ b').$$

In particular given a λ-tridendriform algebra $(A, \prec, \cdot, \succ)$, we may consider A as an associative algebra with the product $* = \prec + \lambda \cdot + \succ$, so $A \otimes A$ has a natural structure of λ-tridendrifrom algebra.

For any λ-tridendriform algebra A, we extend partially the tridendriform structure to $A^+ := \mathbb{K} \bigoplus A$ as follows:

$$a \succ 1_{\mathbb{K}} := 0 =: 1_{\mathbb{K}} \prec a,$$
$$a \cdot 1_{\mathbb{K}} := 0 =: 1_{\mathbb{K}} \cdot a,$$
$$a \prec 1_{\mathbb{K}} := a =: 1_{\mathbb{K}} \succ a,$$

for all $a \in A$. However, it is not possible to extend the operations $\succ$, $\cdot$ and $\prec$ to the whole A^+ in such a way that it becomes a λ-tridendriform algebra, due essentially to the fact that it is impossible to define $1_{\mathbb{K}} \succ 1_{\mathbb{K}}$ and $1_{\mathbb{K}} \prec 1_{\mathbb{K}}$ in a compatible way with the relation $1_{\mathbb{K}} * a = a = a * 1_{\mathbb{K}}$, for all $a \in A^+$. However, we may define a tridendriform algebra structure on $A^+ \otimes A \bigoplus A \otimes A^+$, which coincides with the usual tridendriform structure on $A \otimes A$, just adding the following relations:

$$(a \otimes 1_{\mathbb{K}}) \succ (b \otimes 1_{\mathbb{K}}) := (a \succ b) \otimes 1_{\mathbb{K}},$$
$$(a \otimes 1_{\mathbb{K}}) \cdot (b \otimes 1_{\mathbb{K}}) := (a \cdot b) \otimes 1_{\mathbb{K}},$$
$$(a \otimes 1_{\mathbb{K}}) \prec (b \otimes 1_{\mathbb{K}}) := (a \prec b) \otimes 1_{\mathbb{K}},$$

for all $a, b \in A$.

With the previous definitions, we are able to introduce the notion of λ-tridendriform bialgebra.

Definition 2.3. A λ-*tridendriform bialgebra* is a λ-tridendriform algebra A equipped with a coassociative counital coproduct $\Delta : A^+ \longrightarrow A^+ \otimes A^+$ which is a tridendriform algebra homomorphism on $A^+ \otimes A \bigoplus A \otimes A^+$.

Remark 2.1. It is immediate to check that if $(A, \succ, \cdot, \prec, \Delta, \epsilon)$ is a λ-tridendriform bialgebra, then $(A^+, *, \Delta)$ is a bialgebra in the usual sense.

The proof of the following lemma is straightforward.

Lemma 2.1. *Let* $(A, \cdot, P)$ *be a* λ *Rota-Baxter algebra which admits a coassociative coproduct* $\Delta : A \longrightarrow A \otimes A$ *satisfying that:*

(1) $\Delta(a \cdot b) = \sum (a_{(1)} * b_{(1)}) \otimes (a_{(2)} \cdot b_{(2)})$,
(2) $\Delta(P(a)) = \sum a_{(1)} \otimes P(a_{(2)})$,

where $\Delta(a) = \sum a_{(1)} \otimes a_{(2)}$ *for* $a \in A$ *and* $a * b = P(a) \cdot b + a \cdot P(b) + \lambda\, a \cdot b$. *The associated* λ*-tridendriform algebra* $(A, \succ, \cdot, \prec)$ *equipped with* Δ *is a* λ*-tridendriform bialgebra, where* P *and* $\cdot$ *are extended to* A^+ *in an obvious way.*

3. Tridendriform structure on the space of partitions

Given non-empty ordered partitions $B \in \mathrm{Part}_{[n]}$, $D \in \mathrm{Part}_{[m]}$ and $K \in \mathrm{Part}_{[n+m]}$ such that $K|^{[n]} = B$ and $K|^{[m]+n} = D + n$, the integer $\cap^K_{B,D}$ is the number of blocks K_j such that $K_j \cap [n] \neq \emptyset$ and $K_j \cap [m] + n \neq \emptyset$.

Consider the graded vector space $\bigoplus_{n \geq 1} \mathbb{K}[\mathrm{Part}_{[n]}]$, spanned by all the non-empty ordered partitions of natural numbers. We may extend the constructions of[4] in order to define a λ-tridendriform algebra structure on it, as follows:

$$B \succ D := \sum_{K_r \cap [n] = \emptyset} \lambda^{\cap^K_{B,D}} K,$$

$$B \cdot D := \sum_{\substack{K_r \cap [n] \neq emptyset \\ K_r \cap [m]+n \neq \emptyset}} \lambda^{\cap^K_{B,D} - 1} K,$$

$$B \prec D := \sum_{K_r \cap [m]+n = \emptyset} \lambda^{\cap^K_{B,D}} K,$$

for $B \in \mathrm{Part}_{[n]}$ and $D \in \mathrm{Part}_{[m]}$, where the sum is taken over all partitions $K = (K_1, \dots, K_r) \in \mathrm{Part}_{[n+m]}$ such that $K|^{[n]} = B$ and $K|^{[m]+n} = D + n$.

Example 3.1. For example, let $B = \{(1,3),(4),(2,5)\}$ and $D = \{(1,3),(2)\}$. Then

$$\begin{aligned}
B \succ D &= \{(1,3),(4),(2,5),(6,8),(7)\} + \lambda\{(1,3),(4),(2,5,6,8),(7)\} \\
&+ \{(1,3),(4),(6,8),(2,5),(7)\} + \lambda\{(1,3),(4,6,8),(2,5),(7)\} \\
&+ \{(1,3),(6,8),(4),(2,5),(7)\} + \lambda\{(1,3,6,8),(4),(2,5),(7)\} \\
&+ \{(6,8),(1,3),(4),(2,5),(7)\}, \\
B \cdot D &= \{(1,3),(4),(6,8),(2,5,7)\} + \lambda\{(1,3),(4,6,8),(2,5,7)\} \\
&+ \{(1,3),(6,8),(4),(2,5,7)\} + \lambda\{(1,3,6,8),(4),(2,5,7)\} \\
&+ \{(6,8),(1,3),(4),(2,5,7)\},
\end{aligned}$$

$$B \prec D = \{(6,8),(7),(1,3),(4),(2,5)\} + \lambda\{(6,8),(1,3,7),(4),(2,5)\}$$
$$+\{(6,8),(1,3),(7),(4),(2,5)\} + \lambda\{(6,8),(1,3),(4,7),(2,5)\}$$
$$+\{(6,8),(1,3),(4),(7),(2,5)\} + \lambda\{(1,3,6,8),(7),(4),(2,5)\}$$
$$+\lambda^2\{(1,3,6,8),(4,7),(2,5)\} + \lambda\{(1,3,6,8),(4),(7),(2,5)\}$$
$$+\{(1,3),(6,8),(7),(4),(2,5)\} + \lambda\{(1,3),(6,8),(4,7),(2,5)\}$$
$$+\{(1,3),(6,8),(4),(7),(2,5)\} + \lambda\{(1,3),(4,6,8),(7),(2,5)\}$$
$$+\{(1,3),(4),(6,8),(7),(2,5)\}.$$

Theorem 3.1. *The space $\bigoplus_{n\geq 1} \mathbb{K}[Part_{[n]}]$, equipped with the products $\succ$, $\cdot$ and $\prec$ defined above is a λ-tridendriform algebra.*

Proof. Note first that, given partitions $B \in \mathrm{Part}_{[n]}$, $D \in \mathrm{Part}_{[m]}$ and $E \in \mathrm{Part}_{[p]}$, the sets

$$\mathcal{S}_1 := \{(K,L) \in \mathrm{Part}_{[n+m]} \times \mathrm{Part}_{[n+m+p]} \mid K|^{[n]} = B,$$
$$K|^{[m]+n} = D+n,\ L|^{[n+m]} = K \text{ and } L|^{[p]+n+m} = E+(n+m)\}, \text{ and}$$
$$\mathcal{S}_2 := \{(K,L) \in \mathrm{Part}_{[m+p]} \times \mathrm{Part}_{[n+m+p]} \mid K|^{[m]} = D,$$
$$K|^{[p]+m} = E+m,\ L|^{[n]} = B \text{ and } L|^{[m+p]+n} = K+n\},$$

are both equal to

$$\mathcal{S} := \{L \in \mathrm{Part}_{[n+m+p]} \mid L|^{[n]} = B,\ L|^{[m]+n} = D+n, \text{ and } L|^{[p]+n+m} = E+n+m\}.$$

Consider the bijection $L \mapsto (L|^{[n+m]}, L)$ from $\mathcal{S}$ to $\mathcal{S}_1$. Suppose that $L = (L_1,\dots,L_s)$ and $L|^{[n+m]} = (K_1,\dots,K_r)$, with $K_i = L_{j_i} \cap [n+m]$. We have that:

$$\cap^L_{K,E} = |\{i \in [s] \mid L_i \cap [p]+n+m \neq \emptyset \text{ and } L_i \cap [n+m] \neq \emptyset\}| =$$
$$|\{i \in [s] \mid L_i \cap [n] \neq \emptyset \text{ and } L_i \cap [p]+n+m \neq \emptyset\}| +$$
$$|\{i \in [s] \mid L_i \cap [m]+n \neq \emptyset \text{ and } L_i \cap [p]+n+m \neq \emptyset\}| -$$
$$|\{i \in [s] \mid L_i \cap [n] \neq \emptyset,\ L_i \cap [m]+n \neq \emptyset \text{ and } L_i \cap [p]+n+m \neq \emptyset\}|.$$

and $\cap^K_{B,D} = |\{i \in [s] \mid L_i \cap [n] \neq \emptyset \text{ and } L_i \cap [m] \neq \emptyset\}|$.

So, $\cap^K_{B,D} + \cap^L_{K,E} =$

$$|\{i \in [s] \mid L_i \cap [n] \neq \emptyset \text{ and } L_i \cap [m]+n \neq \emptyset\}| +$$
$$|\{i \in [s] \mid L_i \cap [n] \neq \emptyset \text{ and } L_i \cap [p]+n+m \neq \emptyset\}| +$$
$$|\{i \in [s] \mid L_i \cap [m]+n \neq \emptyset \text{ and } L_i \cap [p]+n+m \neq \emptyset\}| -$$
$$|\{i \in [s] \mid L_i \cap [n] \neq \emptyset\ ,\ L_i \cap [m]+n \neq \emptyset \text{ and } L_i \cap [p]+n+m \neq \emptyset\}|.$$

In the same way we may show that, if $H = L|^{[m+p]+n} - n$, then

$$\begin{aligned}\cap^H_{D,E} + \cap^L_{B,H} = & |\{i \in [s] |\ L_i \cap [n] \neq \emptyset \text{ and } L_i \cap [m] + n \neq \emptyset\}| + \\ & |\{i \in [s] |\ L_i \cap [n] \neq \emptyset \text{ and } L_i \cap [p] + n + m \neq \emptyset\}| + \\ & |\{i \in [s] |\ L_i \cap [m] + n \neq \emptyset \text{ and } L_i \cap [p] + n + m \neq \emptyset\}| - \\ & |\{i \in [s] |\ L_i \cap [n] \neq \emptyset\ ,\ L_i \cap [m] + n \neq \emptyset \text{ and } L_i \cap [p] + n + m \neq \emptyset\}|.\end{aligned}$$

Denote by $\cap^L_{B,D,E}$ the number $\cap^K_{B,D} + \cap^L_{K,E} = \cap^H_{D,E} + \cap^L_{B,H}$.

We show that the conditions 1), 2), 5) and 6) of Definition 2.2 are verified; the other conditions may be easily checked in an analogous way.

In the rest of the proof we assume that $L = (L_1, \ldots, L_s) \in \mathrm{Part}_{[n+m+p]}$ is such that $L|^{[n]} = B$, $L|^{[m]+n} = D+n$ and $L|^{[p]+n+m} = E+n+m$. We denote the restrictions $K = (K_1, \ldots, K_t) = L|^{[n+m]}$ and $H = (H_1, \ldots, H_r) = L|^{[m+p]+n} - n$.

Condition 1) We have that $B \succ (D \succ E) = \sum_{L \in \mathcal{S}} \lambda^{\cap^L_{B,D,E}} L$, where the sum is taken over all L such that $H_r \cap [m] = \emptyset$ and $L_s \cap [n] = \emptyset$. But $H_r \cap [m] = \emptyset$ and $L_s \cap [n] = \emptyset$ mean that $L_s \subseteq [p] + n + m$. We may conclude that:

$$B \succ (D \succ E) = \sum_{L_s \subseteq [p]+n+m} \lambda^{\cap^L_{B,D,E}} L,$$

with no conditions on H. So, $B \succ (D \succ E) = (B * D) \succ E$.

Condition 2) Note that $B \succ (D \prec E) = \sum_{L \in \mathcal{S}} \lambda^{\cap^L_{B,D,E}} L$, where the sum is taken over all L such that $H_r \cap [p] + m = \emptyset$ and $L_s \cap [n] = \emptyset$. But $H_r \cap [p] + m = \emptyset$ implies that $L_s \cap [p] + m + n = \emptyset$, and $L_s \cap [n] = \emptyset$ implies that $L_s \subseteq [m]$. So,

$$B \succ (D \prec E) = \sum_{L_s \subseteq [m]+n} \lambda^{\cap^L_{B,D,E}} L = (B \succ D) \prec E.$$

Condition 5) Computing $B \succ (D \cdot E)$, we get $B \succ (D \cdot E) = \sum_{L \in \mathcal{S}} \lambda^{\cap^L_{B,D,E} - 1} L$, where the sum is taken over all L such that $L_s \cap [n] = \emptyset$ and $H_r \cap [m] + n \neq \emptyset \neq H_r \cap [p] + m$. Since $L_s \cap [n] = \emptyset$, we have that $H_r = L_s$. So, $K_t = L_s \cap [m] + n$, which implies that:

$$B \succ (D \cdot E) = \sum_{\substack{L_s \cap [n] = \emptyset \\ L_s \cap [m]+n \neq \emptyset \neq L_s \cap [p]+n+m}} \lambda^{\cap^L_{B,D,E} - 1} L = (B \succ D) \cdot E.$$

Condition 6) We have that $B \cdot (D \succ E) = \sum_{L \in \mathcal{S}} \lambda^{\cap^L_{B,D,E}-1} L$, where the sum is taken over all L such that $L_s \cap [n] \neq \emptyset \neq L_s \cap [m+p]+n$ and $H_r = L_s \cap [m+p]+n \subseteq [p]+n+m$. Since $L_s \cap [n] \neq \emptyset \neq L_s \cap [m+p]+n$ and $L_s \cap [m+p]+n \subseteq [p]+n+m$, we may conclude that $K_t = L_s \cap [n]$, $K_t \cap [m]+n = \emptyset$ and $L_s \cap ([n+m] \neq \emptyset \neq L_s \cap [p]+n+m$. So,

$$B \cdot (D \succ E) = \sum_{\substack{K_t \cap [m]+n=\emptyset \\ L_s \cap [n+m] \neq \emptyset \neq L_s \cap [p]+n+m}} \lambda^{\cap^L_{B,D,E}-1} L = (B \prec D) \cdot E. \qquad \square$$

We denote by $\mathbf{Part}(\lambda)$ the λ-tridendriform algebra defined on the space $\bigoplus_{n \geq 1} \mathbb{K}[\mathrm{Part}_{[n]}]$.

Let us point out that the associative product $* = \succ + \lambda \cdot + \prec$ associated to the tridendriform structure of $\mathbf{Part}(\lambda)$ does not coincide with the one defined in[14] for $\lambda \neq 1$. Clearly, we may identify an ordered partition $B = (B_1, \ldots, B_r) \in \mathrm{Part}_{[n]}$ with the surjective map $f : \{1 < \cdots < n\} \longrightarrow \{1, \ldots, r\}$ such that $B_i = f^{-1}(i)$ for $1 \leq i \leq r$, so the underlying vector space of $\mathbf{Part}(\lambda)$ is isomorphic to the underlying vector space of **WQSym**. However, the associative product induced by the λ-tridendriform structure of $\mathbf{Part}(\lambda)$ via the isomorphisms gives a formula:

$$f * g = \sum_{h \in H(f,g)} \lambda^{r+s-l(h)} h,$$

for some functions $h : \{1 < \cdots < n+m\} \longrightarrow \{1, \ldots, l(h)\}$, where $f : \{1 < \cdots < n\} \longrightarrow \{1, \ldots, r\}$ and $g : \{1 < \cdots < m\} \longrightarrow \{1, \ldots, s\}$; while the product defined in[14] gives:

$$f *_\lambda g = \sum_{h \in H(f,g)} \lambda^{|sinv(h)|} h,$$

where $sinv(h) := \{(i < j) \in \{1, \ldots, n+m\}^2 \mid h(i) > h(j)\}$ is the set of inversions of h.

The coproduct Δ on $\mathbf{Part}(\lambda)^+$ is given by:

$$\Delta(B) = \sum_{i=0}^{r} \mathrm{std}(B_1, \ldots, B_i) \otimes \mathrm{std}(B_{i+1}, \ldots, B_r),$$

for $B = (B_1, \ldots, B_r)$.

Example 3.2. Let $B = \{(3,4,7),(2,5),(1),(6)\}$, we have that:

$\Delta(B) = \{(3,4,7),(2,5),(1),(6)\} \otimes 1_{\mathbb{K}} + \{(3,4,6),(2,5),(1)\} \otimes \{(1)\}$
$+\{(2,3,5),(1,4)\} \otimes \{(1),(2)\} + \{(1,2,3)\} \otimes \{(2,3),(1),(4)\}$
$+1_{\mathbb{K}} \otimes \{(3,4,7),(2,5),(1),(6)\}$.

Proposition 3.1. *The algebra* ***Part***(λ)*, equipped with* Δ *is a* λ*-tridendriform bialgebra.*

Proof. Let $\overline{\Delta}$ be the reduced coproduct on **Part**(λ), that is

$$\overline{\Delta}(B) := \Delta(B) - (B \otimes 1_{\mathbb{K}} + 1_{\mathbb{K}} \otimes B).$$

Let us denote $\overline{\Delta}(B) = \sum B_{(1)} \otimes B_{(2)}$, for any non-empty ordered partition B.

For $B = (B_1, \ldots, B_p) \in \text{Part}_{[n]}$ and $D = (D_1, \ldots, D_s) \in \text{Part}_{[m]}$, we prove that $\overline{\Delta}(B \succ D) = \sum (B_{(1)} * D_{(1)}) \otimes (B_{(2)} \succ D_{(2)}) + B \otimes D$, which implies that

$$\Delta(B \succ D) = (*\otimes \succ) \circ (id \otimes \tau \otimes id)(\Delta(B) \otimes \Delta(D)),$$

for all partitions B and D, where $\tau(H \otimes K) = K \otimes H$. The proof that $\Delta \circ \cdot = (*\otimes \cdot) \circ (id \otimes \tau \otimes id) \circ (\Delta \otimes \Delta)$ and $\Delta \circ \prec = (*\otimes \prec) \circ (id \otimes \tau \otimes id) \circ (\Delta \otimes \Delta)$ may be obtained applying similar arguments.

Let $K = (K_1, \ldots, K_r)$ be a partition of $[n+m]$ such that $K|^{[n]} = B$, $K|^{[m]+n} = D + n$ and $K_r \subseteq [m] + n$. We have the following possibilities:

(1) There exist integers $1 \leq k \leq p$ and $0 \leq h \leq q - 1$, such that $(K_1, \ldots, K_i)|^{[n]} = (B_1, \ldots, B_k)$ and $(K_1, \ldots, K_i)|^{[m]+n} = (D_1, \ldots, D_h) + n$. Since $K_r \subseteq [m] + n$, the last block of $\text{std}(K_{i+1}, \ldots, K_r)$ is the last block of $\text{std}(D_{h+1}, \ldots, D_s)$.
For $1 \leq i \leq r - 1$, consider the partitions $\text{std}(K_1, \ldots, K_i)$ and $\text{std}(K_{i+1}, \ldots, K_r)$. Note that, as $K|^{[n]} = B$ and $K|^{[m]+n} = D + n$, we may assert that :

$$\begin{aligned} \text{std}(K_1, \ldots, K_i)|^{[n]} &= \text{std}(B_1, \ldots, B_k), \\ \text{std}(K_{i+1}, \ldots, K_{r-1})|^{[n]} &= \text{std}(B_{k+1}, \ldots, B_p), \\ \text{std}(K_1, \ldots, K_i)|^{[m]+n} &= \text{std}(D_1, \ldots, D_h) + n, \\ \text{std}(K_{i+1}, \ldots, K_r)|^{[m]+n} &= \text{std}(D_{h+1}, \ldots, D_s) + n. \end{aligned}$$

(2) There exists $1 \leq h \leq s - 1$ ($h \neq s$ because $K_r \subseteq [m] + n$) such that $(K_1, \ldots, K_i) = (D_1, \ldots, D_h) + n$ and $(K_1, \ldots, K_i)|^{[n]} = 0$.

So, we get that

$$\operatorname{std}(K_1,\dots,K_i) = \operatorname{std}(D_1,\dots,D_h) + n,$$
$$\operatorname{std}(K_{i+1},\dots,K_{r-1})|^{[n]} = B,$$
$$\operatorname{std}(K_{i+1},\dots,K_r)|^{[m]+n} = \operatorname{std}(D_{h+1},\dots,D_s) + n.$$

Conversely, for any pair of integers $0 \leq i \leq p$ and $0 \leq j \leq s-1$, let $\operatorname{std}(B_1,\dots,B_i) \in \operatorname{Part}_{[k]}$ and $\operatorname{std}(D_1,\dots,D_j) \in \operatorname{Part}_{[h]}$.

Suppose that $K \in \operatorname{Part}_{[k+h]}$ and $H \in \operatorname{Part}_{[n+m-k-h]}$ are partitions such that $K|^{[k]} = \operatorname{std}(B_1,\dots,B_i)$, $K|^{[h]+k} = \operatorname{std}(D_1,\dots,D_j)$, $H|^{[n-k]} = \operatorname{std}(B_{i+1},\dots,B_p)$, $H|^{[m-h]+n-k} = \operatorname{std}(D_{j+1},\dots,D_s)$ and the last block of H is contained in $[m-h]+n-k$.

It is easy to see that, in this case, there exists a unique partition $L = (L_1,\dots,L_r) \in \operatorname{Part}_{[n+m]}$ and a unique $1 \leq j \leq r-1$ satisfying that:

(1) $L|^{[n]} = B$ and $L|^{[m]+n} = D+n$,
(2) $\operatorname{std}(L_1,\dots,L_j) = K$ and $\operatorname{std}(L_{j+1},\dots,L_r) = H$.

Moreover, as the last block of H is contained in $[m-h]+n-k$, we get that $L_r \subseteq [m]+n$, which ends the proof. □

Clearly, the tridendriform algebra $\mathbf{Part}(\lambda)$ does not come from a λ-Rota-Baxter structure. We are going to consider a largest class of ordered partitions.

Definition 3.1. For any set X, an *ordered partition* of X is a finite family of subsets $B = \{B_1,\dots,B_p\}$ of X such that $B_i \cap B_j = \emptyset$, for $i \neq j$, and $X = \bigcup_{j=1}^{p} B_j$. The number of blocks p of an ordered partition B is called the length of the partition and is denote $\operatorname{le}(B)$

Let $\operatorname{OPart}(X)$ denote the set of all ordered partitions of X, and let $\mathbf{OPart}$ denote the vector space spanned by all ordered partitions $\bigcup_{n\geq 0}\operatorname{OPart}[n]$ (where $(\emptyset)$ is the unique element of $\operatorname{OPart}[0]$). We want to define a λ-Rota-Baxter algebra structure on $\mathbf{OPart}$.

Definition 3.2. For any pair of non-empty disjoint sets X and Y, let $B \in \operatorname{OPart}(X)$ and $D \in \operatorname{OPart}(Y)$, we define the *shuffle* of B and D as the collection $\operatorname{SH}(B,D)$ of partitions of the disjoint union $X \bigcup Y$ obtained recursively as follows:

(1) If $B = (B_1)$ and $D = (D_1)$, then

$$\operatorname{SH}(B,D) := \{(B_1,D_1); (D_1,B_1); (B_1 \cup D_1)\},$$

(2) If $B = (B_1)$ and $D = (D_1, \ldots, D_r)$, with $r \geq 2$, then

$$\mathrm{SH}(B, D) := \{(E, D_r) \mid \text{with } E \in \mathrm{SH}(B, (D_1, \ldots, D_{r-1}))\} \bigcup \{(D_1, \ldots, D_{r-1}, B_1 \cup D_r); (D, B_1)\},$$

(3) If $B = (B_1, \ldots, B_p)$ and $D = (D_1)$, with $p \geq 2$, then

$$\mathrm{SH}(B, D) := \{(B_1, E) \mid \text{with } E \in \mathrm{SH}((B_2, \ldots, B_p), D_1)\} \bigcup \{(B_1 \cup D_1, B_2, \ldots, B_p); (D_1, B_1, \ldots, B_p)\},$$

(4) If $B = (B_1, \ldots, B_p)$ and $D = (D_1, \ldots, D_r)$, with $p \geq 2$ and $r \geq 2$, then

$$\mathrm{SH}(B, D) := \{(B_1, E) \mid \text{with } E \in \mathrm{SH}((B_2, \ldots, B_p), (D_1, \ldots, D_r))\} \bigcup \{(B_1 \cup D_1, E) \mid \text{with } E \in \mathrm{SH}((B_2, \ldots, B_p), (D_2, \ldots, D_r))\} \bigcup \{(D_1, E) \mid \text{with } E \in \mathrm{SH}(B, (D_2, \ldots, D_r))\},$$

where for any partition $E = (E_1, \ldots, E_l)$ and any subset F_1 such that $F_1 \bigcap (\bigcup_{i=1}^{l} E_i) = \emptyset$, we denote $(F_1, E) := (F_1, E_1, \ldots, E_l)$ and $(E, F_1) := (E_1, \ldots, E_l, F_1)$.

Remark 3.1. Let $B = (B_1, \ldots, B_p)$ and $B' = (B_{p+1}, \ldots, B_{p+r})$. For any $\sigma \in \mathrm{Sh}(p, r)$ we say that a partition $H \in \mathrm{SH}(B, B')$ belongs to $\sigma(B, B')$ if it satisfies one of the following conditions:

(1) $H = (B_{\sigma(1)}, \ldots, B_{\sigma(p+r)})$,
(2) H is obtained from $(B_{\sigma(1)}, \ldots, B_{\sigma(p+r)})$ by joining adjacent blocks $B_{\sigma(i)} \cup B_{\sigma(i+1)}$, for $1 \leq \sigma(i) \leq p$ and $p + 1 \leq \sigma(i + 1) \leq p + r$.

We have that $\mathrm{SH}(B, B')$ is the disjoint union $\bigcup_{\sigma \in Sh(p,r)} \sigma(B, B')$.

Define the product $\cdot_O$ and the operator P_O on **OPart** as follows:

(1) For $B = (B_1, \ldots, B_p) \in \mathrm{OPart}[n]$ and $D = (D_1, \ldots, D_r) \in \mathrm{OPart}[m]$,

$$B \cdot_O D := \sum_{H \in SH(\overline{B}, \overline{D}+n)} \lambda^{p+r-2-le(H)} (H, B_p \cup (D_r + n)),$$

where $\overline{B} = (B_1, \ldots, B_{p-1})$, $\overline{D} = (D_1, \ldots, D_{r-1})$.
(2) For any $B \in \mathrm{OPart}[n]$, $P_O(B) := (B, \emptyset)$.

Example 3.3. Consider the ordered partitions $B = \{(2,4),(1),(3)\}$ and $D = \{(1,2),\emptyset\}$, we have that:

$$B \cdot_O D = \{(2,4),(1),(5,6),(3)\} + \lambda\{(2,4),(1,5,6),(3)\} + \{(2,4),(5,6),(1),(3)\} + \lambda\{(2,4,5,6),(1),(3)\} + \{(5,6),(2,4),(1),(3)\}.$$

Proposition 3.2. *The space **OPart** with the product $\cdot_O$ and the operator P_O is a λ-Rota-Baxter algebra.*

Proof. The associativity of $\cdot_O$ is a straightforward consequence of the associativity of the shuffle (see formula $(*)$) and Remark 3.1. We have just to verify the relationship between $\cdot_O$ and P_O.

Let $B = (B_1, \dots, B_p) \in \mathrm{OPart}[n]$ and $D = (D_1, \dots, D_r) \in \mathrm{OPart}[m]$, then

$$P_O(B) \cdot_O P_O(D) = \sum_{H \in SH(B,D+n)} \lambda^{p+r-le(H)}(H,\emptyset) = P_O\Big(\sum_{H \in SH(B,D+n)} \lambda^{p+r-le(H)} H\Big).$$

So, we need only to compute $\sum_{H \in SH(B,D+n)} \lambda^{p+r-le(H)} H$.

But, for any $H \in SH(B, D+n)$, there exist $\sigma_H \in \mathrm{Sh}(p,r)$ and a family of integers $1 \leq j_1 < \cdots < j_l \leq p+r-1$, with $1 \leq \sigma_H(j_k) \leq p$ and $p+1 \leq \sigma_H(j_k+1) \leq p+r$, such that $H = (H_1, \dots, H_{p+r-k})$ with

$$H_s = \begin{cases} B_{\sigma_H(i)}, & \text{for } \sigma_H(i) \leq p \text{ and } i \notin \{j_1, \dots, j_l\}, \\ D_{\sigma_H(i)-p} + n, & \text{for } \sigma_H(i) > p \text{ and } i \notin \{j_1+1, \dots, j_l+1\}, \\ B_{\sigma_H(i)} \cup (D_{\sigma_H(i+1)-p} + n), & \text{for } i \in \{j_1, \dots, j_l\}. \end{cases}$$

Moreover, any partition $H = (H_1, \dots, H_{p+r-k}) \in \mathrm{SH}(B, D+n)$ fulfills exactly one of the three conditions:

(1) $H_{p+r-k} = B_p$ and $\sigma_H(p+r) = p$,
(2) $H_{p+r-k} = D_r$ and $\sigma_H(p+r) = p+r$ and $j_l < p+r-1$,
(3) $H_{p+r-k} = B_p \cup D_r$, $\sigma_H(p+r) = p+r$, $\sigma_H(p+r-1) = p$ and $j_l = p+r-1$.

Note that:

- H satisfies the first condition if, and only if, $H = (H', B_p)$ with $H' \in \mathrm{SH}(\overline{B}, D+n)$. Moreover, $p+r-\mathrm{le}(H) = p+r-\mathrm{le}(H')$. So, we get that

$$B \cdot_O P_O(D) = \sum_{H \; satisfying \; 1)} \lambda^{p+r-le(H)} H.$$

- H satisfies the second condition if, and only if, $H = (H', D_r + n)$ with $H' \in \mathrm{SH}(B, \overline{D} + n)$. Moreover, $p + r - \mathrm{le}(H) = p + r - \mathrm{le}(H')$. So, we get that

$$P_O(B) \cdot_O D = \sum_{H \ satisfying \ 2)} \lambda^{p+r-le(H)} H.$$

- H satisfies the second condition if, and only if, $H = (H', B_p \cup D_r + n)$ with $H' \in \mathrm{SH}(\overline{B}, \overline{D} + n)$. Moreover, $p + r - \mathrm{le}(H) = p + r - \mathrm{le}(H') - 1$. So, we get that

$$\lambda B \cdot_O D = \sum_{H \ satisfying \ 3)} \lambda^{p+r-le(H)} H.$$

We may conclude that

$$P_O(B) \cdot_O P_O(D) = P_O(B \cdot_O P_O(D) + P_O(B) \cdot_O D) + \lambda P(B \cdot_O D),$$

which ends the proof. □

It is easily seen that the map std which associates to any non-empty ordered partition of a subset $J \subseteq \mathbb{N}$ a non-empty ordered partition of $[|J|]$, extends trivially to a map from the sets of ordered partitions of J to the set of ordered partitions of $[|J|]$.

For instance, $\mathrm{std}((4,7), \emptyset, (5), (1,8), \emptyset)) = ((2,4), \emptyset, (3), (1,5), \emptyset)$.

The coproduct Δ_O on $\mathbf{OPart}^+$ is given by:

$$\Delta_O(B) := \sum_{i=0}^{p} \mathrm{std}(B_1, \ldots, B_i) \otimes \mathrm{std}(B_{i+1}, \ldots, B_p),$$

for $B = (B_1, \ldots, B_p)$. It is immediate to verify that Δ_O is coassociative.

Note that $\Delta_O(\emptyset) = \emptyset \otimes 1_{\mathbb{K}} + 1_{\mathbb{K}} \otimes \emptyset$ and $\Delta_O(1_{\mathbb{K}} = 1_{\mathbb{K}} \otimes 1_{\mathbb{K}}$, so $(\emptyset)$ cannot be identified with $1_{\mathbb{K}}$. The point is that $(\emptyset)$ is the identity element for $\cdot_O$, while $1_{\mathbb{K}}$ is the identity for $*_O$.

We denote by $*_O$ the associative product on $\mathbf{OPart}$ obtained from the λ-tridendriform structure:

$$B *_O D = P_O(B) \cdot_O D + B \cdot_O P_O(D)) + \lambda\ P(B \cdot_O D).$$

Proposition 3.3. *The coproduct Δ_O satisfies the following conditions:*

(1) $\Delta_O(B \cdot_O D) = \sum (B_{(1)} *_O D_{(1)}) \otimes (B_{(2)} \cdot_O D_{(2)})$,
(2) $\Delta_O(P_O(B)) = \sum B_{(1)} \otimes P(B_{(2)})$,

where $\Delta_O(B) = \sum B_{(1)} \otimes B_{(2)}$.

Proof. Note first that

$$B \cdot_O D = (\overline{B} *_O \overline{D}, B_p \cup D_r),$$

for $B = (B_1, \ldots, B_p)$ and $D = (D_1, \ldots D_r)$.

Let $H \in \mathrm{SH}(\overline{B}, \overline{D} + n)$. Recall that H is determined by a permutation $\sigma \in \mathrm{Sh}(p-1, r-1)$ and integers $1 \leq j_1 < \cdots < j_k \leq p-1$ such that $\sigma(j_i) \leq p-1$, $\sigma(j_i + 1) \geq p$, and

$$H = (H_{\sigma(1)}, \ldots, H_{\sigma(j_i)} \cup H_{\sigma(j_i+1)}, \ldots, H_{\sigma(p+r-2)}),$$

where $H_i = B_i$, for $1 \leq i \leq p-1$, and $H_i = D_{i-p-1} + n$, for $p \leq i \leq p+r-2$.

For $1 \leq l \leq p+r-2-k$, there exist $0 \leq l_1 \leq p-1$ and $0 \leq l_2 \leq r-1$ such that

$$\bigcup_{i=1}^{l} H_i = (\bigcup_{i=1}^{l_1} B_i) \cup (\bigcup_{i=1}^{l_2} D_i + n),$$

and $0 \leq t \leq k$ such that $i_t \leq l_1 < i_t + 1$.

Moreover (see Section 1), we have that there exist unique $\sigma_{(1)} \in \mathrm{Sh}(l_1, l_2)$, $\sigma_{(2)} \in \mathrm{Sh}(p-1-l_1, r-1-l_2)$ and $\delta \in \mathrm{Sh}(l_1 + l_2, p+r-2-l_1-l_2)$ such that:

$$\sigma = \delta^{-1} \cdot (\sigma_{(1)} \times \sigma_{(2)}).$$

It is easy to verify that:

- $(H_1, \ldots, H_l) \in \sigma_{(1)}((B_1, \ldots, B_{l_1}), (D_1, \ldots, D_{l_2}) + n)$ is obtained by applying $\sigma_{(1)}$ and joining the blocks $H_{j_i} = B_{\sigma_{(1)}(j_i)} \cup (D_{\sigma_{(1)}(j_i - p)} + n$, for $0 \leq i \leq t$.
- $(H_{l+1}, \ldots, H_{p+r-2-k}) \in \sigma_{(2)}((B_{l_1+1}, \ldots, B_{p-1}), (D_{l_2+1}, \ldots, D_{r-1}) + n)$ is obtained by applying $\sigma_{(2)}$ and joining the blocks $H_{j_i} = B_{\sigma_{(2)}(j_i)} \cup (D_{\sigma_{(2)}(j_i - p)} + n$, for $t+1 \leq i \leq k$.

Conversely, for any pair of integers $0 \leq l_1 \leq p-1$ and $0 \leq l_2 \leq r-1$, and any pair of permutations $\sigma_{(1)} \in \mathrm{Sh}(l_1, l_2)$ and $\sigma_{(2)} \in \mathrm{Sh}(p-1-l_1, r-1-l_2)$, let

$$\delta^{-1}(i) = \begin{cases} i, & \text{for } 1 \leq i \leq l_1 \\ p-1+i-l_1, & \text{for } l_1+1 \leq i \leq l_1+l_2 \text{ or } p+l_2 \leq i \leq p+r-2 \\ i-l_2, & \text{for } l_1+l_2+1 \leq i \leq p+l_2-1. \end{cases}$$

Let $\sigma = \delta^{-1} \cdot (\sigma_{(1)}) \times \sigma_{(2)} \in \mathrm{Sh}(p-1, q-1)$ and $l = l_1 + l_2$. If

- $H_1 \in \mathrm{SH}((B_1, \ldots, B_{l_1}), (D_1, \ldots, D_{l_2}))$ is determined by $\sigma_{(1)}$ and integers $1 \leq i_1 < \cdots < i_t \leq l$,
- $H_2 \in \mathrm{SH}((B_{l_1+1}, \ldots, B_{p-1}), (D_{l_2+1}, \ldots, D_{r-1}))$ is determined by $\sigma_{(2)}$ and integers $1 \leq j_1 < \cdots < j_s \leq p+r-2-l$,

then $H = (H_1, H_2) \in \mathrm{SH}(\overline{B}, \overline{D})$ is the unique element determined by the permutation σ and the integers $i_1, \ldots, i_t, j_1 + l_1, \ldots, j_s + l_1$.

So,

$$\Delta_O(B \cdot_O D) = \sum (\overline{B}_{(1)} *_O \overline{D}_{(1)}) \otimes (\overline{B}_{(2)} *_O \overline{D}_{(2)}, B_p \cup D_r) =$$

$$\sum (B_{(1)} *_O D_{(1)}) \otimes (B_{(2)} \cdot_O D_{(2)}),$$

which proves the first condition.

The second condition is evident. □

By Lemma 2.1, we get that, for any $\lambda \in \mathbb{K}$, **OPart** has a natural structure of λ-tridendriform bialgebra, which we denote by $\mathbf{OPart}(\lambda)$.

Remark 3.2. It is easy to verify that $\mathbf{Part}(\lambda)$ is a sub-λ-tridendriform bialgebra of $\mathbf{OPart}(\lambda)$, for all $\lambda \in \mathbb{K}$. However, the Rota-Baxter structure of $\mathbf{OPart}(\lambda)$ does not restrict to $\mathbf{Part}(\lambda)$.

4. Tridendriform algebra structure on the maps between finite sets

Let $\mathbb{K}[\mathcal{F}]$ be the vector space spanned by $\bigcup_{n \geq 1} \mathcal{F}_n$.

To any ordered partition $B = (B_1, \ldots, B_p) \in \mathrm{OPart}[n]$ we may associate a map $f_B \in \mathcal{F}_n$ by setting $f_B(i) := j$ whenever $i \in B_j$.

This application defined a surjective linear homomorphism ϑ : **OPart** $\longrightarrow \mathbb{K}[\mathcal{F}]$. Note that for any ordered partition $B = (B_1, \ldots, B_p)$, the image under ϑ of any partition of the form $(B, \emptyset, \ldots, \emptyset)$ is always f_B.

The map $f \mapsto B_f := (f^{-1}(1), \dots, f^{-1}(r))$, for $f : [n] \longrightarrow [r]$ and $f^{-1}(r) \neq \emptyset$ is a section of ϑ. So, we may consider $\mathbb{K}[\mathcal{F}]$ as a subspace of **OPart**.

The following result is immediate to check.

Lemma 4.1. *Let $f \in \mathcal{F}_n$ and $g \in \mathcal{F}_m$ be two maps, then the elements $P_O(B_f) \cdot_O B_g$, $B_f \cdot_O B_g$ and $B_f \cdot_O P_O(B_g)$ belong to the image of $\mathbb{K}[\mathcal{F}]$ in **OPart**.*

Note that for any $f \in \mathcal{F}_n$, the image in $\mathbb{K}[\mathcal{F}]$ of $P_O(\varphi(f))$ coincides with f. So, the restriction of the Rota-Baxter operator P_O to $\mathbb{K}[\mathcal{F}]$ is the identity homomorphism. However, Lemma 4.1 implies that $\mathbb{K}[\mathcal{F}]$ is a sub-λ-tridendriform algebra of $\mathbf{OPart}(\lambda)$, denoted $\mathbb{K}[\mathcal{F}](\lambda)$. So, we get inclusions of λ-tridendriform algebras:

$$\mathbf{Part}(\lambda) \hookrightarrow \mathbb{K}[\mathcal{F}](\lambda) \hookrightarrow \mathbf{OPart}(\lambda).$$

In fact, the λ-tridendriform bialgebra $\mathbf{Part}(\lambda)$ coincides with the one defined in terms of surjective maps in.[15]

Note that the coproduct Δ_O does not restrict to $\mathbb{K}[\mathcal{F}]$.

References

1. Marcelo Aguiar. PrePoisson algebras. *Lett. Math. Phys.*, 54(4):263–277, 2000.
2. Marcelo Aguiar and Walter Moreira. Combinatorics of the free Baxter algebra. *Elect. J. of Combinatorics*, 13(1):–, 2006.
3. Glenn Baxter. An analytic problem whose solution follows from a simple algebraic identity. *Pacific J. Math.*, 10:731–742, 1960.
4. Emily Burgunder and María Ronco. Tridendriform structures and combinatorial Hopf algebras. *J. Algebra*, 324:2860–2883, 2010.
5. Pierre Cartier. On the structure of free Baxter algebras. *Adv. in Math.*, 9:253–265, 1972.
6. Kurush Ebrahimi-Fard. Loday-type algebras and the Rota-Baxter relation. *Lett. Math. Phys.*, 61(2):139–147, 1963.
7. Kurush Ebrahimi-Fard and Li Guo. Free Rota-Baxter algebras and rooted trees. *J. Algebra and its Applications*, 7:167–194, 2008.
8. Li Guo and William Keigher. Baxter algebras and shuffle products. *Adv. in Math.*, 150(1):117–149, 2000.
9. Thomas Lam and Paavlo Pylyavskyy. Combinatorial Hopf algebras and k-homology of Grassmannians. *Intern. Math. Res. Notices*, 207(2):544–565, 2006.
10. Jean-Louis Loday. *Dialgebras*, volume 1763 of *Lecture Notes in Maths.*, pages 7–66. Springer, Berlin, 2001.
11. Jean-Louis Loday and María Ronco. *Trialgebras and families of polytopes*, volume 346 of *Contemporary Mathematics*, pages 369–398. American Mathematical Society, Providence, R.I., 2004.

12. Jean-Christophe Novelli and Jean-Yves Thibon. Polynomial realizations of some trialgebras. In *Proceedings Formal Power Series and Algebraic Combinatorics, San Diego, California*, volume 2006, pages +12, 2006.
13. Jean-Christophe Novelli and Jean-Yves Thibon. Hopf algebras and dendriform structures arising from parking functions. *Fund. Math.*, 193(3):189–241, 2007.
14. Jean-Christophe Novelli, Jean-Yves Thibon, and Lauren Williams. Combinatorial Hopf algebras, non-commutative Hall-Littlewood functions and permutation tableaux. *Adv. in Math.*, 224:1311–1348, 2010.
15. Patricia Palacios and María Ronco. Weak Bruhat order on the set of faces of the permutahedra. *J. Algebra*, 299(2):648–678, 2006.
16. Gian-Carlo Rota. Baxter algebras and combinatorial identities i, ii. *Bull. Amer. Math. Soc.*, 75:325–329, ibid. 330–334, 1969.
17. Louis Solomon. A Mackey formula in the group ring of a Coxeter group. *J. Algebra*, 41(2):255–268, 1976.

Generalized Disjunctive Languages and Universal Algebra

Yun Liu
Department of Mathematics, Yuxi Normal University,
Yuxi, Yunnan, 653100, P. R. China
E-mail: slliuyun@sina.com

In this paper, we investigate the algebraic properties of some classes of generalized disjunctive languages by using tools coming from universal algebra, including syntactic monoid characterizations and the relations and hierarchy of these classes of languages.

Keywords: Generalized disjunctive langauge; syntactic monoid; formal language; universal algebra.

1. Introduction

Let A be a nonempty set called an *alphabet* whose elements are called *letters*. Finite sequences of elements of A are called *words* over A. Let A^* be the set of all words, which is a monoid under the concatenation operation of two words and the empty sequence is the neutral element called the empty word and denoted by 1. The monoid A^* is called the *free monoid* on A. Let $A^+ = A^* \setminus \{1\}$. If $w = a_1 a_2 \cdots a_n$ is a word with $a_i \in A$, then n is called the *length* of w and denoted by $lg(w)$.

$x \in A^*$ is called a *prefix* (resp. *suffix*) of $y \in A^*$, if there exists a $u \in A^*$ such that $y = xu$ (resp. $y = ux$). $x \in A^*$ is called an *infix* (or a *factor*) of $y \in A^*$ if there exist $u, v \in A^*$ such that $y = uxv$. A prefix (resp. suffix, infix) x of y is called *proper*, if $x \neq y$.

Usually, subsets of A^* are called *languages* over A. Remark that such concept are defined on free monoids A^*. We can also define languages on the free objects of some other algebraic systems. Usually, subsets of free monoids are called $*$-*languages* whereas subsets of free semigroups are called $+$-*languages*. Generally speaking, $*$-languages and $+$-languages have parallel theories. But in some occasion, such as dealing with varieties of languages, distinguishing between $*$-languages and $+$-languages is necessary.

In this paper, unless otherwise stated, all languages we mentioned are $*$-languages. The corresponding results for $+$-languages can be similarly obtained.

Let C be a nonempty language over A. C is called a *code* over A if any word $w \in A^*$ has at most one C-factorization. That is

$$x_1x_2\cdots x_m = y_1y_2\cdots y_n, \quad x_i, y_j \in C, i = 1, 2, \ldots, m, j = 1, 2, \ldots, n$$

implies $m = n$ and $x_i = y_i$, $i = 1, 2, \ldots, n$.

Some important classes of codes are listed as follows. Let C be a nonempty langauge over A. C is called

(1) a *prefix code* if $x, xy \in C$ implies $y = 1$;
(2) a *suffix code* if $x, yx \in C$ implies $y = 1$;
(3) a *bifix code* if it is both prefix and suffix;
(4) an *infix code* if $y, xyz \in C$ implies $x = z = 1$;
(5) an *outfix code* if $xz, xyz \in C$ implies $y = 1$;
(6) a *p-infix code* if $y, xyz \in C$ implies $z = 1$;
(7) a *s-infix code* if $y, xyz \in C$ implies $x = 1$.

It can be easily shown that all prefix (resp. suffix, bifix, infix, outfix, p-infix, s-infix) codes other than $\{1\}$ is a code. So for convenience, we treat the singleton $\{1\}$ as a code in this paper.

Let L be a language over A. The submonoid (resp. subsemigroup) generated by L is denoted by L^* (resp. L^+).

A language $L \subseteq A^*$ is called *dense* if any word $w \in A^*$ is a factor of some word in L (or equivalently L intersects with all ideals of A^*).

For a language L over A, the congruence P_L defined by

$$P_L = \{(x, y) \in A^* \times A^* \mid \text{for all } u, v \in A^*, uxv \in L \text{ if and only if } uyv \in L\}$$

is called the *syntactic congruence* of L. And the natural homomorphism $P_L^\natural$ is call the *syntactic homomorphism* of L and denoted by φ_L. The quotient monoid A^*/P_L is called the *syntactic monoid* of L and denoted by $\mathbf{M}(L)$. For any word u, we often use $[u]_L$ to denote the P_L-class containing u.

Let L be a language over A. Then L is said to be *regular* if the syntactic monoid of L is finite (that is, there is only finitely many P_L-classes). It is well known that regular languages play a critical role in theoretical computer science and studied by many authors in literature, which have many equivalent definitions and hence have many different names such as *recognizable languages*, *rational languages*, etc. Regular languages have remarkable algebraic and combinatorial properties and various applications. For theoretical and applied requirements, many kinds of generalizations of regular languages and some relevant non-regular languages are also studied in literature including disjunctive languages and their generalizations.

A language L over A is said to be *disjunctive* if each P_L-class contains exactly one element. Disjunctive languages have been investigated by many authors in literature.[4,5,9–11,16,18–21,23,24]

There are many kinds of generalizations of disjunctive languages. A langauge L over A is said to be

(1) *f-disjunctive*,[6] if each P_L-class is finite;

(2) *nd-disjunctive*,[15] if each P_L-class is not dense;

(3) *ni-disjunctive*,[15] if each P_L-class is not an ideal;

(4) *regular disjunctive*, if each P_L-class is regular;

(5) *(prefix, suffix, bifix, infix, outfix, p-infix, s-infix, etc)code disjunctive*, if each P_L-class is a (prefix, suffix, bifix, infix, outfix, p-infix, s-infix, etc) code;

(6) *relatively f-disjunctive*[7] (resp. *relatively disjunctive*[7]) if there exists a dense language D such that for all $u \in A^*$, $|[u]_L \cap D| < \infty$ (resp. $|[u]_L \cap D| \leq 1$).

In this paper, we mainly study the algebraic properties of these classes of generalized disjunctive languages including syntactic monoids characterizations in Section 2 and the relations and hierarchy of these classes of languages in Section 3.

In the following discussion, $\mathbb{N}$ (resp. $\mathbb{N}^0$) stands for the set of all positive (resp. nonnegative) integers. And $\mathfrak{P}(S)$ represents the power set of a set S.

2. K-Disjunctive languages and universal algebra

In this section, we characterize the syntactic monoids of some classes of generalized disjunctive languages mentioned in the previous section by using universal algebra.

Let $\mathscr{A}$ be a given class of alphabets. We call

$$\mathbf{L} = \{(A, \mathbf{L}(A^*)) \mid A \in \mathscr{A}\}$$

a *language class* (over $\mathscr{A}$), where $\mathbf{L}(A^*)$ is a family of languages over A, that is $\mathbf{L}(A^*) \subseteq \mathfrak{P}(A^*)$. Elements of $\mathbf{L}(A^*)$ are called $\mathbf{L}$ *languages* over A. $\mathbf{L}$ can also been treated as a mapping from $\mathscr{A}$ to $\bigcup_{A\in\mathscr{A}} \mathfrak{P}(\mathfrak{P}(A^*))$, called a *language map* in some literatures. We also denote $\bigcup_{A\in\mathscr{A}} \mathbf{L}(A^*)$ by $\mathbf{L}$ and call its elements $\mathbf{L}$ *languages* if no ambiguity arises.

We usually denote by $\mathbf{D}$ (resp. $\mathbf{D}_f$, $\mathbf{D}_r$, $\mathbf{D}_{rf}$, $\mathbf{D}_{nd}$, $\mathbf{D}_{ni}$, $\mathbf{D_R}$, $\mathbf{D_C}$, $\mathbf{D_P}$, $\mathbf{D_S}$, $\mathbf{D_B}$, $\mathbf{D_I}$, $\mathbf{D_O}$, $\mathbf{D_{PI}}$, $\mathbf{D_{SI}}$) the class of all (resp. f-, r-, rf-, nd-, ni-, R-,

C-, P-, S-, B-, I-, O-, PI-, SI-) disjunctive languages. In fact, most of them can be defined by the following more general manner.

Definition 2.1. Let $\mathbf{K}$ be a language class. A congruence ρ on A^* is said to be $\mathbf{K}$-*disjunctive*, if every ρ-class is a $\mathbf{K}$ language. A language is said to be $\mathbf{K}$-*disjunctive language*, if the syntactic congruence P_L is a $\mathbf{K}$-disjunctive congruence. Denote by $\mathbf{D_K}$ the language class of all $\mathbf{K}$-disjunctive languages (that is $\mathbf{D_K}(A^*)$ contains all $\mathbf{K}$-disjunctive languages of A^*). Notice that if $\mathbf{K}(A^*)$ contains all singleton of A^*, then the class of $\mathbf{K}$-disjunctive language contains all disjunctive languages and can be treat as a generalized disjunctive language.

All classes of the generalized disjunctive languages mentioned above except $\mathbf{D}_r$ and $\mathbf{D}_{rf}$ can be defined by this manner.

Syntactic monoids of f-, r-, rf-, nd-, ni-disjunctive, have been characterized by many authors.[7,14,15,17] Here we list some results as follows which will be used later:

Proposition 2.1 (cf. 15). *Let $L \subseteq A^*$. Then L is nd-disjunctive if and only if its syntactic monoid contains no minimal ideal.*

Proposition 2.2 (cf. 15). *Let $L \subseteq A^*$. Then the following statements are equivalent:*

(1) *L is an ni-disjunctive language;*
(2) *Both L and its complement L^c are dense;*
(3) *$\mathbf{M}(L)$ contains no zero element;*
(4) *A^* contains either no dense P_L-class, or at least two such classes.*

Proposition 2.3 (cf. 7). *Let L be a language over A. Then the following properties are equivalent:*

(1) *$L \in \mathbf{D}_{rf}$;*
(2) *$\mathbf{M}(L)$ contains no finite ideal;*
(3) *If A^* contains dense P_L-classes then it contains infinitely many such classes;*
(4) *$L \in \mathbf{D}_r$.*

We now study the syntactic monoids of the above mentioned generalized disjunctive languages which has not been discussed in literature.

For prefix, suffix, bifix, infix, outfix, p-infix and s-infix disjunctive languages, we can characterize their syntactic monoids by using the following general manner coming from universal algebra.

Let Σ be a special alphabet, elements in which are called *variables* and words over which are called *expressions.* A homomorphism from Σ^* to a monoid M is said to be an *evaluation mapping.* Usually, we use $\alpha(x_1, x_2, \ldots, x_n)$ to represent an expression α, where $\alpha \subseteq \{x_1, x_2, \ldots, x_n\}^*$, $x_1, x_2, \ldots, x_n \in \Sigma$. Use $\alpha(\varphi(x_1), \varphi(x_2), \ldots, \varphi(x_n))$ to represent $\varphi(\alpha(x_1, x_2, \ldots, x_n))$, where φ is an evaluation mapping. Let $\alpha = \alpha(x_1, x_2, \ldots, x_n)$, $\beta = \beta(x_1, x_2, \ldots, x_n)$, $\gamma = \gamma(x_1, x_2, \ldots, x_n)$, $\delta = \delta(x_1, x_2, \ldots, x_n)$ be expressions over Σ, $\Lambda \notin \Sigma^*$ be a symbol called a *language variable* (which stands for a language in the following implications). A language $L \subseteq A^*$ is said to *satisfy* the following implication

$$\alpha, \beta \in \Lambda \ \Rightarrow \ \gamma = \delta,$$

if

$$(\forall w_1, w_2, \ldots, w_n \in A^*) \ \alpha(w_1, w_2, \ldots, w_n), \beta(w_1, w_2, \ldots, w_n) \in L$$

implies

$$\gamma(w_1, w_2, \ldots, w_n) = \delta(w_1, w_2, \ldots, w_n)$$

A language class $\mathbf{L}$ is said to be *defined by* the following set

$$\Gamma = \{\alpha_i, \beta_i \in \Lambda \ \Rightarrow \ \gamma_i = \delta_i\}_{i \in I}$$

of implications, denoted by

$$\mathbf{L} = [\Gamma] = [\{\alpha_i, \beta_i \in \Lambda \ \Rightarrow \ \gamma_i = \delta_i\}_{i \in I}],$$

if L is an $\mathbf{L}$ language if and only if L satisfies all implications in Γ.

A monoid M is said to *satisfy* the following implication

$$\alpha = \beta \ \Rightarrow \ \gamma = \delta,$$

if

$$(\forall w_1, w_2, \ldots, w_n \in M) \ \alpha(w_1, w_2, \ldots, w_n) = \beta(w_1, w_2, \ldots, w_n)$$

implies

$$\gamma(w_1, w_2, \ldots, w_n) = \delta(w_1, w_2, \ldots, w_n).$$

A class of monoids $\mathcal{M}$ is said to be *defined by* the following set

$$\Gamma = \{\alpha_i = \beta_i \ \Rightarrow \ \gamma_i = \delta_i\}_{i \in I}$$

of implications, denoted by

$$\mathcal{M} = [\Gamma] = [\{\alpha_i = \beta_i \ \Rightarrow \ \gamma_i = \delta_i\}_{i \in I}],$$

if $M \in \mathcal{M}$ if and only if M satisfies all implications in Γ.

A congruence ρ on a monoid M is called an 1-*free congruence*, if $1\rho = \{1\}$. A language $L \subseteq A^*$ is called a 1-*free language*, if $[1]_L = \{1\}$. Denote by $\mathbf{1_F}$ the language class of all 1-free languages.

Following our convention, we usually discuss $*$-languages. While the $*$-languages involved in the following theorem are similar to $+$-languages. We call a language class $\mathbf{L}$ a $+$-*like language class*, if for any alphabet A, $\{\{1\}\} \subseteq \mathbf{L}(A^*) \subseteq \mathfrak{P}(A^+) \cup \{\{1\}\}$. For instance, the class of [prefix, suffix, bifix, infix, outfix, p-infix, s-infix] codes are all $+$-like language classes.

Let $\mathcal{M}$ be a class of monoids. Define

$$\mathbb{L}(\mathcal{M}) = \{(A, \mathcal{M}(A^*)) \mid A \in \mathscr{A}, \mathcal{M}(A^*) = \{L \subseteq A^* \mid \mathbf{M}(L) \in \mathcal{M}\}\} .$$

Theorem 2.1. *Let* $\mathbf{K} = [\{\alpha_i, \beta_i \in \Lambda \;\Rightarrow\; \gamma_i = 1\}_{i \in I}]$ *be a* $+$-*like language class. Then a congruence* ρ *on* A^* *is a* $\mathbf{K}$-*disjunctive congruence if and only if* $1\rho = \{1\}$ *and*

$$A^*/\rho \in [\{\alpha_i = \beta_i \;\Rightarrow\; \gamma_i = 1\}_{i \in I}].$$

In particular, $L \subseteq A^*$ *is a* $\mathbf{K}$-*disjunctive language if and only if* $[1]_L = \{1\}$ *and*

$$\mathbf{M}(L) \in [\{\alpha_i = \beta_i \;\Rightarrow\; \gamma_i = 1\}_{i \in I}].$$

Therefore,

$$\mathbf{D_K} = \mathbb{L}[\{\alpha_i = \beta_i \;\Rightarrow\; \gamma_i = 1\}_{i \in I}] \cap \mathbf{1_F}.$$

Proof. Let $M = A^*/\rho$, $\alpha_i = \alpha_i(x_1, x_2, \ldots, x_{n_i})$, $\beta_i = \beta_i(x_1, x_2, \ldots, x_{n_i})$, $\gamma_i = \gamma_i(x_1, x_2, \ldots, x_{n_i})$.

Necessity. Since $\mathbf{K}$ is a $+$-like language class, the only $\mathbf{K}$ language containing 1 is the singleton set $\{1\}$. Thus $1\rho = \{1\}$.

Next we prove M satisfies the implications $\alpha_i = \beta_i \;\Rightarrow\; \gamma_i = 1$, $i \in I$. Suppose that

$$\alpha_i(m_1, m_2, \ldots, m_{n_i}) = \beta_i(m_1, m_2, \ldots, m_{n_i}),$$

where $m_1, m_2, \ldots, m_{n_i} \in M$. Let $w_j \in A^*$ satisfying $\rho^\natural(w_j) = m_j$, $j = 1, 2, \ldots, n_i$. Then we have

$$\begin{aligned}
\rho^\natural(\alpha_i(w_1, w_2, \ldots, w_{n_i})) &= \alpha_i(\rho^\natural(w_1), \rho^\natural(w_2), \ldots, \rho^\natural(w_{n_i})) \\
&= \alpha_i(m_1, m_2, \ldots, m_{n_i}) \\
&= \beta_i(m_1, m_2, \ldots, m_{n_i}) \\
&= \beta_i(\rho^\natural(w_1), \rho^\natural(w_2), \ldots, \rho^\natural(w_{n_i})) \\
&= \rho^\natural(\beta_i(w_1, w_2, \ldots, w_{n_i})).
\end{aligned}$$

Hence $\alpha_i(w_1, w_2, \ldots, w_{n_i})$ and $\beta_i(w_1, w_2, \ldots, w_{n_i})$ are in the same ρ-class. Then by the fact that every ρ-class is a **K** language, and **K** satisfies $\alpha_i, \beta_i \in \Lambda \;\Rightarrow\; \gamma_i = 1$, we know that $\gamma_i(w_1, w_2, \ldots, w_{n_i}) = 1$. Hence

$$\gamma_i(m_1, m_2, \ldots, m_{n_i}) = \rho^\natural(\gamma_i(w_1, w_2, \ldots, w_{n_i})) = 1.$$

That is M satisfies the implications $\alpha_i = \beta_i \;\Rightarrow\; \gamma_i = 1$, $i \in I$. Thus,

$$M \in [\{\alpha_i = \beta_i \;\Rightarrow\; \gamma_i = 1\}_{i \in I}].$$

Sufficiency. Let C be a ρ-class. Suppose that

$$\alpha_i(w_1, w_2, \ldots, w_{n_i}), \beta_i(w_1, w_2, \ldots, w_{n_i}) \in C,$$

where $w_1, w_2, \ldots, w_{n_i} \in A^*$. Let $m_j = \rho^\natural(w_j)$, $j = 1, 2, \ldots, n_i$. Then we have

$$\begin{aligned}
\alpha_i(m_1, m_2, \ldots, m_{n_i}) &= \alpha_i(\rho^\natural(w_1), \rho^\natural(w_2), \ldots, \rho^\natural(w_{n_i})) \\
&= \rho^\natural(\alpha_i(w_1, w_2, \ldots, w_{n_i})) \\
&= \rho^\natural(\beta_i(w_1, w_2, \ldots, w_{n_i})) \\
&= \beta_i(\rho^\natural(w_1), \rho^\natural(w_2), \ldots, \rho^\natural(w_{n_i})) \\
&= \beta_i(m_1, m_2, \ldots, m_{n_i}).
\end{aligned}$$

Then by the fact that M satisfies $\alpha_i = \beta_i \;\Rightarrow\; \gamma_i = 1$, we know that

$$\gamma_i(m_1, m_2, \ldots, m_{n_i}) = 1.$$

Notice that $1\rho = \{1\}$, we have $\gamma_i(w_1, w_2, \ldots, w_{n_i}) = 1$. That is C satisfies implications $\alpha_i, \beta_i \in \Lambda \;\Rightarrow\; \gamma_i = 1$, $i \in I$. Thus C is a **K** language, which implies that ρ is a **K**-disjunctive congruence. □

Many **K**-disjunctive language classes can be characterized by the above theorem. For instance, we have

Corollary 2.1. *In the following table, we list the syntactic characterizations of some generalized disjunctive languages defined by given classes of codes.*

	$\mathbf{K}$	$\mathbf{D_K}$
prefix code	$\mathbf{P} = [x, xy \in \Lambda \;\Rightarrow\; y = 1]$	$\mathbf{D_P} = \mathbb{L}[x = xy \;\Rightarrow\; y = 1] \cap \mathbf{1_F}$
suffix code	$\mathbf{S} = [x, yx \in \Lambda \;\Rightarrow\; y = 1]$	$\mathbf{D_S} = \mathbb{L}[x = yx \;\Rightarrow\; y = 1] \cap \mathbf{1_F}$
bifix code	$\mathbf{B} = \mathbf{P} \cap \mathbf{S}$	$\mathbf{D_B} = \mathbf{D_P} \cap \mathbf{D_S}$
infix code	$\mathbf{I} = [y, xyz \in \Lambda \;\Rightarrow\; x = z = 1]$	$\mathbf{D_I} = \mathbb{L}[y = xyz \;\Rightarrow\; x = z = 1] \cap \mathbf{1_F}$
outfix code	$\mathbf{O} = [xz, xyz \in \Lambda \;\Rightarrow\; y = 1]$	$\mathbf{D_O} = \mathbb{L}[xz = xyz \;\Rightarrow\; y = 1] \cap \mathbf{1_F}$
p-infix code	$\mathbf{PI} = [y, xyz \in \Lambda \;\Rightarrow\; z = 1]$	$\mathbf{D_{PI}} = \mathbb{L}[y = xyz \;\Rightarrow\; z = 1] \cap \mathbf{1_F}$
s-infix code	$\mathbf{SI} = [y, xyz \in \Lambda \;\Rightarrow\; x = 1]$	$\mathbf{D_{SI}} = \mathbb{L}[y = xyz \;\Rightarrow\; x = 1] \cap \mathbf{1_F}$

Similar results can also be given for some other classes of codes such as the so-called shuffle codes including n-prefix codes $\mathbf{P}_n$, n-suffix codes $\mathbf{S}_n$, n-infix codes $\mathbf{I}_n$, n-outfix codes $\mathbf{O}_n$ and hypercodes $\mathbf{H}$. While all kinds of intercodes including comma-free codes can not be used to construct generalized disjunctive languages in this manner. Since such codes contain primitive words only, but for any congruence, it is impossible that every congruence class contains only primitive words.

A class of monoids closed under the three operations: homomorphism, submonoid and direct product are called *variety*, which is equivalent to say that this class of monoids can be defined by identities by a well-known Birkhoff Theorem in universal algebra (cf. 1 or 2). The langauge class $\mathbb{L}(\mathcal{M})$ defined by a variety $\mathcal{M}$ has good algebraic closure properties (which is closed under boolean operation, inverse homomorphism, left and right quotients by Eilenberg's Correspondence Theorem[3]).

The class of monoids defined by implications is called a *quasivariety*. It can be easily checked that any quasivariety is closed under isomorphism, submonoid and direct product but not necessary closed under homomorphism, which cause the language class $\mathbb{L}(\mathcal{M})$ defined by a quasivariety $\mathcal{M}$ possibly has less algebraic closure properties. In fact, it is easy to check that the language class $\mathbf{D_P}$, $\mathbf{D_S}$, $\mathbf{D_B}$, $\mathbf{D_I}$, $\mathbf{D_O}$, $\mathbf{D_{PI}}$, $\mathbf{D_{SI}}$ is a *anti-AFL*. That is it is not closed under the operations of union, concatenation, "+", 1-free homomorphism, inverse homomorphism, intersection with regular languages. Up to now, we do not know that a class of monoids $\mathcal{M}$ being closed under submonoid and direct product can bring $\mathbb{L}(\mathcal{M})$ what kinds of closure properties. So, unlike varieties, we do not know language classes defined by quasivarieties have what kinds of common properties.

For syntactic monoids of regular disjunctive languages, we have the following characterization.

A monoid M is said to be *residually finite*,[12] if any pair of distinct elements of M can be separated by a finite-indexed congruence on M, that is to say, for any $x, y \in M$, $x \neq y$, there exists a congruence ρ on M with finite index such that $(x, y) \notin \rho$. It is easy to show that, a monoid is residually finite if and only if it is isomorphic to a subdirect product of some finite monoids.

Let M be a monoid. $x \in M$ is said to be *uniformly residually finite*, if there is a congruence ρ on M with finite index such that for any $y \in M$, $x \neq y$, $(x, y) \notin \rho$ (That is $x\rho = \{x\}$). M is said to be *uniformly residually finite*, if every element of M is uniformly residually finite.

A monoid M_1 is said to *divide* a monoid M_2, denoted by $M_1 \mid M_2$, if M_1 is a homomorphic image of some submonoid of M_2.

Lemma 2.1 (cf. 12). *Let L be a language over A, M a monoid. Then $\mathbf{M}(L) \mid M$ if and only if there exist a homomorphism φ from A^* to M and $P \subseteq M$ such that $L = \varphi^{-1}(P)$.*

Lemma 2.2. *Let M be a finitely generated monoid, A a finite alphabet, $\varphi : A^* \to M$ a surjective homomorphism. Then for any $x \in M$, $\varphi^{-1}(x)$ is regular if and only if x is uniformly residually finite.*

Proof. Let $L = \varphi^{-1}(x)$. Then there is a surjective homomorphism ψ from M to $\mathbf{M}(L)$ such that $\varphi_L = \psi \circ \varphi$ (cf. 12, Chapter 6, Lemma 5.1). Hence

$$\begin{aligned}\psi^{-1}\psi(x) &= \psi^{-1}\psi(\varphi(L))\\ &= \psi^{-1}\varphi_L(L)\\ &= \varphi\varphi^{-1}\psi^{-1}\varphi_L(L)\\ &= \varphi\varphi_L^{-1}\varphi_L(L)\\ &= \varphi(L)\\ &= \{x\}.\end{aligned}$$

Let $\rho = \operatorname{Ker}\psi$. Then we have $x\rho = \{x\}$. If L is regular, then $\mathbf{M}(L)$ is finite, which implies the index of ρ is finite. Therefore, x is uniformly residually finite.

Conversely, if x is uniformly residually finite, then there is a congruence ρ on M with finite index such that $x\rho = \{x\}$. Let $\psi = \rho^{\natural}$, then we have $\psi^{-1}\psi(x) = \{x\}$. Hence $\varphi^{-1}\psi^{-1}\psi\varphi(L) = L$. Then by Proposition 2.1, $\mathbf{M}(L)$ is a homomorphic image of M/ρ. Therefore, $\mathbf{M}(L)$ is finite, that is L is regular. □

By the definition of uniformly residual finiteness and Lemma 2.2, we have:

Theorem 2.2. *Let M be a finitely generated monoid. Then the following statements are equivalent.*

(1) *M is uniformly residually finite;*

(2) *For any finite alphabet A and any surjective homomorphism φ from A^* to M, $Ker\varphi$ is a regular disjunctive congruence on A^*;*

(3) *There exist a finite alphabet A and a surjective homomorphism φ from A^* to M, such that $Ker\varphi$ is a regular disjunctive congruence on A^*;*

(4) *S is isomorphic to a subdirect product of a family $\{S_i\}_{i\in I}$ of finite monoids satisfying the following condition: for any $x \in S$, there exists $i \in I$ such that $\pi_i^{-1}\pi_i(x) = \{x\}$.*

Corollary 2.2. *$L \subseteq A^*$ is a regular disjunctive language if and only if $\mathbf{M}(L)$ is uniformly residually finite.*

Remark 2.1. Clearly, every uniformly residually finite monoid is residually finite. While the converse is not necessarily true. For instance, let $A = \{a, b\}$, $L = \{w \in A^* \mid w_a = w_b\}$. Direct computations show that L is not a regular disjunctive language. Hence $\mathbf{M}(L) \simeq \mathbb{Z}$ is not uniformly residually finite. On the other hand, since for any $m, n \in \mathbb{Z}$, $m > n$, we have $(m, n) \notin \rho_k$, where $k = m - n + 1$ and $\rho_k = \{(x, y) \in \mathbb{Z} \times \mathbb{Z} \mid x \equiv y \pmod{k}\}$ is a congruence with finite index. Therefore, $\mathbb{Z}$ is a residually finite monoid.

Remark 2.2. It is easy to show that the class of all residually finite monoids is a so-called semivariety, that is, it is closed under isomorphism, submonoid and direct product. While the class of all uniformly residually finite monoids is not a semivariety, it is closed under isomorphism, submonoid and finite direct product but not closed under arbitrary direct product.

Regular disjunctive languages can be treat as a common generalization of regular languages and disjunctive languages. Characterizing their syntactic monoids may be a valuable work.

3. Generalized disjunctive hierarchy

We now discuss the relationship among some classes of generalized disjunctive languages mentioned in the previous section. The main result is Theorem 3.1. For this purpose, we will make some preparations first.

For any $L \subseteq A^*$, let

$$\begin{aligned} P(L) &= \{u \in A^+ \mid L \cap uA^+ \neq \emptyset\}; \\ S(L) &= \{u \in A^+ \mid L \cap A^+u \neq \emptyset\}; \\ I(L) &= \{u \in A^* \mid L \cap A^*uA^* \neq \emptyset\}. \end{aligned}$$

A language L over A is said to be *overlap-free* if $P(u) \cap S(v) = \emptyset$ for any $u, v \in L$.

Definition 3.1 (cf. 22). *For a nonempty language $L \subseteq A^+$ and any $w \in A^*$, a factorization of w on A^**

$$w = x_1y_1x_2y_2 \cdots x_ny_nx_{n+1}$$

such that $y_i \in L$ *and* $I(x_j) \cap L = \emptyset$, $i = 1, 2, \ldots, n$, $j = 1, 2, \ldots, n+1$, $n \geq 0$, *is called an* L-representation *of* w, *and* $(x_1, x_2, \ldots, x_{n+1})$ *is called the coefficients of the representation. If every word* w *has a unique* L*-representation, then we call* L *a* solid code.

Proposition 3.1 (cf. 8,22). *Let* $L \subseteq A^+$ *and* $L \neq \emptyset$. *Then* L *is a solid code if and only if* L *is an overlap-free infix code.*

Definition 3.2 (cf. 13). *Let* S *be a solid code and* w *a word over* A. *Then* w *has a unique* S*-representation* $w = t_1 s_1 t_2 s_2 \cdots t_l s_l t_{l+1}$. *We call the above* l *the* S-length *of* w *and denote it by* $l_S(w)$. *The number* $\max(lg(t_1), lg(t_2), \ldots, lg(t_{l+1}))$ *associated to the representation is called the* S-height *of* w *and denoted by* $h_S(w)$. *If* $l_S(w) \geq 2$, *the number* $\max(lg(t_2), \ldots, lg(t_l))$ *is called the* inner S-height *of* w *and denoted by* $h_S^*(w)$, *if* $l_S(w) < 2$, *then let* $h_S^*(w) = 0$. *We also denote* t_1 *and* t_{l+1} *by* $p_S(w)$ *and* $s_S(w)$ *respectively. Let* $m_S(w) = \max(r(l_S(w)), r(h_S(w))) = r(\max(l_S(w), h_S(w)))$ *and* $m_S^*(w) = \max(r(l_S(w)), r(h_S^*(w))) = r(\max(l_S(w), h_S^*(w)))$, *where*

$$r(n) = \begin{cases} 2^{\lceil \log_2 n \rceil} & \text{if } n > 0; \\ 0 & \text{if } n = 0. \end{cases}$$

That is, $r(n)$ *is the least number in* $\{0, 2^0, 2^1, \ldots\}$ *which is not less than* n. *For the sake of simplicity, we use* $l(w)$, $h(w)$, $h^*(w)$, $m(w)$, $m^*(w)$, $p(w)$ *and* $s(w)$ *to represent* $l_S(w)$, $h_S(w)$, $h_S^*(w)$, $m_S(w)$, $m_S^*(w)$, $p_S(w)$ *and* $s_S(w)$ *respectively if the solid code* S *is known in the context. Now, by means of the solid code* S, *we are ready to construct the following language* $L = L(S)$ *which will be used in the proof of the main result of this section. For any* $p, s \in A^*$, $l \in \{0, 1, \ldots\}$, $m, m^* \in M = \{0, 2^0, 2^1, \ldots\}$, $m, m^* \geq r(l)$, *let*

$$U(l, m) = \{w \in A^* \mid l(w) = l \text{ and } m(w) = m\};$$
$$V(l, m, m^*, p, s) = \{w \in U(l, m) \mid p(w) = p, s(w) = s \text{ and } m^*(w) = m^*\};$$
$$L = L(S) = \bigcup_{m \in M} U(m, m) = \{1\} \cup \bigcup_{l=0}^{\infty} U(2^l, 2^l).$$

For the sake of simplicity, for any word w, *we also denote* $U(l(w), m(w))$ *and* $V(l(w), m(w), m^*(w), p(w), s(w))$ *by* $U(w)$ *and* $V(w)$ *respectively.*

Then we have:

Proposition 3.2 (cf. 13). *Let* S *be a solid code,* $L = L(S)$. *Then for any* $w \in A^*$ *we have*

(1) $V(w) \subseteq [w]_L \subseteq U(w)$.

(2) $[w]_L$ *is an infix code.*

(3) *If* $p(w) = s(w) = 1$, *then* $[w]_L = V(w)$. *In particular,* S *is a* P_L-*class.*

Proposition 3.3 (cf. 13). *Let* S *be a solid code over* A, $L = L(S)$. *Then* S *is finite if and only if* L *is f-disjunctive.*

Proposition 3.4. *Let* S *be a solid code over* A, $L = L(S)$. *Then* S *is regular if and only if* L *is regular disjunctive.*

Proof. By Proposition 3.2(3), S is a P_L-class. Therefore, if L is regular disjunctive, then S must be regular.

Conversely, suppose that S is regular. Since S is a P_L-class, for any $n \in \mathbb{N}^0$, $x_i \in A^*$, $i = 1, 2, \ldots, n+1$, $x_1 S x_2 S \cdots x_n S x_{n+1}$ must be contained in a P_L-class. Moreover, by the regularity of S, we know that $x_1 S x_2 S \cdots x_n S x_{n+1}$ is regular. For any $n \in \mathbb{N}^0$, let $A^{\leq n}$ denote the set of all words with length $\leq n$. By Proposition 3.2(1), we have $[w]_L \subseteq U(w) \subseteq (A^{\leq m} S)^l A^{\leq m}$ for any $w \in A^*$, $l_S(w) = l$, $m_S(w) = m$. Then $[w]_L$ is a finite union of the form $x_1 S x_2 S \cdots x_l S x_{l+1}$, where $x_i \in A^{\leq m}$. Thus $[w]_L$ is regular, which implies L is regular disjunctive. □

Definition 3.3. Let L be a nonempty language in A^+, $w \in A^*$. An L-representation

$$w = x_1 y_1 x_2 y_2 \cdots x_n y_n x_{n+1}$$

of w is said to be *maximal*, if any infix of w properly containing y_i does not belong to L.

Proposition 3.5. *If* O *is an overlap-free language over* A, *then every word over* A *has a unique maximal* O-*representation.*

Proof. Suppose that the proposition is not true. Let $w \in A^*$ be a word with minimal length among the words with at least two maximal O-representations. Let

$$w = x_1 y_1 x_2 y_2 \cdots x_n y_n x_{n+1},$$

$$w = x'_1 y'_1 x'_2 y'_2 \cdots x'_{n'} y'_{n'} x'_{n'+1}$$

be two distinct maximal O-representations of w.

By the minimality of $lg(w)$, we have $x_1 \neq x'_1$. Suppose, without loss of generality, that $lg(x_1) < lg(x'_1)$. Then by $y_1, y'_1 \in O$ and O is overlap-free, we know that y'_1 is a proper infix of y_1, which contradicts the maximality of the representation

$$w = x'_1 y'_1 x'_2 y'_2 \cdots x'_{n'} y'_{n'} x'_{n'+1}.$$

Therefore, every word has a unique maximal O-representation. □

Remark 3.1. The converse of the above proposition is not necessarily true. For example, let $O = A^*$. Then every word w over A has a unique maximal O-representation, i.e. $w = w$. Clearly, O is not overlap-free. So, it is quite different from the corresponding property of solid codes (cf. Proposition 3.1).

By Proposition 3.5, we have the following definition.

Definition 3.4. Let O be an overlap-free language over A. Then the l in the unique maximal O-representation

$$w = x_1 y_1 x_2 y_2 \cdots x_l y_l x_{l+1}$$

is called the O-length of w, denoted by $l_O(w)$.

Proposition 3.6. *Let O be an overlap-free prefix code or suffix code over A,*

$$U_O(n) = \{w \in A^* \mid l_O(w) = n\}, \quad n \in \mathbb{N}^0,$$

$$L = L_O = \bigcup_{n=0}^{\infty} U_O(2^n).$$

Then

(1) *For any $w \in A^*$, $l_O(w) = l$, we have $[w]_L \subseteq U_O(l)$;*
(2) *L is an nd-disjunctive language over A.*

Proof. We need only prove the case of that O is a prefix code.

(1) Let $w = x_1 y_1 x_2 y_2 \cdots x_l y_l x_{l+1}$ be the maximal O-representation of w. For any $w' \in [w]_L$, let $l_O(w') = l'$. Suppose that $l > l'$. Let r be a positive integer satisfying $l - l' < 2^{r-1}$ and u an arbitrary word in O. Then by the fact that O is an overlap-free prefix code, we obtain that

$$u^{2^r - l} w = \overbrace{1u1u1 \cdots 1u}^{2^r - l \ \ u} x_1 y_1 x_2 y_2 \cdots x_l y_l x_{l+1}$$

is the maximal O-representation of $u^{2^r-l}w$. Hence $l_O(u^{2^r-l}w) = 2^r$, which implies that $u^{2^r-l}w \in L$. Similarly, $l_O(u^{2^r-l}w') = 2^r - (l - l')$. Since $2^{r-1} <$

$2^r - (l - l') < 2^r$, we have $u^{2^r - l} w' \notin L$, This contradicts $(w, w') \in P_L$. Thus $l \leq l'$. Symmetrically, one can prove that $l' \leq l$. Thus $l = l'$, that is $w' \in U_O(l)$.

(2) is a direct consequence of (1). □

Lemma 3.1. *Let ρ be a congruence on A^*, $w \in A^*$. If $w\rho$ is finite, then $w\rho$ is an infix code.*

Proof. Suppose that $x, uxv \in w\rho$, where $u, v, x \in A^*$. Then for any $n \in \mathbb{N}$, $u^n x v^n \in w\rho$. If $uv \neq 1$, then $w\rho$ is infinite. Thus, if $w\rho$ is finite, then $u = v = 1$. That is $w\rho$ is an infix code. □

Lemma 3.2. *Every infix code is not dense.*

Proof. Let L be an infix code over A, $w \in L$, $a \in A$. Then for any $u, v \in A^*$, $uwav \notin L$. Thus, L is not dense. □

Lemma 3.3 (cf. 18). *If $|A| = 1$, then a language over A is disjunctive if and only if it is not regular.*

Theorem 3.1. (1) *If $|A| = 1$, then*

$$\mathbf{D}(A^*) = \mathbf{D}_f(A^*) = \mathbf{D}_\mathbf{I}(A^*) = \mathbf{D}_{nd}(A^*) = \mathbf{D}_{rf}(A^*) = \mathbf{D}_{ni}(A^*) \setminus \mathbf{R}(A^*).$$

(2) *If $|A| \geq 2$, then*

$$\mathbf{D}(A^*) \subsetneq \mathbf{D}_f(A^*) \subsetneq \mathbf{D}_\mathbf{I}(A^*) \subsetneq \mathbf{D}_{nd}(A^*) \subsetneq \mathbf{D}_{rf}(A^*) \subsetneq \mathbf{D}_{ni}(A^*) \setminus \mathbf{R}(A^*).$$

For regular disjunctive languages, we have

(3) $\mathbf{D}_f \cup \mathbf{R} \subseteq \mathbf{D}_\mathbf{R}$.

(4) *If $|A| = 1$, then every language over A is regular disjunctive.*

(5) *If $|A| \geq 2$, $\mathbf{D}_\mathbf{R}(A^*)$ is not comparable with $\mathbf{D}_\mathbf{I}(A^*)$, $\mathbf{D}_{nd}(A^*)$, $\mathbf{D}_{rf}(A^*)$, $\mathbf{D}_{ni}(A^*)$ under set inclusion. (The relationship among these language classes is shown in Figure 1)*

Proof. By the definitions of these classes of generalized disjunctive languages and Proposition 2.1, 2.2, 2.3, Lemma 3.1 and 3.2 we have

$$\mathbf{D}(A^*) \subseteq \mathbf{D}_f(A^*) \subseteq \mathbf{D}_\mathbf{I}(A^*) \subseteq \mathbf{D}_{nd}(A^*) \subseteq \mathbf{D}_{rf}(A^*) \subseteq \mathbf{D}_{ni}(A^*) \setminus \mathbf{R}(A^*),$$

(1) By Lemma 3.3, when the alphabet A contains only one letter, a language over A is either regular or disjunctive. Hence, $\mathbf{D}_{ni}(A^*) \setminus \mathbf{R}(A^*) \subseteq \mathbf{D}(A^*)$. Thus we have

$$\mathbf{D}(A^*) = \mathbf{D}_f(A^*) = \mathbf{D}_{nd}(A^*) = \mathbf{D}_{rf}(A^*) = \mathbf{D}_{ni}(A^*) \setminus \mathbf{R}(A^*).$$

(2) It can be easily shown that $L_0 = \bigcup_{n=0}^{\infty} A^{2^n} \in \mathbf{D}_f(A^*) \setminus \mathbf{D}(A^*)$, which shows the inclusion $\mathbf{D}(A^*) \subseteq \mathbf{D}_f(A^*)$ is proper. By the proof of (5) below, we know that all other inclusions of (2) is also proper.

(3) Since finite language is trivially regular, we have $\mathbf{D}_f \subseteq \mathbf{D_R}$. The inclusion $\mathbf{R} \subseteq \mathbf{D_R}$ is a consequence of Corollary 2.2.

(4) is a consequence of (3) and Lemma 3.3.

(5) Let $a, b \in A$, $D = \{a^n b^n \mid n \in \mathbb{N}\}$, $E = \{a^{2^n} \mid n \in \mathbb{N}^0\}$,

$L_1 = L(S)$, where S is an infinite regular solid code, the operator L follows Definition 3.2,

$L_2 = L(S')$, where S' is a non-regular solid code, operator L has the same meaning as the above statement,

$L_3 = \{w \in A^* \mid w_a = 2^n, n \in \mathbb{N}^0\}$,

$L_4 = L_D = \{w \in A^* \mid l_D(w) = 2^n, n \in \mathbb{N}^0\}$,

$L_5 = EbA^*$,

$L_6 = \{w \in A^* \mid w_a = w_b\}$,

$L_7 = E \cup A(A^2)^*$,

$L_8 = D \cup A(A^2)^*$,

$L_9 = (A^2)^*$,

$L_{10} = E$,

$L_{11} = \{a\}$,

$L_{12} = D$.

(i) By Proposition 3.2(2), 3.3 and 3.4, we have $L_1 \in (\mathbf{D_I}(A^*) \cap \mathbf{D_R}(A^*)) \setminus \mathbf{D}_f(A^*)$, $L_2 \in \mathbf{D_I}(A^*) \setminus \mathbf{D_R}(A^*)$.

(ii) Direct computation shows that

$$\mathbf{M}(L_3) = \{C_n \mid n \in \mathbb{N}^0\},$$

where $C_n = \{w \in A^* \mid w_a = n\}$. Clearly, C_n is non-dense regular language, but not an infix code, $n \in \mathbb{N}^0$. Thus $L_3 \in (\mathbf{D}_{nd}(A^*) \cap \mathbf{D_R}(A^*)) \setminus \mathbf{D_I}(A^*)$.

(iii) Notice that D is an overlap-free bifix code, then by Proposition 3.6(2), L_4 is an nd-disjunctive language over A.

Next we show that D is a P_{L_4}-class. For any $u, v \in A^*$, $x \in D$, we have $l_D(uxv) = l_D(u) + l_D(v) + 1$. Hence for any $x, y \in D$, $uxv \in L_4$ if and only if $uyv \in L_4$. That is D is contained in a P_{L_4}-class.

Let $w \in D$, $w' \in A^*$, $(w, w') \in P_{L_4}$. Then by Proposition 3.6(1), $l_D(w') = l_D(w) = 1$. Let $w' = uxv$ be the maximal L_4-representation of w', $x \in D$, $u, v \in A^*$. Then we have $uv \in b^*a^*$ and $(x, w') \in P_{L_4}$. If $u \notin a^*$, then $u = b^i a^j$ for some $i \geq 1$, $j \geq 0$. Then $l_D(aba^i w') = l_D(aba^i b^i a^j xv) = 3$ and $l_D(aba^i x) = 2$. Hence $aba^i w' \notin L_4$ while $aba^i x \in L_4$, which contradicts

$(x, w') \in P_{L_4}$. Thus $u \in a^*$. Similarly, $v \in b^*$. Therefore, $uv \in a^*b^*$, which together with $uv \in b^*a^*$ implies $u = v = 1$, and hence $w' = x \in D$.

Now we have proved that D is a P_{L_4}-class. Since D is neither a regular language nor an infix code over A, we have $L_4 \in \mathbf{D}_{nd}(A^*) \setminus (\mathbf{D}_{\mathbf{R}}(A^*) \cup \mathbf{D}_{\mathbf{I}}(A^*))$.

(iv) It is easy to show that

$$\mathbf{M}(L_5) = \{a^n bA^* \mid n \in \mathbb{N}^0\} \cup \{\{a^n\} \mid n \in \mathbb{N}^0\}.$$

Hence L_5 is regular disjunctive. Since there exist infinitely many dense P_{L_5}-class $a^n bA^*$, $n \in \mathbb{N}$, by Theorem 2.3 and the definition of nd-disjunctive languages, L_5 is an rf-disjunctive language but not an nd-disjunctive language. Thus $L_5 \in (\mathbf{D}_{rf}(A^*) \cap \mathbf{D}_{\mathbf{R}}(A^*)) \setminus \mathbf{D}_{nd}(A^*)$.

(v) $\mathbf{M}(L_6) = \{C_n \mid n \in \mathbb{Z}\}$, where $C_n = \{w \in A^* \mid w_a = w_b + n\}$. It is easy to check that C_n is a dense non-regular language. Hence L_6 is neither nd-disjunctive nor regular disjunctive. While by Theorem 2.3, L_6 is rf-disjunctive. Thus $L_6 \in \mathbf{D}_{rf}(A^*) \setminus (\mathbf{D}_{\mathbf{R}}(A^*) \cup \mathbf{D}_{nd}(A^*))$.

(vi) $\mathbf{M}(L_7) = \{C_1, C_2\} \cup \{\{a^n\} \mid n \in \mathbb{N}^0\}$, where $C_1 = (A^* \setminus a^*) \cap (A^2)^*$, $C_2 = (A^* \setminus a^*) \cap A(A^2)^*$. Clearly, L_7 is regular disjunctive but not regular. By Proposition 2.2 and 2.3, $L_7 \in \mathbf{D}_{ni}(A^*) \setminus \mathbf{D}_{rf}(A^*)$. Thus $L_7 \in (\mathbf{D}_{ni}(A^*) \cap \mathbf{D}_{\mathbf{R}}(A^*)) \setminus (\mathbf{R}(A^*) \cup \mathbf{D}_{rf}(A^*))$.

(vii) $\mathbf{M}(L_8) = \{C_1, C_2\} \cup \{D_n \mid n \in \mathbb{Z}\}$, where $C_1 = (A^* \setminus a^*b^*) \cap (A^2)^*$, $C_2 = (A^* \setminus a^*b^*) \cap A(A^2)^*$, $D_n = \{a^i b^j \mid i, j \in \mathbb{N}^0, i - j = n\}$. Clearly, both C_1 and C_2 are dense regular language. And it is easy to check that D_n is neither regular nor dense. Thus $L_8 \in \mathbf{D}_{ni}(A^*) \setminus (\mathbf{D}_{\mathbf{R}}(A^*) \cup \mathbf{D}_{rf}(A^*))$.

(viii) L_9 is obviously a regular language. Moreover, both L_9 and its complement are dense. Thus $L_9 \in \mathbf{R}(A^*) \cap \mathbf{D}_{ni}(A^*)$.

(ix) By $\mathbf{M}(L_{10}) = \{A^* \setminus a^*\} \cup \{\{a^n\} \mid n \in \mathbb{N}^0\}$, we have L_{10} is regular disjunctive but not regular. Moreover, since there is only one dense $P_{L_{10}}$-class, $L_{10} \notin \mathbf{D}_{ni}$. Thus $L_{10} \in \mathbf{D}_{\mathbf{R}}(A^*) \setminus (\mathbf{D}_{ni}(A^*) \cup \mathbf{R}(A^*))$.

(x) Since L_{11} is regular but not dense, we have $L_{11} \in \mathbf{R}(A^*) \setminus \mathbf{D}_{ni}(A^*)$.

(xi) $\mathbf{M}(L_{12}) = \{A^* \setminus a^*b^*\} \cup \{D_n \mid n \in \mathbb{Z}\}$, where $D_n = \{a^i b^j \mid i, j \in \mathbb{N}^0, i - j = n\}$. Since there is only one dense $P_{L_{12}}$-class, $L_{12} \notin \mathbf{D}_{ni}$. Since D_n is not regular, $n \in \mathbb{Z}$, we have $L_{12} \notin \mathbf{D}_{\mathbf{R}}$. Thus $L_{12} \notin \mathbf{D}_{ni}(A^*) \cup \mathbf{D}_{\mathbf{R}}(A^*)$.

(i)–(xi) not only proves that all inclusions $\mathbf{D}_f(A^*) \subsetneq \mathbf{D}_{\mathbf{I}}(A^*) \subsetneq \mathbf{D}_{nd}(A^*) \subsetneq \mathbf{D}_{rf}(A^*) \subsetneq \mathbf{D}_{ni}(A^*) \setminus \mathbf{R}(A^*)$ are proper, but also shows that $\mathbf{D}_{\mathbf{R}}(A^*)$ is not comparable with $\mathbf{D}_{\mathbf{I}}(A^*)$, $\mathbf{D}_{nd}(A^*)$, $\mathbf{D}_{rf}(A^*)$ and $\mathbf{D}_{ni}(A^*)$ under set inclusion (cf. Figure 1). □

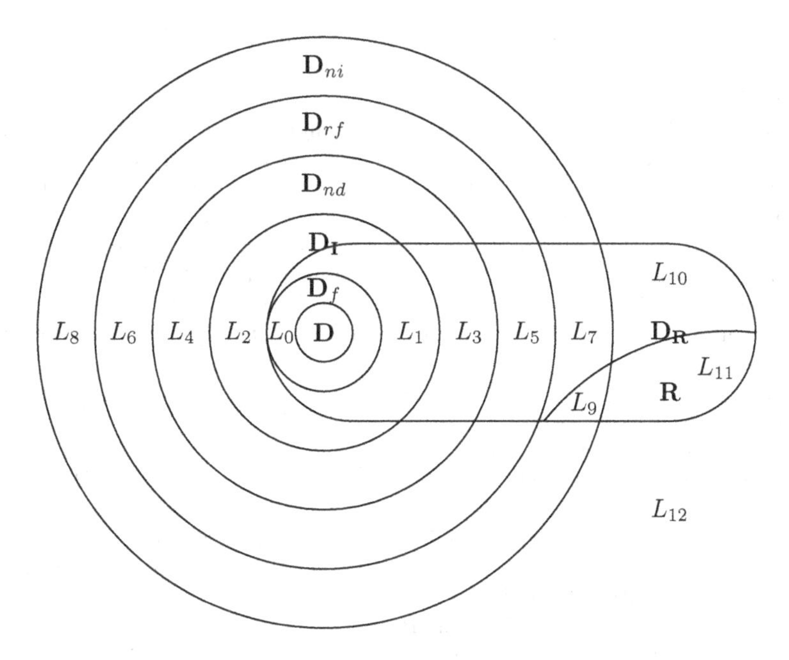

Fig. 1. Generalized disjunctive hierarchy

Acknowledgements

The research is supported by National Natural Science Foundation of China (Grant No. 11101354) and Natural Science Foundation of Yunnan Province, China (Grant No. 2008ZC162M).

References

1. Burris, S. N. and Sankappanavar, H. P., A Course in Universal Algebra, Graduate Texts in Mathematics, No. 78, Springer-Verlag, 1981.
2. Cohn, P. M., Universal Algebra, D. Reidel Publishing Co., Dordrecht, Revised Ed., 1981.
3. Eilenberg, S., Automata, Languages and Machines, Vol. B, Academic Press, 1976.
4. Guo, Y. Q., Li, L. and Xu, G. W., On the Disjunctive Structure of Dense Languages, Scientia Sinica, Ser. A, Vol. 28, No. 12 (1985), 1233–1238.
5. Guo, Y. Q., Xu, G. W. and Thierrin, G., Disjunctive Decomposition of Languages, Theoretical Computer Science, Vol. 46, No. 1 (1986), 47–51.
6. Guo, Y. Q., Shyr, H. J. and Thierrin, G., F-Disjunctive Languages, International Journal of Computer Mathematics, Vol. 18 (1986), 219–237.
7. Guo, Y. Q., Reis, C. M. and Thierrin, G., Relatively f-Disjunctive Languages, Semigroup Forum, Vol. 37 (1988), 289–299.

8. Jiang, Z. H., On Strong Dense Languages and an Equivalent Description of the GRT Problem (Chinese), Acta Mathematica Sinica, Vol. 32, No. 3, (1989), 307–315.
9. Jiang, S. W., Shyr, H. J. and Yu, S. S., Completely Disjunctive Languages, Proceedings of the Second International Mathematical Mini Conference, Budapest (1988), Periodica Polytechnica, Vol. 19, No. 1–2 (1991), 101–110.
10. Katsura, M. and Shyr, H. J., Decompositions of Languages into Disjunctive Outfix Codes, Semigroups, Theory and Application, Proceedings, Oberwolfach, (1986), edited by H. Jurgensen, G. Lallement, H. J. Weinert, Lecture Notes in Mathematics 1320, Springer-Verlag, Berlin, (1988), 172–175.
11. Katsura, M., Disjunctive splittability of languages, Semigroup Forum, Vol. 37 (1988), 201–213.
12. Lallement, G., Semigroups and Combinatorial Applications, Wiley Interscience, New York, 1979.
13. Liu, Y., Guo, Y. Q. and Tsai, Y. S., Solid Codes and the Uniform Density of fd-Domains, Science in China, Ser. A., Vol. 50, No. 7 (2007), 1026–1034.
14. Liu, Y. and Guo, Y. Q., Semigroups with I-quasi Length and Syntactic Semigroups, Communications in Algebra, Vol. 39 (2011), 1069–1081.
15. Mu, L. W., Some Kinds of Generalized Disjunctive Languages, Dissertation of Harbin Institute of Technology, 1988.
16. Reis, C. M. and Shyr, H. J., Some Properties of Disjunctive Languages on a Free Monoid, Information and Control, Vol. 37, No. 3 (1978), 334–344.
17. Reis, C. M., F-Disjunctive Congruences and a Generalization of Monoids with Length, Semigroup Forum, Vol. 41 (1990), 291–306.
18. Shyr, H. J., Disjunctive Languages on a Free Monoid, Information and Control, Vol. 34 (1977), 123–129.
19. Shyr, H. J. and Thierrin, G., Disjunctive Languages and Codes, Fundamentals of Computation Theory, Proceeding of the 1977 Inter. FCT-Conference, Poznan, Poland, Lecture Notes in Computer Science, No. 56, Springer-Verlag (1977), 171–176.
20. Shyr, H. J., Ordered Catenation and Regular Free Disjunctive Languages, Information and Control, Vol. 46, No. 3 (1980), 257–269.
21. Shyr, H. J., A Characterization of Dense Languages, Semigroup Forum, Vol. 30 (1984), 237–240.
22. Shyr, H. J. and Yu, S. S., Solid Codes and Disjunctive Domains, Semigroup Forum, Vol. 41, No. 1 (1990), 23–37.
23. Shyr, H. J., Free Monoids and Languages, Third Edition, Hon Min Book Company, Taichung, Taiwan, 2001.
24. Xu, G. W., Guo, Y. Q. and Li, L., The Union Decomposition of a Dense Language into Disjunctive Languages (in Chinese), Acta Math Sinica, Vol. 29, No. 2 (1986), 184–188.

Koszul Duality of the Category of Trees and Bar Constructions for Operads

Muriel Livernet

Université Paris 13, CNRS, UMR 7539 LAGA,
99 avenue Jean-Baptiste Clément,
93430 Villetaneuse, France
**E-mail: livernet@math.univ-paris13.fr*

In this paper, we study a category of trees $\mathcal{T}_I$ and prove that it is a Koszul category. Consequences are the interpretation of the reduced bar construction of operads of Ginzburg and Kapranov as the Koszul complex of this category, and the interpretation of operads up to homotopy as a functor from the minimal resolution of $\mathcal{T}_I$ to the category of graded vector spaces. We compare also three different bar constructions of operads. Two of them have already been compared by Shnider-Von Osdol and Fresse.

Keywords: Operads; bar construction; monad; Koszul duality.

Introduction

The bar construction is an old machinery that applies to different objects, such as algebras, monads[14] or categories.[16] Ginzburg and Kapranov[11] built a bar construction B^{GK} for operads, as an analogue of the bar construction for algebras. Except that, *stricto sensu*, this bar construction is not the exact analogue of the bar construction for algebras. C. Rezk,[19] S. Shnider and D. Von Osdol,[21] and B. Fresse[7] considered another bar construction for operads, denoted by B°, the one viewing an operad $\mathcal{P}$ as a monoid in the monoidal (non-symmetric) category of symmetric sequences, with the plethysm as monoidal structure. B. Fresse proved that the associated complexes of the two bar constructions are related by an explicit quasi-isomorphism, improving the result by S. Shnider and D. Von Osdol who proved that the two complexes have isomorphic homology. This explicit morphism enables also B. Fresse to work over any commutative ring.

The purpose of this paper is first to give an interpretation of the original bar construction B^{GK} of Ginzburg and Kapranov. In this process, we view an operad as a left module over the category of trees $\mathcal{T}_I$ as in[11] and build the bar construction B of this category. Note that the idea of using categories of trees for operads, or PROPS, goes back to Boardman and Vogt.[3] The crucial point is that the category $\mathcal{T}_I$ is Koszul and it is immediate to see that the original bar construction of Ginzburg and Kapranov is precisely the 2-sided Koszul complex of the category with coefficients in the left-module $\mathcal{P}$ and the unit right module. The bar construction of the category of trees is the same as the bar construction of the monad (or triple) arising from the adjunction

$$\text{Operads} \rightleftarrows \text{Symmetric Sequences},$$

where the left adjoint to the augmentation ideal functor is the free operad functor. Symmetric sequences are also known as species, terminology used in the present paper, or $\mathbb{S}$-modules.

We prove that the inclusion $B^{GK} \to B$ factors through the quasi-isomorphism described by Fresse $B^{GK} \to B^\circ$, where B° denotes the bar construction with respect to the monoidal structure. We describe explicitly the quasi-isomorphism $B^\circ \to B$.

Note that S. Shnider and D. Von Osdol[21] interpret the bar construction B° as a bar construction of a category, which is not the same as our category of trees. The one used by the authors is the usual category associated to an operad or a PROP (see e.g.[15]).

The advantage of proving that the category $\mathcal{T}_I$ is Koszul is that we can provide a smaller resolution of the category $\mathcal{T}_I$ than the usual cobar-bar resolution, inspired by the work of B. Fresse.[8] We prove that this resolution yields the definition of operads up to homotopy recovering the original definition given by P. Van der Laan in his PhD thesis.[23]

The plan of the paper is the following one. In Section 1, we study the category of trees $\mathcal{T}_I$, define the two-sided bar construction B, the Koszul complex K and prove that the category is Koszul (Theorem 1.8). In Section 2, we compare the three bar constructions B, B^{GK} and B° and prove the main Theorem 2.6: the factorization of $B^{GK} \to B$ through the levelization morphism of B. Fresse. Section 3 is devoted to operads up to homotopy.

Notation. We work over a field $\mathbf{k}$ of any characteristic.
• The category of differential graded $\mathbf{k}$-vector spaces is denoted by dgvs. An object in this category is often called a *complex*.
• The symmetric group acting on n elements is denoted by S_n.

• Let F, G be subsets of a set E. The notation $F \sqcup G = E$ means that $\{F, G\}$ forms a partition of E, that is, $F \cup G = E$ and $F \cap G = \emptyset$.
• To any set E, one associates the vector space $\mathbf{k}[E]$ spanned by E.

1. The tree category is Koszul

1.1. *The tree category $\mathcal{T}_I$*

Definition 1.1. A *tree* is a non-empty connected oriented graph t with no loops, with the property that at each vertex v, there is at least one incoming edge and exactly one outgoing edge. The target of the incoming edges at v is v and the source of the outgoing edge of v is v. We allow some edges to have no sources and these edges are called the leaves of the tree t. The other ones are called the internal edges of t and we denote by E_t the set of internal edges of t. We denote by V_t the set of vertices of t and by $\mathrm{In}(v)$ the set of incoming edges (leaves or internal edges) at the vertex v. A tree is *reduced* if for every $v \in V_t$ one has $|\mathrm{In}(v)| > 1$.

Let I be a finite set. An I-tree is a tree such that there is a bijection between the set of its leaves and I. The objects of the category $\mathcal{T}_I$ are the isomorphism classes of reduced I-trees. One can find a detailed account on this category in the book in progress of B. Fresse (see [6, AppendixA]). Note that the set of objects of $\mathcal{T}_I$ is finite. Let t be a tree in $\mathcal{T}_I$ and E be a set of internal edges which can be empty. The tree t/E is the tree obtained by contracting the edges $e \in E$. For a given pair of trees (t, s) the set of morphisms $\mathcal{T}_I(t, s)$ is a point if there is $E \subset E_t$ such that $s = t/E$ and is empty if not. Note that if there is $E \subset E_t$ such that $s = t/E$ then E is unique. The category $\mathbf{k}\mathcal{T}_I$ is the $\mathbf{k}$-linear category spanned by $\mathcal{T}_I$: it has the same set of objects and has for morphisms $\mathbf{k}\mathcal{T}_I(t, s) = \mathbf{k}[\mathcal{T}_I(t, s)]$. When $s = t/E$ we denote again by E the basis of the one dimensional vector space $\mathbf{k}\mathcal{T}_I(t, t/E)$.

Definition 1.2. A *left $\mathcal{T}_I$-module* is a covariant functor $\mathcal{T}_I \to$ dgvs and a *right $\mathcal{T}_I$-module* is a contravariant functor $\mathcal{T}_I \to$ dgvs. To any left $\mathcal{T}_I$-module L and right $\mathcal{T}_I$-module R, we associate the differential graded vector space

$$R \otimes_{\mathcal{T}_I} L = \bigoplus_{t \in \mathcal{T}_I} R(t) \otimes L(t)/\sim$$

with $f^*(x) \otimes y \sim x \otimes f_*(y)$ whenever $f \in \mathcal{T}_I(t, s), x \in R(s)$ and $y \in L(t)$.

Recall that the Yoneda Lemma implies the functorial equivalences

$$\mathbf{k}\mathcal{T}_I(-, s) \otimes_{\mathcal{T}_I} L \cong L(s) \quad \text{and} \quad R \otimes_{\mathcal{T}_I} \mathbf{k}\mathcal{T}_I(t, -) \cong R(t).$$

1.2. *Bar construction for the category $\mathcal{T}_I$*

1.2.1. *Bar construction*

The *bar construction*, or standard complex in the terminology of Mitchell,[16] of the $\mathbf{k}$-linear category $\mathbf{k}\mathcal{T}_I$ is a simplicial bifunctor $\mathcal{T}_I^{op} \times \mathcal{T}_I \to$ dgvs defined by

$$B_n(\mathcal{T}_I, \mathcal{T}_I, \mathcal{T}_I)(t,s) = \bigoplus_{s_0,\cdots,s_n \in \mathcal{T}_I} \mathbf{k}\mathcal{T}_I(s_0,s) \otimes \mathbf{k}\mathcal{T}_I(s_1,s_0) \otimes \cdots \otimes \mathbf{k}\mathcal{T}_I(s_n,s_{n-1}) \otimes \mathbf{k}\mathcal{T}_I(t,s_n)$$

with the simplicial structure given by

for $0 \le i \le n$,

$$\begin{aligned} d_i : B_n(\mathcal{T}_I, \mathcal{T}_I, \mathcal{T}_I)(t,s) &\to B_{n-1}(\mathcal{T}_I, \mathcal{T}_I, \mathcal{T}_I)(t,s) \\ a_0 \otimes \cdots \otimes a_{n+1} &\mapsto a_0 \otimes \cdots \otimes a_i a_{i+1} \otimes \cdots \otimes a_{n+1}, \end{aligned}$$

for $0 \le j \le n$,

$$\begin{aligned} s_j : B_n(\mathcal{T}_I, \mathcal{T}_I, \mathcal{T}_I)(t,s) &\to B_{n+1}(\mathcal{T}_I, \mathcal{T}_I, \mathcal{T}_I)(t,s) \\ a_0 \otimes \cdots \otimes a_{n+1} &\mapsto a_0 \otimes \cdots \otimes a_j \otimes 1 \otimes a_{j+1} \otimes \cdots \otimes a_{n+1}, \end{aligned}$$

To this simplicial bifunctor is associated the usual complex $(B_n, d = \sum_{i=0}^n (-1)^i d_i)_{n \ge 0}$. If $s = t/E$ then the complex B_n simplifies as

$$B_n(\mathcal{T}_I, \mathcal{T}_I, \mathcal{T}_I)(t, t/E) = \bigoplus_{E_0 \sqcup \cdots \sqcup E_{n+1} = E} \mathbf{k}(E_0, \cdots, E_{n+1})$$

with

$$d(E_0, \cdots, E_{n+1}) = \sum_{i=0}^n (-1)^i (E_0, \cdots, E_i \sqcup E_{i+1}, \cdots, E_{n+1}).$$

This complex is augmented by letting $B_{-1}(\mathcal{T}_I, \mathcal{T}_I, \mathcal{T}_I)(t,s) = \mathbf{k}\mathcal{T}_I(t,s)$. From B. Mitchell[16] this complex is acyclic and $B_*(\mathcal{T}_I, \mathcal{T}_I, \mathcal{T}_I) \to \mathbf{k}\mathcal{T}_I$ is a free resolution of bifunctors.

1.2.2. *Resolution of left and right $\mathcal{T}_I$-modules and* Tor *functors*

Let L be a left $\mathcal{T}_I$-module and R be a right $\mathcal{T}_I$-module. The left $\mathcal{T}_I$-module $B(\mathcal{T}_I, \mathcal{T}_I, \mathcal{T}_I) \otimes_{\mathcal{T}_I} L$ is denoted by $B(\mathcal{T}_I, \mathcal{T}_I, L)$, the right $\mathcal{T}_I$-module $R \otimes_{\mathcal{T}_I} B(\mathcal{T}_I, \mathcal{T}_I, \mathcal{T}_I)$ is denoted by $B(R, \mathcal{T}_I, \mathcal{T}_I)$ and the differential graded vector space $R \otimes_{\mathcal{T}_I} B(\mathcal{T}_I, \mathcal{T}_I, \mathcal{T}_I) \otimes_{\mathcal{T}_I} L$ is denoted by $B(R, \mathcal{T}_I, L)$.

From B. Mitchell one gets that $B(\mathcal{T}_I, \mathcal{T}_I, L)$ is a free resolution of L in the category of left $\mathcal{T}_I$-modules and $B(R, \mathcal{T}_I, \mathcal{T}_I)$ is a free resolution of R in the category of right $\mathcal{T}_I$-modules. Consequently,

$$H_n B(R, \mathcal{T}_I, L) = \mathrm{Tor}_n^{\mathcal{T}_I}(R, L).$$

The Yoneda Lemma implies that the complex computing the Tor functor has the following form:

$$B_n(R, \mathcal{T}_I, L) = \bigoplus_{(t, E \subset E_t)} \bigoplus_{E_1 \sqcup \cdots \sqcup E_n = E} R(t/E) \otimes \mathbf{k}(E_1, \cdots, E_n) \otimes L(t),\ n \geq 1$$

$$B_0(R, \mathcal{T}_I, L) = \bigoplus_{t \in \mathcal{T}_I} R(t) \otimes L(t).$$

with the differential given by

$$\begin{aligned} d(x \otimes (E_1, \cdots, E_n) \otimes y) = {} & (E_1)^*(x) \otimes (E_2, \cdots, E_n) \otimes y \\ & + \sum_{i=1}^{n-1} (-1)^i x \otimes (E_1, \cdots, E_i \sqcup E_{i+1} \cdots, E_n) \otimes y \\ & + (-1)^n x \otimes (E_1, \cdots, E_{n-1}) \otimes (E_n)_*(y), \end{aligned}$$

where $(E_1)^* = R(E_1 : t/(E \setminus E_1) \to t/E)$ and $(E_n)_* = L(E_n : t \to t/E_n)$.

1.2.3. *Normalized bar complex*

Because B_* is a simplicial bifunctor, one can mod out by the degeneracies to get the normalized bar complex of the category

$$N_n(\mathcal{T}_I, \mathcal{T}_I, \mathcal{T}_I)(t, t/E) = \bigoplus_{\substack{E_0 \sqcup \cdots \sqcup E_{n+1} = E, \\ E_i \neq \emptyset \text{ for } 1 \leq i \leq n}} \mathbf{k}(E_0, \cdots, E_{n+1}),$$

as well as the normalized bar complexes with coefficients $N_*(R, \mathcal{T}_I, \mathcal{T}_I)$, $N_*(\mathcal{T}_I, \mathcal{T}_I, L)$ and $N_*(R, \mathcal{T}_I, L)$. Furthermore for any left $\mathcal{T}_I$-module L and right $\mathcal{T}_I$-module R one has quasi-isomorphisms

$$B_*(\mathcal{T}_I, \mathcal{T}_I, L) \to N_*(\mathcal{T}_I, \mathcal{T}_I, L) \to L,$$

in the category of left $\mathcal{T}_I$-modules and quasi-isomorphisms

$$B_*(R, \mathcal{T}_I, \mathcal{T}_I) \to N_*(R, \mathcal{T}_I, \mathcal{T}_I) \to R,$$

in the category of right $\mathcal{T}_I$-modules and quasi-isomorphisms in dgvs

$$B(R, \mathcal{T}_I, L) \to N(R, \mathcal{T}_I, L).$$

Since $N_*(R, \mathcal{T}_I, \mathcal{T}_I)$ is a free right $\mathcal{T}_I$-module and $N_*(\mathcal{T}_I, \mathcal{T}_I, L)$ is a free left $\mathcal{T}_I$-module one can either use the bar complex or the normalized bar complex in the sequel, as free resolutions of R or L.

1.3. *The Koszul complex of the category $\mathcal{T}_I$*

1.3.1. *The Koszul complex*

Notation 1.3. For any tree t we denote by b_t the left and right $\mathcal{T}_I$-module which sends t to $\mathbf{k}$ and $s \neq t$ to 0. If t is the corolla c_I we use the notation b_I instead of b_{c_I}.

Let $E = \{e_1, \cdots, e_n\}$ be a finite set with n elements. Let $\mathbf{k}[E]$ be the n-dimensional vector space spanned by E. The vector space $\Lambda^n(\mathbf{k}[E])$ is a one dimensional vector space. Let $e_1 \wedge \cdots \wedge e_n$ be a basis.

The Koszul complex of the category $\mathcal{T}_I$ is a bifunctor

$$K(\mathcal{T}_I, \mathcal{T}_I, \mathcal{T}_I) : \mathcal{T}_I^{op} \times \mathcal{T}_I \to \text{dgvs}.$$

For any pair of trees (t, s), if there is no E such that $s = t/E$ we let $K(\mathcal{T}_I, \mathcal{T}_I, \mathcal{T}_I)(t, s) = 0$. If $s = t/E$ we let

$$K(\mathcal{T}_I, \mathcal{T}_I, \mathcal{T}_I)(t, t/E) = \bigoplus_{F \sqcup G \subset E} \mathbf{k}\mathcal{T}_I(t/(F \sqcup G), t/E) \otimes \Lambda^{|G|}(\mathbf{k}[G]) \otimes \mathbf{k}\mathcal{T}_I(t, t/F)$$

For any $F \sqcup (G = \{e_1, \cdots, e_g\}) \sqcup H = E$ we define

$$d(H \otimes e_1 \wedge \cdots \wedge e_g \otimes F) = \sum_{i=1}^{g} (-1)^{i-1} H \cup \{e_i\} \otimes e_1 \wedge \cdots \wedge \hat{e}_i \wedge \cdots \wedge e_g \otimes F +$$
$$\sum_{i=1}^{g} (-1)^i H \otimes e_1 \wedge \cdots \wedge \hat{e}_i \wedge \cdots \wedge e_g \otimes F \cup \{e_i\}.$$

Lemma 1.4. *The map d satisfies $d^2 = 0$.*

Proof. The map d splits into two parts $d_l + d_r$.

One has $d_l d_r + d_r d_l = 0$: if x_i denotes the element $e_1 \wedge \cdots \wedge \hat{e}_i \wedge \cdots \wedge e_g$ and $x_{i,j}$, $i < j$ denotes the element $e_1 \wedge \cdots \wedge \hat{e}_i \wedge \cdots \wedge \hat{e}_j \wedge \cdots \wedge e_g$ then

$$\begin{aligned}
&(d_l d_r + d_r d_l)(H \otimes e_1 \wedge \cdots \wedge e_g \otimes F) \\
&= d_l\Big(\sum_{i=1}^{g} (-1)^i H \otimes x_i \otimes F \cup \{e_i\}\Big) \\
&\quad + d_r\Big(\sum_{j=1}^{g} (-1)^{j-1} H \cup \{e_j\} \otimes x_j \otimes F\Big)
\end{aligned}$$

$$
\begin{aligned}
&= \sum_{j<i} (-1)^{i+j-1} H \cup \{e_j\} \otimes x_{j,i} \otimes F \cup \{e_i\} \\
&\quad + \sum_{j>i} (-1)^{i+j} H \cup \{e_j\} \otimes x_{i,j} \otimes F \cup \{e_i\} \\
&\quad + \sum_{i<j} (-1)^{i+j-1} H \cup \{e_j\} \otimes x_{i,j} \otimes F \cup \{e_i\} \\
&\quad + \sum_{i>j} (-1)^{i+j-2} H \cup \{e_j\} \otimes x_{j,i} \otimes F \cup \{e_i\} = 0.
\end{aligned}
$$

Let V be an n-dimentional vector space. Let $\mathcal{V} = \{v_1, \cdots, v_n\}$ be a basis of V with a given order $v_1 < \cdots < v_n$. Recall that the Koszul complex $\Lambda(V) \otimes S(V)$ has the following differential

$$
d(x_1 \wedge \cdots \wedge x_p \otimes y_1 \cdots y_q) = \sum_{i=1}^{p} (-1)^i x_1 \wedge \cdots \wedge \hat{x_i} \wedge \cdots \wedge x_p \otimes x_i y_1 \cdots y_q.
$$

This complex splits into subcomplexes

$$
(\Lambda(V) \otimes S(V), d) = \bigoplus_{\emptyset \neq W \subset \mathcal{V}} (C_*^W(V), d_W)
$$

where

$$
C_p^W(V) = \bigoplus_{\{x_1 < \cdots < x_p ; y_1 \leq \cdots \leq y_q\} = W} k[x_1 \wedge \cdots \wedge x_p \otimes y_1 \cdots y_q].
$$

For $F \subset E$ we let V_F be the vector space with basis $\mathcal{V}_F = \{e_k, e_k \notin F\}$. The map d_l corresponds to the differential $d_{\mathcal{V}_F}$ of $C_*^{\mathcal{V}_F}(V_F)$ and $d_l^2 = 0$. The same is true for d_r, with $\mathcal{V}_H$. □

Note that the Koszul complex is augmented by letting $K_{-1}(\mathcal{T}_I, \mathcal{T}_I, \mathcal{T}_I)(t,s) = \mathbf{k}\mathcal{T}_I(t,s)$.

1.3.2. *The Koszul complex of the category $\mathcal{T}_I$ with coefficients*

Let L be a left $\mathcal{T}_I$-module and R be a right $\mathcal{T}_I$-module. The left $\mathcal{T}_I$-module $K(\mathcal{T}_I, \mathcal{T}_I, \mathcal{T}_I) \otimes_{\mathcal{T}_I} L$ is a free left $\mathcal{T}_I$-module denoted by $K(\mathcal{T}_I, \mathcal{T}_I, L)$. The right $\mathcal{T}_I$-module $R \otimes_{\mathcal{T}_I} K(\mathcal{T}_I, \mathcal{T}_I, \mathcal{T}_I)$ is a free right $\mathcal{T}_I$-module denoted by $K(R, \mathcal{T}_I, \mathcal{T}_I)$. The differential graded vector space $R \otimes_{\mathcal{T}_I} K(\mathcal{T}_I, \mathcal{T}_I, \mathcal{T}_I) \otimes_{\mathcal{T}_I} L$ is denoted by $K(R, \mathcal{T}_I, L)$.

Let t be a tree in $\mathcal{T}_I$ and $s = t/E$ for a given $E \subset E_t$. The right $\mathcal{T}_I$-module $K(b_s, \mathcal{T}_I, \mathcal{T}_I)$ has the following form

$$K(b_s, \mathcal{T}_I, \mathcal{T}_I)(t) = \bigoplus_{G \sqcup F = E} \Lambda^{|G|}(\mathbf{k}[G]) \otimes \mathbf{k}[F].$$

From the previous proof, one gets that it corresponds to a summand of the Koszul complex $\Lambda(\mathbf{k}[E]) \otimes S(\mathbf{k}[E])$. If E is non empty, this complex is acyclic (see e.g.[20]) and if E is empty it is $\mathbf{k}$ in degree 0. As a consequence, we have

Theorem 1.5. *The augmentation $\epsilon : K(b_s, \mathcal{T}_I, \mathcal{T}_I) \to b_s$ is a quasi-isomorphism, thus $K(b_s, \mathcal{T}_I, \mathcal{T}_I)$ is a free resolution of b_s in the category of right $\mathcal{T}_I$-modules.*

The augmentation $\epsilon : K(\mathcal{T}_I, \mathcal{T}_I, b_t) \to b_t$ is a quasi-isomorphism, thus $K(\mathcal{T}_I, \mathcal{T}_I, b_t)$ is a free resolution of b_t in the category of left $\mathcal{T}_I$-modules.

1.4. *The category $\mathcal{T}_I$ is Koszul*

The aim of this section is to prove that the homology of the complex $N(b_s, \mathcal{T}_I, b_t)$ is concentrated in top degree with value $K(b_s, \mathcal{T}_I, b_t)$ which amounts to say that the category $\mathcal{T}_I$ is Koszul.

Lemma 1.6. *The map $\kappa : K(\mathcal{T}_I, \mathcal{T}_I, \mathcal{T}_I) \to B(\mathcal{T}_I, \mathcal{T}_I, \mathcal{T}_I)$ defined by*

$$\kappa(t, t/E)(H \otimes e_1 \wedge \cdots \wedge e_n \otimes F) = \sum_{\sigma \in S_n} \epsilon(\sigma) H \otimes (e_{\sigma(1)}, \cdots, e_{\sigma(n)}) \otimes F$$

is a natural transformation of bifunctors. For any right $\mathcal{T}_I$-module R the induced map $R \otimes_{\mathcal{T}_I} \kappa$ commutes with the augmentation maps $K(R, \mathcal{T}_I, \mathcal{T}_I) \to R$ and $B(R, \mathcal{T}_I, \mathcal{T}_I) \to R$.

For any left $\mathcal{T}_I$-module L the induced map $\kappa \otimes_{\mathcal{T}_I} L$ commutes with the augmentation maps.

Proof. The only thing we need to prove is that κ commutes with the differentials. One has

$$\begin{aligned}
& d\kappa(H \otimes e_1 \wedge \cdots \wedge e_n \otimes F) \\
& = \sum_{\sigma \in S_n} \epsilon(\sigma)(H \cup e_{\sigma(1)}) \otimes (e_{\sigma(2)}, \cdots, e_{\sigma(n)}) \otimes F \\
& \quad + \sum_{i=1}^{n-1} (-1)^i \sum_{\sigma \in S_n} \epsilon(\sigma) H \otimes (e_{\sigma(1)}, \cdots, e_{\sigma(i)} \cup e_{\sigma(i+1)}, \cdots, e_{\sigma(n)}) \otimes F \\
& \quad + (-1)^n \sum_{\sigma \in S_n} \epsilon(\sigma) H \otimes (e_{\sigma(1)}, \cdots, e_{\sigma(n-1)}) \otimes F \cup e_{\sigma(n)}.
\end{aligned}$$

The middle term vanishes. For the first term, we split the sum over S_n into sums over $\sigma \in S_n$ such that $\sigma(1) = i$. Such a σ is a composite $\tau\rho$ with τ having i as fixed point and with ρ being the cycle $1 \to i \to i-1 \to \cdots \to 2 \to 1$. Hence $\epsilon(\sigma) = \epsilon(\tau)(-1)^{i-1}$. Thus the first term writes

$$\sum_{i=1}^{n}(-1)^{i-1} \sum_{\tau \in S_n, \tau(i)=i} \epsilon(\tau) H \cup e_i \otimes (e_{\tau(1)}, \cdots, \hat{e}_i, \cdots, e_{\tau(n)}) \otimes F.$$

For the last term, we split the sum over S_n into sums over $\sigma \in S_n$ such that $\sigma(n) = i$. Such a σ is a composite $\tau\eta$ with τ having i as fixed point and with η being the cycle $i \to i+1 \to \cdots \to n \to i$. Hence $\epsilon(\sigma) = \epsilon(\tau)(-1)^{n-i}$. Thus the last term writes

$$\sum_{i=1}^{n}(-1)^{i} \sum_{\tau \in S_n, \tau(i)=i} \epsilon(\tau) H \otimes (e_{\tau(1)}, \cdots, \hat{e}_i, \cdots e_{\tau(n)}) \otimes F \cup e_i.$$

As a consequence $d\kappa = \kappa d$. □

Proposition 1.7. *The morphisms of right $\mathcal{T}_I$-modules*

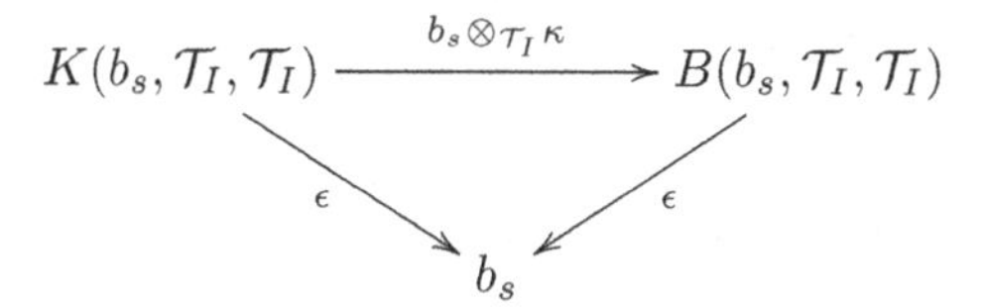

are quasi-isomorphisms.

Proof. This is a direct corollary of Theorem 1.5 □

Theorem 1.8. *The category $\mathcal{T}_I$ is Koszul.*

Proof. Because we have quasi-isomorphisms of free left modules

$$K(b_s, \mathcal{T}_I, \mathcal{T}_I) \to B(b_s, \mathcal{T}_I, \mathcal{T}_I) \to N(b_s, \mathcal{T}_I, \mathcal{T}_I),$$

we have quasi-isomorphisms of differential graded vector spaces

$$K(b_s, \mathcal{T}_I, b_t) \to B(b_s, \mathcal{T}_I, b_t) \to N(b_s, \mathcal{T}_I, b_t).$$

If $s = t/E$ with $E = \{e_1, \cdots, e_n\}$ then $N(b_s, \mathcal{T}_I, b_t)$ is bounded with top degree n. Namely

$$N_n(b_s, \mathcal{T}_I, b_t) = \bigoplus_{\sigma \in S_n} k[(e_{\sigma(1)}, \cdots, e_{\sigma(n)})],$$

whereas $K(b_s, \mathcal{T}_I, b_t)$ is concentrated in degree n, of dimension one with basis $e_1 \wedge \cdots \wedge e_n$. Since $K(b_s, \mathcal{T}_I, b_t) \to N(b_s, \mathcal{T}_I, b_t)$ is a quasi-isomorphism, one gets that K is precisely the homology of N. □

Corollary 1.9. *For any right $\mathcal{T}_I$-module R and left $\mathcal{T}_I$-module L the morphisms*

$$\kappa : K(R, \mathcal{T}_I, L) \to B(R, \mathcal{T}_I, L) \quad and$$

$$\overline{\kappa} : K(R, \mathcal{T}_I, L) \to N(R, \mathcal{T}_I, L)$$

are quasi-isomorphisms and

$$\mathrm{Tor}_*^{\mathcal{T}_I}(R, L) = H_*(K(R, \mathcal{T}_I, L)).$$

Proof. It is enough to prove that $\overline{\kappa}$ is a quasi-isomorphism. Let us consider the filtration by the number of internal vertices

$$F_p(N) = \bigoplus_{\substack{E:t\to s, |E|\leq p \\ E=E_1\sqcup\cdots\sqcup E_n}} R(s) \otimes \mathbf{k}(E_1, \cdots, E_n) \otimes L(t),$$

$$F_p(K) = \bigoplus_{n\leq p} K_n(R, \mathcal{T}_I, L),$$

which are subcomplexes of $N(R, \mathcal{T}_I, L)$ and $K(R, \mathcal{T}_I, L)$ respectively. One has, as complexes,

$$F_p(N)/F_{p-1}(N) = \bigoplus_{E:t\to s, |E|=p, n\leq p} R(s) \otimes N_n(b_s, \mathcal{T}_I, b_t) \otimes L(t),$$

$$F_p(K)/F_{p-1}(K) = \bigoplus_{E:t\to s, |E|=p} R(s) \otimes K_p(b_s, \mathcal{T}_I, b_t) \otimes L(t).$$

From Theorem 1.8, the map $F_p(K)/F_{p-1}(K) \to F_p(N)/F_{p-1}(N)$ is a quasi-isomorphism. Since $F_0(K) = \oplus_s R(s) \otimes L(s) = F_0(N)$, then for every p, the map $F_p(K) \to F_p(N)$ is a quasi-isomorphism. This yields the result. □

1.5. *Bibliographical remarks*

There are other ways to prove that the category $\mathcal{T}_I$ is a Koszul category, using different points of view. One point of view is to consider operads as algebras over a coloured operad. This has been done by P. Van der Laan in[22] where he proved that this coloured operad is Koszul.

Another point of view is to use incidence algebras of a poset as considered by C. Cibils,[4] for the category $\mathcal{T}_I$ forms a poset. Saying that the category is Koszul amounts to say that its incidence algebra is Koszul.

Moreover, the Koszul complex we have described in Section 1.3 coincides with the Koszul complex of the incidence algebra as presented by D. Woodcock.[24] If the poset is Cohen-Macaulay (see e.g.[18] or[24]), then its incidence algebra is Koszul. In our case the poset is Cohen-Macaulay because every interval is boolean: given a tree t and a tree $s = t/E$ where E is a set of edges of t, then the interval $[s, t]$ is the product of $|E|$ linear intervals of length 1, each of them corresponding to a contraction along a specific edge $e \in E$. I thank the anonymous referee for drawing this to my attention.

2. Comparison of three different types of bar constructions for an operad

The aim of this section is to compare different kinds of bar constructions for operads, depending on the way we consider operads, either as left $\mathcal{T}_I$-modules, or algebras over the free operad monad, or monoids in the monoidal category of species.

Section 2.1 is an attempt to generalize the bar construction in a framework that applies to all the cases considered in the paper. Section 2.2 shows that an operad $\mathcal{P}$ can be considered as a left $\mathcal{T}_I$-module, leading to the bar construction $B(R, \mathcal{T}_I, \mathcal{P})$, for $R = \mathcal{T}_I$ or R a right $\mathcal{T}_I$-module. Section 2.3 defines the free operad functor $\mathcal{F}$, yielding to the Godemont/May bar construction $B(\mathcal{R}, \mathcal{F}, \mathcal{P})$ for an $\mathcal{F}$-functor $\mathcal{R}$. We prove in Proposition 2.6 that to any right $\mathcal{T}_I$-module R is associated an $\mathcal{F}$-functor $\pi_I(R)$ such that $B(\pi_I(R), \mathcal{F}, \mathcal{P}) = B(R, \mathcal{T}_I, \mathcal{P})$. In Section 2.4, we recall the bar construction $B^\circ(R, \mathcal{P}, L)$ of an operad $\mathcal{P}$ with coefficients in a right $\mathcal{P}$-module R and left $\mathcal{P}$-module L obtained by viewing an operad as a monoid in the monoidal category of species. In Section 2.5, we recall the original reduced bar construction B^{GK} given by Ginzburg and Kapranov, which coincides with the Koszul complex $K(b_I, \mathcal{T}_I, \mathcal{P})$ introduced in Section 1. We recall the levelization morphism defined by B. Fresse from B^{GK} to B°. The last Section 2.6 is devoted to the factorization of $\bar{\kappa} : K(b_I, \mathcal{T}_I, \mathcal{P}) \to N(b_I, \mathcal{T}_I, \mathcal{P})$, introduced in Section 1.4, through the levelization morphism.

2.1. *Principle of the bar construction with coefficients*

Section 2.3 of the book of M. Markl, S. Shnider and J. Stasheff[13] can serve as our definition of two-sided bar construction. The idea is to work in a "context" for which any object X admits the notions of left X-modules and right X-modules, as

(a) A **k**-algebra X with its usual notions of left X-module and right X-module;

(b) a linear category X where left X-modules and right X-modules are covariant and contravariant functors $X \to$ dgvs;
(c) A monoid X in a monoidal category $(\mathcal{C}, \otimes, I)$ where left X-modules L and right X-modules R are objects in $\mathcal{C}$ together with maps $X \otimes L \to X$ and $R \otimes X \to X$ commuting with the monoid structure of X;
(d) A monad $X : \mathcal{T} \to \mathcal{T}$ where left modules L and right modules R are functors $L : \mathcal{D} \to \mathcal{T}$ and $R : \mathcal{T} \to \mathcal{E}$ together with natural transformations $\rho : XL \Rightarrow L$ and $\lambda : RX \Rightarrow R$ commuting with the monad structure.

Note that the last example is very close to a monoid X in a monoidal category except that L and R are not objects in the same category as X. Certainly the right notion in order to unify all the examples enumerated above is to start with a monoidal category $\mathcal{C}$, left and right module categories $\mathcal{L}$ and $\mathcal{R}$ (see[17] for the definition), and pick a monoid X in $\mathcal{C}$ and left module $L \in \mathcal{L}$ and right module $R \in \mathcal{R}$.

In this context, the above examples resume to

(a) The category $\mathcal{C}$ is the category of **k**-modules with the tensor product as monoidal structure and $\mathcal{L} = \mathcal{C} = \mathcal{R}$.
(b) The category $\mathcal{C}$ is the category of bifunctors $X^{op} \times X \to$ dgvs with the tensor product defined in Section 1. The category $\mathcal{L}$ is the category of covariant functors $X \to$ dgvs and $\mathcal{R}$ the one of contravariant functors.
(c) The category $\mathcal{C}$ is the monoidal category $(\mathcal{C}, \otimes, I)$ and $\mathcal{L} = \mathcal{C} = \mathcal{R}$
(d) The category $\mathcal{C}$ is the category of endo-functors of the category $\mathcal{T}$, with composition as monoidal structure. The category $\mathcal{L}$ is the category of functors from $\mathcal{D}$ to $\mathcal{T}$ and the category $\mathcal{R}$ is the category of functors from $\mathcal{T}$ to $\mathcal{E}$.

Definition 2.1. We say that a simplicial complex $B_*(R, \mathcal{P}, L)$ endowed with an augmentation $\epsilon : B(R, \mathcal{P}, L) \to B_{-1}(R, \mathcal{P}, L)$ satisfies the *principle of the simplicial bar construction with coefficients*, if

- $\forall n$, $B_n(R, \mathcal{P}, \mathcal{P})$ is a free right $\mathcal{P}$-module and $\epsilon : B(R, \mathcal{P}, \mathcal{P}) \to B_{-1}(R, \mathcal{P}, \mathcal{P}) = R$ is a quasi-isomorphism.
- $\forall n$, $B_n(\mathcal{P}, \mathcal{P}, L)$ is a free left $\mathcal{P}$-module and $\epsilon : B(\mathcal{P}, \mathcal{P}, L) \to B_{-1}(\mathcal{P}, \mathcal{P}, L) = L$ is a quasi-isomorphism.

Since we are working in linear categories, the normalized complex $N_*(R, \mathcal{P}, L)$ makes sense and we say that it satisfies the *principle of the bar construction with coefficients*, if it satifies the properties analogous to the ones stated above. More generally a complex $K_*(R, \mathcal{P}, L)$ satisfies the *principle of the bar construction with coefficients*, if it satisfies these properties.

The result of Section 1 can be summed up in the following proposition.

Proposition 2.2. *The standard resolution $B_*(R, \mathcal{T}_I, L)$ satisfies the principal of the simplicial bar construction. The normalized complex $N_*(R, \mathcal{T}_I, L)$ satisfies the principle of the bar construction as does the Koszul complex $K_*(R, \mathcal{T}_I, L)$.*

2.2. *Operads as left $\mathcal{T}_I$-modules*

In this section, we recall that operads can be considered as left $\mathcal{T}_I$-modules as presented in [11, Section 1.2].

Definition 2.3. Let Bij be the category whose objects are finite sets, possibly empty, and morphisms are bijections. A *vector species* is a contravariant functor $\mathcal{M} : \mathrm{Bij} \to \mathrm{dgvs}$. An *operad* is a vector species $\mathcal{P}$ together with partial composition maps

$$\circ_i : \mathcal{P}(I) \otimes \mathcal{P}(J) \to \mathcal{P}(I \setminus \{i\} \sqcup J), \ \forall i \in I,$$

and unit $\mathbf{k} \to \mathcal{P}(\{x\})$ satisfying functoriality, associativity and unit axioms. A *connected operad* is an operad $\mathcal{P}$ such that $\mathcal{P}(\emptyset) = 0$ and $\mathcal{P}(\{x\}) = \mathbf{k}$. We denote by $\mathcal{O}\mathrm{p}$ the category of connected operads. Given a connected operad $\mathcal{P}$, the species

$$\begin{cases} \overline{\mathcal{P}}(I) = 0, & \text{if } |I| \leq 1, \\ \overline{\mathcal{P}}(I) = \mathcal{P}(I), & \text{if } |I| > 1, \end{cases}$$

is called the *augmentation ideal* of the operad. A species $\mathcal{M}$ satisfying $\mathcal{M}(I) = 0$ for $|I| \leq 1$ is called connected. We denote by $\mathcal{S}\mathrm{p}$ the category of connected species. Hence, the augmentation ideal of a connected operad forms a functor $\mathcal{O}\mathrm{p} \to \mathcal{S}\mathrm{p}$. A species $\mathcal{M}$ satisfying $\mathcal{M}(\emptyset) = 0$ and $\mathcal{M}(\{x\}) = \mathbf{k}$ is called augmented. We denote by $\mathcal{S}\mathrm{p}_+$ the category of augmented species. Given a connected species $\mathcal{M}$ one can build an augmented species $\mathcal{M}_+$ by adding $\mathbf{k}$ in arity one. The induced functor $\mathcal{S}\mathrm{p} \to \mathcal{S}\mathrm{p}_+$ is an equivalence of category, the inverse functor being $\mathcal{M} \mapsto \overline{\mathcal{M}}$. The composition of the functors $\mathcal{O}\mathrm{p} \to \mathcal{S}\mathrm{p} \to \mathcal{S}\mathrm{p}_+$ is the forgetful functor.

Let t be a tree in $\mathcal{T}_I$ and let $\mathcal{M}$ be a vector species. The graded vector space $\mathcal{M}(t)$ is defined by

$$\mathcal{M}(t) = \bigotimes_{v \in V_t} \mathcal{M}(\mathrm{In}(v)).$$

When $\mathcal{P}$ is an operad, this definition extends to morphisms in $\mathcal{T}_I$ so that one gets a functor $\mathcal{P}:\mathcal{T}_I\to$ dgvs, as follows. Let $e\in E_t$ be an internal edge of t going from w to v. By reordering the terms in the tensor product one gets

$$\mathcal{P}(t)=\mathcal{P}(\mathrm{In}(v))\otimes\mathcal{P}(\mathrm{In}(w))\otimes\underbrace{\otimes_{z\in V_t\setminus\{v,w\}}\mathcal{P}(\mathrm{In}(z))}_{X_{v,w}}$$

and $\mathcal{P}(t\to t/e):\mathcal{P}(t)\to\mathcal{P}(t/e)$ is defined as

$$\circ_e\otimes X_{v,w}:\mathcal{P}(\mathrm{In}(v))\otimes\mathcal{P}(\mathrm{In}(w))\otimes X_{v,w}\to\mathcal{P}(\mathrm{In}(v)\setminus\{e\}\sqcup\mathrm{In}(w))\otimes X_{v,w}.$$

Iterating the process, and because of the axioms of the operad, to any $E\subset E_t$, is associated a well defined map $\mathcal{P}(t\to t/E):\mathcal{P}(t)\to\mathcal{P}(t/E)$. Consequently $\mathcal{P}$ is a left $\mathcal{T}_I$-module. In the sequel we will use the notation E_* for the map $\mathcal{P}(t\to t/E)$.

In the sequel we will consider the two-sided bar construction $B(\mathcal{T}_I,\mathcal{T}_I,\mathcal{P})$ and $B(R,\mathcal{T}_I,\mathcal{P})$ for $\mathcal{P}$ an operad considered as a left $\mathcal{T}_I$-module and R a right $\mathcal{T}_I$-module.

2.3. *Two-sided bar construction from the free operad functor*

In [14, chapter 9], P. May defines $B_*(R,C,X)$ for any monad C, a C-algebra X and a C-functor R to be RC^nX in degree n with the obvious faces and degeneracies corresponding to the C-structure, which satisfies the principle of the simplicial bar construction. The idea generalizes the Godement resolution associated to a triple and constructions used by J. Beck. P. May applied this simplicial resolution to the operad $\mathcal{C}_n$ of little n-cubes. C. Berger and I. Moerdijk[2] compare this construction for operads with the Boardman-Vogt W construction.

In this section, we use this construction and compare it to the bar construction for the category $\mathcal{T}_I$, in the spirit of E. Getzler and M. Kapranov in [10, 2.17].

Let $C:\mathcal{C}\to\mathcal{C}$ be a monad with structural maps $\mu:C^2\to C$ and $\eta:\mathrm{id}_\mathcal{C}\to C$. A *$C$-functor* R is a functor $R:\mathcal{C}\to\mathcal{D}$ together with a natural transformation $\lambda:RC\Rightarrow R$ satisfying the following identities

$$\lambda\circ R\eta=\mathrm{id}:R\Rightarrow R$$
$$\lambda\circ R\mu=\lambda\circ\lambda C:RC^2\Rightarrow R$$

Definition 2.4. The augmentation ideal functor $\mathcal{O}\mathrm{p} \to \mathcal{S}\mathrm{p}$ admits a left adjoint functor, the *free operad functor*

$$\begin{array}{rl} \mathcal{S}\mathrm{p} \to & \mathcal{O}\mathrm{p} \\ \mathcal{M} \mapsto \mathcal{F}\mathcal{M} : I \mapsto & \bigoplus_{t \in \mathcal{T}_I} \mathcal{M}(t) \end{array}$$

The partial composition maps $\circ_i : \mathcal{F}(\mathcal{M})(I) \otimes \mathcal{F}(\mathcal{M})(J) \to \mathcal{F}(\mathcal{M})(I \setminus \{i\} \sqcup J)$ correspond to the grafting of the root of a tree $s \in \mathcal{T}_J$ on the leave i of a tree $t \in \mathcal{T}_I$. When $|I| = 1$ we let $\mathcal{F}(\mathcal{M})(I) = \mathbf{k}$.

An element in $\mathcal{M}(t) \subset \mathcal{F}(\mathcal{M})(I)$ writes (t, E_t, m_t). There is an injection of species $\mathcal{M} \to \mathcal{F}(\mathcal{M})$ where the map $\mathcal{M}(I) \to \mathcal{F}(\mathcal{M})(I)$ sends m to $(c_I, \emptyset, m) \in \mathcal{M}(c_I)$, then identifying $\mathcal{M}(I)$ with $\mathcal{M}(c_I)$.

The equivalence of categories between $\mathcal{S}\mathrm{p}$ and $\mathcal{S}\mathrm{p}_+$ implies that the forgetful $\mathcal{O}\mathrm{p} \to \mathcal{S}\mathrm{p}_+$ admits a left adjoint, namely $\mathcal{M} \mapsto \mathcal{F}(\overline{\mathcal{M}})$ that we will denote also by $\mathcal{F}$ in the sequel.

2.3.1. *The two-sided bar construction*

The above adjunction yields a monad on $\mathcal{S}\mathrm{p}_+$ denoted also by $\mathcal{F}$. The tripleability Theorem implies that $\mathcal{F}$-algebras are exactly connected operads [9, Theorem 1.2]. We denote by $\mathcal{F}^{(n)}$ the n-th iteration of $\mathcal{F}$. An element in $\mathcal{F}^{(n)}(\mathcal{P})(I)$ writes $(t; E_1, \cdots, E_n, p_t)$ with $t \in \mathcal{T}_I$, $E_1 \sqcup \cdots \sqcup E_n = E_t$ and $p_t \in \mathcal{P}(t)$. The counit ϵ of the adjunction corresponds to the composition in the left $\mathcal{T}_I$-module $\mathcal{P}$, namely

$$\begin{array}{rrcl} \epsilon : & \mathcal{F}(\mathcal{P}) & \to & \mathcal{P} \\ & (t, E_t, p_t) & \mapsto & (E_t)_*(p_t) \end{array}$$

where $(E_t)_*(p_t)$ is in the component $\mathcal{P}(c_I)$ of $\mathcal{F}(\mathcal{P})(I)$ that we identify with $\mathcal{P}(I)$. The two-sided bar construction $B_n(\mathcal{F}, \mathcal{F}, P)(I)$ is the simplicial differential graded vector space $\mathcal{F}^{n+1}(\mathcal{P})(I)$ with faces $d_i : \mathcal{F}^{(n+1)}(\mathcal{P})(I) \to \mathcal{F}^{(n)}(\mathcal{P})(I)$ defined by

$$\begin{array}{ll} d_i(t; E_0, \cdots, E_n, p_t) = (t; E_0, \cdots, E_i \cup E_{i+1}, \cdots, E_n, p_t), & 0 \le i \le n-1, \\ d_n(t; E_0, \cdots, E_n, p_t) = \qquad (t/E_n; E_0, \cdots, E_{n-1}, (E_n)_*(p_t)). & \end{array}$$

As a consequence, comparing with the construction in 1.2, we have

Proposition 2.5. *The two-sided bar construction* $B(\mathcal{F}, \mathcal{F}, \mathcal{P})(I)$ *coincides with* $B(\mathcal{T}_I, \mathcal{T}_I, \mathcal{P})(c_I) = B(b_I, \mathcal{T}_I, \mathcal{P})$.

2.3.2. *Right $\mathcal{T}_I$-modules and $\mathcal{F}$-functors*

Let $R : \mathcal{T}_I \to$ dgvs be a right $\mathcal{T}_I$-module. The functor

$$\begin{array}{rcl} \pi_I(R) : \mathcal{S}\mathrm{p}_+ & \to & \mathrm{dgvs} \\ \mathcal{M} & \mapsto & \bigoplus_{t \in \mathcal{T}_I} \mathcal{M}(t) \otimes R(t) \end{array}$$

determines an $\mathcal{F}$-functor. In order to define the structural map $\lambda : \pi_I(R)\mathcal{F} \Rightarrow \pi_I(R)$ one needs to describe, for any $\mathcal{M} \in \mathcal{S}\mathrm{p}_+$, the map $\lambda_{\mathcal{M}} : \pi_I(R)\mathcal{F}(\mathcal{M}) \to \pi_I(R)(\mathcal{M})$. The vector space $\pi_I(R)\mathcal{F}(\mathcal{M})$ is the direct summand of the vector spaces $\mathcal{F}(\mathcal{M})(t) \otimes R(t)$, for $t \in \mathcal{T}_I$. An element in $\mathcal{F}(\mathcal{M})(t)$ writes $(t', E_1, E_2, m_{t'})$ with $t'/E_2 = t$, $E_1 \sqcup E_2 = E_{t'}$ and $m_{t'} \in \mathcal{M}(t')$. The map $\lambda_{\mathcal{M}}$ assigns the element $m_{t'} \otimes (E_2)^*(r_t) \in \mathcal{M}(t') \otimes R(t')$ to the element $(t', E_1, E_2, m_{t'}) \otimes r_t \in \mathcal{F}(\mathcal{M})(t) \otimes R(t)$.

As an example, the $\mathcal{F}$-functor $\pi_I(b_I)$, where b_I has been defined in 1.3, is the functor $\mathcal{M} \mapsto \mathcal{M}(I)$ with structural map

$$\begin{array}{rcl} \lambda_{\mathcal{M}} : \mathcal{F}(\mathcal{M})(I) & \to & \mathcal{M}(I) \\ (t, E_t, m_t) & \mapsto & \begin{cases} 0, & \text{if } t \neq c_I \Leftrightarrow E_t \neq \emptyset, \\ m_{c_I}, & \text{if } t = c_I. \end{cases} \end{array}$$

Comparing with the construction in Section 1.2, one gets easily

Proposition 2.6. *Let R be a right $\mathcal{T}_I$-module and let $\mathcal{P}$ be an operad. The two-sided bar construction $B(\pi_I(R), \mathcal{F}, \mathcal{P})$ coincides with $B(R, \mathcal{T}_I, \mathcal{P})$.*

2.4. *The bar construction with respect to the monoidal structure* $\circ$

As pointed out in the introduction, C. Rezk, S. Shnider, D. Von Osdol and B. Fresse have considered a bar construction for operads related to the fact that operads are monoids in the monoidal category of species, adapting the usual bar construction for algebras. Though the category of species is not monoidal symmetric and the monoidal structure is left distributive with respect to the coproduct but not right distributive, one can still perform the bar construction and then define cohomology theories. For the reader interested by this aspect, we refer to the paper by H.-J. Baues, M. Jibladze and A. Tonks.[1]

The category of augmented species admits a monoidal structure given by

$$(\mathcal{M} \circ \mathcal{N})(J) = \bigoplus_{J_1 \sqcup \cdots \sqcup J_r = J} \mathcal{M}(\{1, \cdots, r\}) \otimes_{S_r} \mathcal{N}(J_1) \otimes \cdots \mathcal{N}(J_r),$$

with unit

$$\mathcal{I}(J) = \begin{cases} \mathbf{k}, & \text{if } |J| = 1, \\ 0, & \text{if } |J| \neq 1. \end{cases}$$

A connected operad as defined in definition 2.3 is exactly a monoid in the monoidal category of augmented species $(\mathcal{S}\mathrm{p}_+, \circ, \mathcal{I})$. Let $\mathcal{P}$ be a connected operad. In the sequel we will use the notation $\mathcal{P}(n)$ for $\mathcal{P}(\{1, \cdots, n\})$ and $u(v_1, \cdots, v_k)$ for the image of the element $u \otimes v_1 \otimes \cdots \otimes v_k \in \mathcal{P}(k) \otimes \mathcal{P}(I_1) \otimes \cdots \otimes \mathcal{P}(I_k)$ under the structure map $\mathcal{P} \circ \mathcal{P} \to \mathcal{P}$.

There exists a simplicial bar construction, a normalized bar construction, and construction with coefficients related to the monoidal structure.

Definition 2.7 (B. Fresse[7]). *Let $\mathcal{P}$ be an operad, let R be a right $\mathcal{P}$-module, that is, a species together with a right action $R \circ \mathcal{P} \to R$ satisfying the usual associativity and unit condition of a right module, and let L be a left $\mathcal{P}$-module. The bar construction with coefficients R and L is the simplicial species*

$$B_n^\circ(R, \mathcal{P}, L) = R \circ \underbrace{\mathcal{P} \circ \cdots \circ \mathcal{P}}_{n \text{ terms}} \circ L$$

where faces d_i are induced either by the multiplication $\gamma_{\mathcal{P}} : \mathcal{P} \circ \mathcal{P} \to \mathcal{P}$ or by the left and right action and where degeneracies are induced by the unit map $\mathcal{I} \to \mathcal{P}$. Modding out by the degeneracies, one gets the normalized bar complex $N^\circ(R, \mathcal{P}, L)$.

Theorem 2.8 (B. Fresse[7]). *The simplicial complex $B_*^\circ(R, \mathcal{P}, L)$ satisfies the principle 2.1 of the simplicial bar construction with coefficients.*

As pointed out by B. Fresse, $N_*^\circ(R, \mathcal{P}, \mathcal{P})$ is a free resolution of the right $\mathcal{P}$-module R, but $N_*^\circ(\mathcal{P}, \mathcal{P}, L)$ is not. So $N_*^\circ(R, \mathcal{P}, \mathcal{P})$ satisfies the "right" principle of the bar construction only.

The species $\mathcal{I}$ is a right and left module for any connected operad $\mathcal{P}$, using the augmentation map $\epsilon : \mathcal{P} \to \mathcal{I}$:

$$\mathcal{I} \circ \mathcal{P} \xrightarrow{\mathrm{id} \circ \epsilon} \mathcal{I} \circ \mathcal{I} \xrightarrow{\gamma_{\mathcal{I}}} \mathcal{I} \qquad \text{and} \qquad \mathcal{P} \circ \mathcal{I} \xrightarrow{\epsilon \circ \mathrm{id}} \mathcal{I} \circ \mathcal{I} \xrightarrow{\gamma_{\mathcal{I}}} \mathcal{I}.$$

In the sequel we will be interested by the bar construction $B^\circ_*(\mathcal{I},\mathcal{P},\mathcal{I})$ with coefficients in the $\mathcal{P}$-module $\mathcal{I}$ and its normalized complex $N^\circ_*(\mathcal{I},\mathcal{P},\mathcal{I})$. An element in $B^\circ_n(\mathcal{I},\mathcal{P},\mathcal{I}) = \mathcal{P}^{\circ n}$ is represented by a tree with n levels as in [7, Section 4.3.1]. As an example the tree

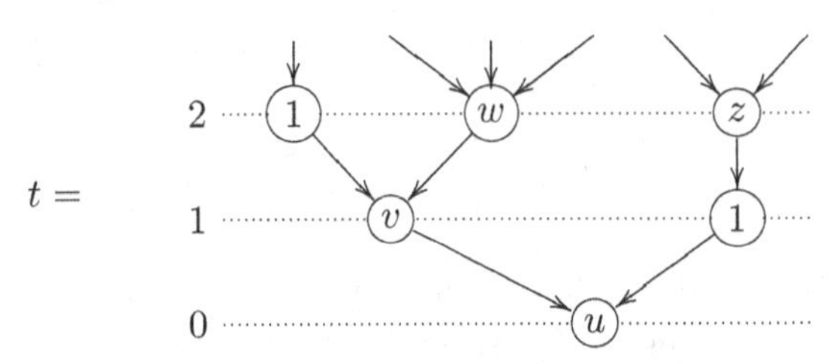

lives in $\mathcal{P}\circ\mathcal{P}\circ\mathcal{P} = B^\circ_3(\mathcal{I},\mathcal{P},\mathcal{I})$ and has 3 levels. The differential of t is a sum of trees with 2 levels in $\mathcal{P}\circ\mathcal{P}$:

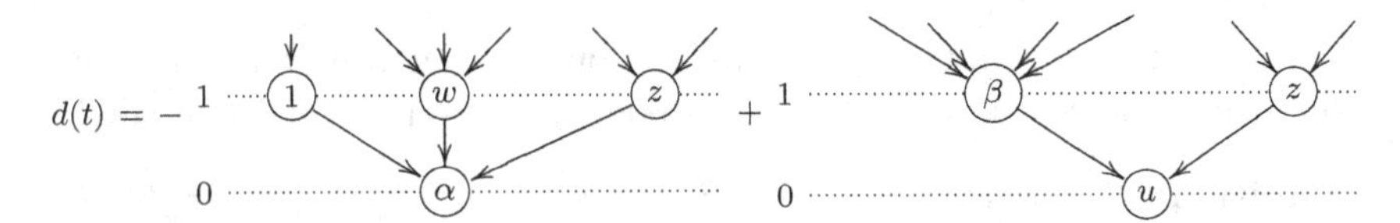

with $\alpha = u(v,1)\in\mathcal{P}(3)$ and $\beta = v(1,w)\in\mathcal{P}(4)$.

Note that an element in $N^\circ_n(\mathcal{I},\mathcal{P},\mathcal{I})$ is represented by a tree with n levels with the condition that at each level there is at least one vertex labelled by an element in $\mathcal{P}(r), r\geq 2$. For instance the tree with 3 levels

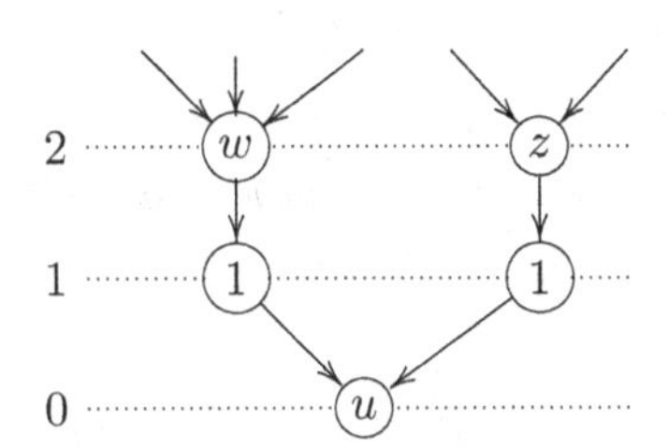

is zero in $N^\circ_3(\mathcal{I},\mathcal{P},\mathcal{I})$.

2.5. *The classical bar construction of operads, and the levelization morphism*

Ginzburg and Kapranov[11] introduced the reduced bar construction, based on partial compositions, as defined in definition 2.3. The classical bar construction $B^{GK}(\mathcal{P})$ of an operad $\mathcal{P}$ is the cofree cooperad generated by $\Sigma\tilde{\mathcal{P}}$ with unique coderivation extending the partial composition on $\mathcal{P}$ (see [9,

Section 2]). It has a description in terms of trees and it is graded by the number of vertices of the trees. Indeed, one has, for any finite set I

$$\bar{B}_n^{GK}(\mathcal{P})(I) = K_{n-1}(b_I, \mathcal{T}_I, \mathcal{P}).$$

B. Fresse in [7, Section 4.1] builds also a complex $B^{GK}(R,\mathcal{P},L) = R \circ \bar{B}^{GK}(\mathcal{P}) \circ L$ and proves that it satisfies the principal 2.1 of the bar construction with coefficients. He builds the levelization morphism

$$\Phi(R,\mathcal{P},L) : B^{GK}(R,\mathcal{P},L) \to N^\circ(R,\mathcal{P},L)$$

and proves that it is a quasi-isomorphism. In particular, for any finite set I, the quasi-isomorphism

$$\Phi(\mathcal{I},\mathcal{P},\mathcal{I})(I)_{n+1} : K_n(b_I,\mathcal{T}_I,\mathcal{P}) = B_{n+1}^{GK}(\mathcal{I},\mathcal{P},\mathcal{I})(I) \to N_{n+1}^\circ(\mathcal{I},\mathcal{P},\mathcal{I})(I)(1)$$

is described as follows.

Let $t \in \mathcal{T}_I$ be a tree with n internal edges : $e_1, \cdots, e_n$. The source of an internal edge is the adjacent vertex closest to the leaves of the tree and its target is the adjacent vertex closest to the root of the tree. The set of internal edges of a tree t is partially ordered: let e and f be internal edges, $e \leq f$ if there is a path from a leaf of t to the root of t meeting f before e. As an example the following figure

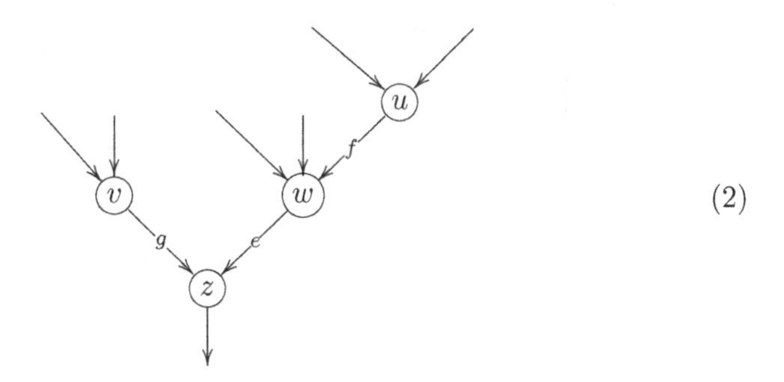

(2)

represents an element in $\mathcal{P}(t)$, where $\{e,f,g\}$ is the set of internal edges of t, with the partial order $e \leq f$. The source of g, e and f are v, w and u respectively. The target of g and e is z and the target of f is w.

Let $e_1\wedge\cdots\wedge e_n \otimes p_t$ be an element in $\mathbf{k}(e_1\wedge\cdots\wedge e_n)\otimes\mathcal{P}(t) \subset K_n(b_I,\mathcal{T}_I,\mathcal{P})$. The levelization morphism associates to this element a sum of trees with $(n+1)$-levels. The set $\{e_1,\cdots,e_n\}$ is partially ordered because it is the set of internal edges of a tree $t \in \mathcal{T}_I$. The set $\{1,\cdots,n\}$ is totally ordered as a subset of $\mathbb{N}$. To any order-preserving bijection $\sigma : \{e_1, \cdots, e_n\} \to \{1, \cdots, n\}$ one associates the level tree t_σ where the source of e_i is placed at level $\sigma(e_i)$, and where we complete the tree by adding vertices labelled by 1 in $\mathcal{P}(1)$. The resulting element in $\mathcal{P}^{\circ(n+1)}$ is denoted by $\sigma(p_t)$. The signature of σ, denoted by $\epsilon(\sigma)$ is the signature of the permutation $i \mapsto \sigma(e_i)$. The levelization morphism is defined by the following formula

$$\Phi(\mathcal{I}, \mathcal{P}, \mathcal{I})_{n+1}(I)(e_1 \wedge \cdots \wedge e_n \otimes p_t) = \sum_{\substack{\sigma:\{e_1,\cdots,e_n\}\to\{1,\cdots,n\} \\ \text{order-preserving}}} \epsilon(\sigma)\sigma(p_t) \quad (3)$$

Example 2.9. As an example we compute the levelization morphism associated to the element $e \wedge f \wedge g \otimes p_t$ of figure (2). The order-preserving maps involved in the formula (3) are $\sigma_1 : (e, f, g) \mapsto (1, 2, 3)$, $\sigma_2 : (e, f, g) \mapsto (2, 3, 1)$ and $\sigma_3 : (e, f, g) \mapsto (1, 3, 2)$.

$\Phi(e \wedge f \wedge g \otimes p_t) =$

$\sigma_1(p_t)$ + $\sigma_2(p_t)$

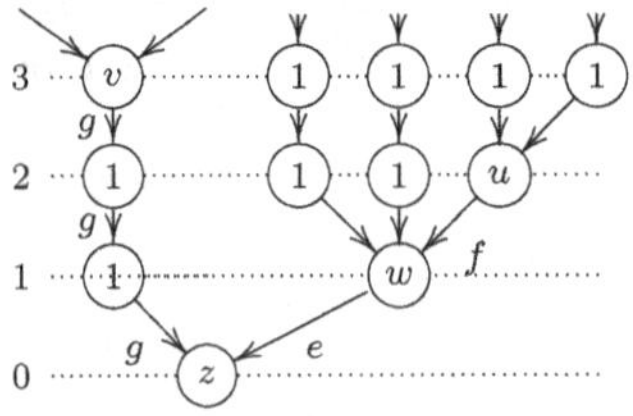

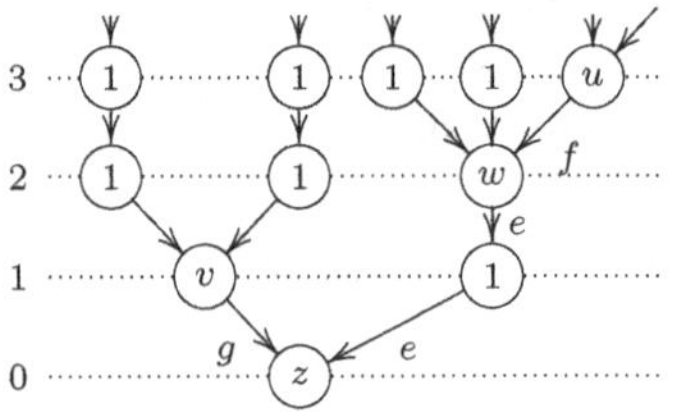

− $\sigma_3(p_t)$

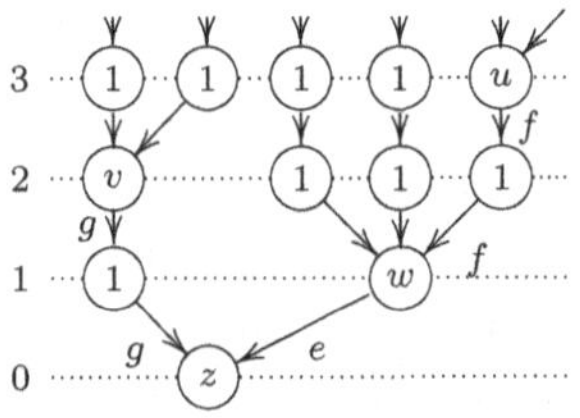

2.6. *The factorization of $\overline{\kappa} : K(b_I, \mathcal{T}_I, \mathcal{P}) \to N(b_I, \mathcal{T}_I, \mathcal{P})$*

We have seen in corollary 1.9 that $\overline{\kappa}$ is a quasi-isomorphism and that $B(b_I, \mathcal{T}_I, \mathcal{P})$ is identified with $B(\pi_I(b_I), \mathcal{F}, \mathcal{P})$ in Proposition 2.6.

The aim of this section is to prove that there exists a map, which will turn out to be a quasi-isomorphism:

$$\overline{\psi} : N_*^\circ(\mathcal{I},\mathcal{P},\mathcal{I})(I) \to N_{*-1}(b_I,\mathcal{T}_I,\mathcal{P}) = N_{*-1}(\pi_I(b_I),\mathcal{F},\mathcal{P}),$$

such that $\overline{\psi}\Phi(\mathcal{I},\mathcal{P},\mathcal{I})(I) = \overline{\kappa}$, that is, the following diagram is commutative

$$\begin{array}{ccc} & K_{*-1}(b_I,\mathcal{T}_I,\mathcal{P}) & \\ {}^{\Phi(\mathcal{I},\mathcal{P},\mathcal{I})(I)}\swarrow & & \searrow^{\overline{\kappa}} \\ N_*^\circ(\mathcal{I},\mathcal{P},\mathcal{I})(I) & \xrightarrow[\overline{\psi}]{} & N_{*-1}(b_I,\mathcal{T}_I,\mathcal{P}) \end{array}$$

We start with the description of a map

$$\psi : B_{n+1}^\circ(\mathcal{I},\mathcal{P},\mathcal{I})(I) = \underbrace{\mathcal{P}\circ\cdots\circ\mathcal{P}}_{n+1 \text{ terms}}(I) \to B_n(b_I,\mathcal{T}_I,\mathcal{P}).$$

An element p_t in $B_{n+1}^\circ(\mathcal{I},\mathcal{P},\mathcal{I})(I)$ is represented by a tree with $n+1$-levels, counted from 0 to n with vertices labelled by elements in $\mathcal{P}$. Such a level tree has subtrees of the form

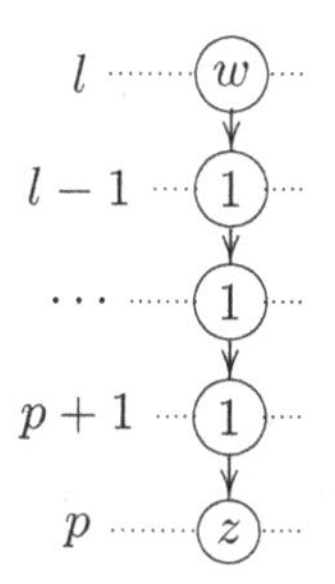

with $w \in \mathcal{P}(x), x \geq 2$ and $z \in \mathcal{P}(y), y \geq 2$. We define the *level-edge set* $N(p_t)$ as $e \in N(p_t)$ if and only if there is a sequence of consecutive edges in p_t, $e = \{e_1 > \cdots > e_k\}$ such that the source of e_1 lives in $\mathcal{P}(x)$ with $x \geq 2$, the target of e_k lives in $\mathcal{P}(y)$ with $y \geq 2$ and all other sources and targets lives in $\mathcal{P}(1)$. The source of e is the source of e_1 and the target of e is the target of e_k. The levels of the source and target of e are denoted by $s(e)$ and $t(e)$ respectively. The previous figure shows an element e in $N(p_t)$ such that $s(e) = l$ and $t(e) = p$. The idea underlying the definition of $N(p_t)$ is that we don't want to consider vertices labelled by $1 \in \mathcal{P}(1)$. For $1 \leq i \leq n$, let

$$N_i(p_t) = \{e \in N(p_t) | t(e) < i \leq s(e)\}.$$

One has $N(p_t) = \cup_{1\leq i\leq n} N_i(p_t)$, for $0 \leq t(e) < n$ and $1 \leq s(e) \leq n$. Note that this decomposition is not necessarily a partition of $N(p_t)$ as we will see in Example 2.10. Let t be a level tree and $p_t \in \otimes_{v\in V_t}\mathcal{P}(\mathrm{In}(v))$. By forgetting the units, we denote by $r(t)$ the associated rooted tree and by $r(p_t)$ the associated element in $\mathcal{P}(r(t))$. In the sequel the level-edge set $N(p_t)$ is written according to its decomposition $N(p_t) = (N_1(p_t), \cdots, N_n(p_t))$.

Example 2.10. The associated reduced tree to any of the trees of Example 2.9 is the tree

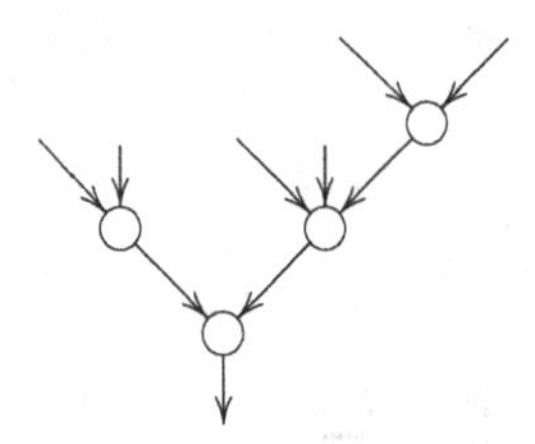

and the associated element $r(p_t)$ is the tree of figure (2). The set of level-edges of

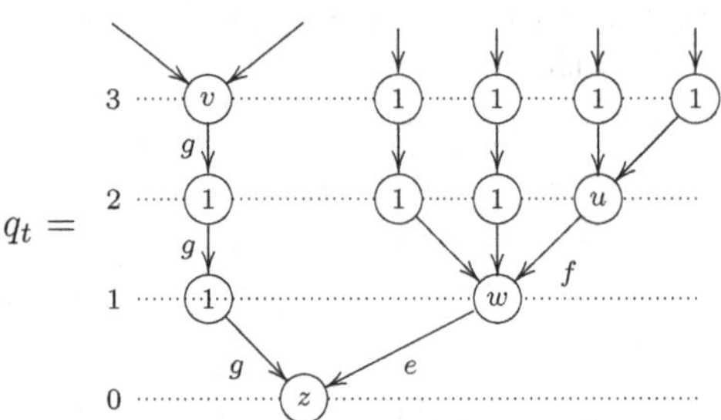

is $N(q_t) = \{e, f, g\}$ with $N_1(q_t) = \{e, g\}, N_2(q_t) = \{f, g\}$ and $N_3(q_t) = \{g\}$.

Definition 2.11. Let p_t be a level tree and N its associated level-edge set. For $\sigma \in S_n$, we define $E_1^\sigma = N_{\sigma(1)}$ and $E_i^\sigma = N_{\sigma(i)} \setminus \{\underset{1\leq j\leq i-1}{\cup} N_{\sigma(j)}\} = N_{\sigma(i)} \setminus \{\underset{1\leq j\leq i-1}{\cup} E_j^\sigma\}$. The map ψ is defined as follows

$$\begin{array}{rcl}\psi : B^\circ_{n+1}(\mathcal{I}, \mathcal{P}, \mathcal{I})(I) \to & & B_n(b_I, \mathcal{T}_I, \mathcal{P}) \\ p_t & & \mapsto \sum_{\sigma\in S_n} \epsilon(\sigma)(E_1^\sigma, \cdots, E_n^\sigma) \otimes r(p_t),\end{array}$$

Example. The computation of $\psi(q_t)$ for the tree q_t of Example 2.10 gives

$$\begin{aligned}\psi(q_t) = (&\underbrace{(\{e,g\}, f, \emptyset)}_{\sigma=(123)} - \underbrace{(\{f,g\}, e, \emptyset)}_{\sigma=(213)} + \underbrace{(\{f,g\}, \emptyset, e)}_{\sigma=(231)} \\ &- \underbrace{(g, f, e)}_{\sigma=(321)} + \underbrace{(g, e, f)}_{\sigma=(312)} - \underbrace{(\{e,g\}, \emptyset, f)}_{\sigma=(132)}) \otimes r(p_t).\end{aligned}$$

Lemma 2.12. *The map ψ induces a well-defined map*

$$\overline{\psi} : N^{\circ}_{n+1}(\mathcal{I},\mathcal{P},\mathcal{I})(I) \to N_n(b_I,\mathcal{T}_I,\mathcal{P}),$$

which commutes with the differentials.

Proof. Assume $p_t = s_j(q)$ with $s_j : B^{\circ}_n(\mathcal{I},\mathcal{P},\mathcal{I})(I) \to B^{\circ}_{n+1}(\mathcal{I},\mathcal{P},\mathcal{I})(I)$ being the degeneracy map sending $\mathcal{I}\circ\mathcal{P}^{\circ n}\circ\mathcal{I}$ to $\mathcal{I}\circ\mathcal{P}^{\circ j}\circ\mathcal{I}\circ\mathcal{P}^{\circ n-j}\circ\mathcal{I}$. If $1\leq j\leq n-1$, then the vertices of the tree t at level j are all labelled by 1. Consequently, $N_j(p_t) = N_{j+1}(p_t)$ and using the transposition $(j\,j+1)$ one gets that $\psi(p_t) = 0$. If $j=0$, then $N_1(p_t) = \emptyset$ and the composite of ψ with the projection $B_n(b_I,\mathcal{T}_I,\mathcal{P}) \to N_n(b_I,\mathcal{T}_I,\mathcal{P})$ is zero. If $j=n$ then $N_n(p_t) = \emptyset$ and the composite of ψ with the projection $B_n(b_I,\mathcal{T}_I,\mathcal{P}) \to N_n(b_I,\mathcal{T}_I,\mathcal{P})$ is zero.

In order to prove that for every x in $N^{\circ}_{n+1}(\mathcal{I},\mathcal{P},\mathcal{I})(I)$ one has $\overline{\psi}(dx) = d\overline{\psi}(x)$, it is enough to prove the equality for a representative p_t of x in $B^{\circ}_n(\mathcal{I},\mathcal{P},\mathcal{I})(I)$, such that $N_j(p_t)\neq\emptyset, \forall j$. To keep track of the levels we write such an element $(N_1,\cdots,N_n,p_t)$, where we consider $p_t\in\mathcal{P}(r(t))$, forgetting the units. On the one hand the differential is given by

$$d(N_1,\cdots,N_n,p_t) = \sum_{i=1}^{n}(-1)^i\left(N_1,\cdots,\hat{N_i},\cdots,N_n,(N_i\setminus\cup_{j\neq i}N_j)_*(r(p_t))\right).$$

Identifying permutations in S_{n-1} with permutations σ in S_n such that $\sigma(n)=i$ one gets

$$\begin{aligned}&\overline{\psi}d(N_1,\cdots,N_n,r(p_t))\\ &=\sum_{i=1}^{n}(-1)^n\sum_{\sigma\in S_n|\sigma(n)=i}\epsilon(\sigma)(E_1^{\sigma},\cdots,E_{n-1}^{\sigma})\otimes(N_i\setminus\cup_{j\neq i}N_j)_*(p_t).\end{aligned}$$

On the other hand one has

$$\begin{aligned}d\overline{\psi}(N_1,\cdots,N_n,p_t) &= \sum_{i=1}^{n}(-1)^i d_i(\sum_{\sigma\in S_n}\epsilon(\sigma)(E_1^{\sigma},\cdots,E_n^{\sigma})\otimes p_t)\\ &= (-1)^n\sum_{\sigma\in S_n}\epsilon(\sigma)(E_1^{\sigma},\cdots,E_{n-1}^{\sigma})\otimes(E_n^{\sigma})_*(p_t),\end{aligned}$$

for, regrouping permutations by pairs (σ,τ) such that $\sigma(i)=k, \sigma(i+1)=l$ and $\tau(i)=l, \tau(i+1)=k$, one gets $d_i(\sum_{\sigma\in S_n}(E_1^{\sigma},\cdots,E_n^{\sigma})\otimes r(p_t))=0$ when $1\leq i<n$.

Furthermore $E_n^\sigma = N_{\sigma(n)} \setminus \cup_{j\neq\sigma(n)} N_j$ implies

$$d\overline{\psi}(N_1,\cdots,N_n,p_t)$$
$$=\sum_{i=1}^{n}(-1)^n \sum_{\sigma\in S_n|\sigma(n)=i} \epsilon(\sigma)(E_1^\sigma,\cdots,E_{n-1}^\sigma)\otimes(N_i\setminus\cup_{j\neq i}N_j)_*(p_t).$$

The two expressions coincide. □

Theorem 2.13. *The map $\overline{\kappa} : K(b_I,\mathcal{T}_I,\mathcal{P}) \to N(b_I,\mathcal{T}_I,\mathcal{P})$ factorizes through $N_*^\circ(\mathcal{I},\mathcal{P},\mathcal{I})(I)$, and the following diagram*

$$\begin{array}{ccc} & K_{*-1}(b_I,\mathcal{T}_I,\mathcal{P}) & \\ {}^{\Phi(\mathcal{I},\mathcal{P},\mathcal{I})(I)}\swarrow & & \searrow^{\overline{\kappa}} \\ N_*^\circ(\mathcal{I},\mathcal{P},\mathcal{I})(I) & \xrightarrow[\overline{\psi}]{} & N_{*-1}(b_I,\mathcal{T}_I,\mathcal{P}) \end{array}$$

is commutative. Consequently $\overline{\psi}$ is a quasi-isomorphism.

Proof. The symbol $[k]$ denotes the set $\{1,\cdots,k\}$. Recall that

$$\overline{\psi}\Phi(e_1\wedge\cdots\wedge e_n\otimes p_t)=\overline{\psi}\Big(\sum_{\substack{f:\{e_1,\cdots,e_n\}\to\{1,\cdots,n\}\\ \text{order-preserving}}}\epsilon(f)f(p_t)\Big).$$

We prove the Theorem by induction on n. If $n=1$ it is obvious. Assume the result is true for any tree with $n-1$ internal edges. Let p_t be a tree with n internal edges $E=\{e_1,\cdots,e_n\}$. One can re-order the internal edges so that, there is a chain of consecutive edges from the root to a leaf $a_1<\cdots<a_p$ such that $a_p=e_n$ and if $p>1$, then $a_{p-1}=e_{n-1}$. By convention, if $p=1$, we let $a_0=\emptyset$.

Let $\tilde{p}_t$ be the tree obtained from p_t by removing the edge e_n. It has exactly $n-1$ internal edges $\tilde{E}=\{e_1,\cdots,e_{n-1}\}$. Let $f:\tilde{E}\to[n-1]$ be an order-preserving map. By convention $f(\emptyset)=0$. For $f(a_{p-1})<i\leq n$, let us define

$$\begin{aligned} f^i:\{e_1,\cdots,e_n\}\to\ & [n] \\ e_j,\ j<n \ \mapsto\ & \begin{cases} f(e_j) & \text{if } f(e_j)<i \\ f(e_j)+1 & \text{if } f(e_j)\geq i \end{cases} \\ e_n\ \mapsto\ & i \end{aligned}$$

The map f^i is an order-preserving bijection. One has $\epsilon(f^n)=\epsilon(f)$. As a consequence $\epsilon(f^i)=(-1)^{n-i}\epsilon(f)$, for $f^i=(i\cdots n)f^n$ where $(i\cdots n)$ denotes the cycle $i\to i+1\to\cdots\to n\to i$ in S_n.

Furthermore, to any order-preserving bijection $\tau : E \to [n]$ there exists a unique $f : \tilde{E} \to [n-1]$ and a unique i with $f(a_{p-1}) < i \leq n$ such that $f^i = \tau$.

Consequently

$$\Phi(e_1 \wedge \cdots \wedge e_n \otimes p_t) = \sum_{\substack{f:\tilde{E}\to[n-1] \\ \text{order-preserving}}} \sum_{i=f(a_{p-1})+1}^{n} (-1)^{n-i}\epsilon(f) f^i(p_t).$$

In order to evaluate $\overline{\psi}$ on the above expression, one needs to express $N_k^{f^i}(p_t)$ in terms of the $N_j^f(\tilde{p}_t)$'s. Because of the choices of the level for $f^i(p_t)$, one has, for $f(a_{p-1}) < i \leq n$,

$$N_k^{f^i}(p_t) = \begin{cases} N_k^f(\tilde{p}_t), & \text{if } k \leq f(a_{p-1}) < i, \\ N_k^f(\tilde{p}_t) \cup e_n, & \text{if } f(a_{p-1}) < k \leq i, \\ N_{k-1}^f(\tilde{p}_t), & \text{if } i < k \leq n. \end{cases} \tag{4}$$

Note that if $i = n$ the second equality reads $N_n^{f^n}(p_t) = \{e_n\}$ since $N_n^f = \emptyset$.

For example, if $n = 4$ and $f(a_{p-1}) = 1$, writing the sets N^{f^i} as $(N_1^{f^i}, N_2^{f^i}, N_3^{f^i}, N_4^{f^i})$ one gets

$$\begin{aligned} N^{f^2} &= \ (N_1^f, N_2^f \cup e_4, N_2^f, N_3^f), \\ N^{f^3} &= (N_1^f, N_2^f \cup e_4, N_3^f \cup e_4, N_3^f), \\ N^{f^4} &= \ (N_1^f, N_2^f \cup e_4, N_3^f \cup e_4, e_4). \end{aligned}$$

Recall that

$$\bar{\psi}(f(p_t)) = \sum_{\sigma \in S_n} \epsilon(\sigma)(E_1^{f,\sigma}, \cdots, E_n^{f,\sigma}) \otimes p_t, \text{ with } E_k^{f,\sigma} = N_{\sigma(k)}^f \setminus \bigcup_{i<k} N_{\sigma(i)}^f.$$

Let $\sigma \in S_n$, $j = f(a_{p-1})$ and $j < i \leq n$. By relations (4), the set N^{f^i} decomposes as

$$N^{f^i} = (N_1^f, \cdots, N_j^f, N_{j+1}^f \cup \{e_n\}, \cdots, N_i^f \cup \{e_n\}, N_i^f, N_{i+1}^f, \cdots, N_{n-1}^f).$$

Firstly, if $\sigma^{-1}(i) = k < \sigma^{-1}(i+1) = l$ then the sequence $(E_1^{f^i,\sigma}, \cdots, E_n^{f^i,\sigma})$ satisfies $E_l^{f^i,\sigma} = \emptyset$ and vanishes in $N_n(b_I, \mathcal{T}_I, \mathcal{P})$. Hence we only need to consider the elements $\sigma \in S_n$ such that $\sigma^{-1}(i+1) < \sigma^{-1}(i)$. In that case $E_{\sigma^{-1}(i)}^{f^i,\sigma} = \{e_n\}$.

Secondly, for $j+1 \leq r \leq i-1$, if $\sigma^{-1}(r) = k < \sigma^{-1}(i) = l$ then the sequence $(E_1^{f^i,\sigma}, \cdots, E_n^{f^i,\sigma})$ satisfies $E_l^{f^i,\sigma} = \emptyset$ and vanishes in $N_n(b_I, \mathcal{T}_I, \mathcal{P})$.

As a consequence, we only need to consider the elements $\sigma \in S_n$ such that

$$\sigma^{-1}(i+1) < \sigma^{-1}(i) < \{\sigma^{-1}(j+1), \cdots, \sigma^{-1}(i-1)\}.$$

Note that if $i = n$ the latter condition writes

$$\sigma^{-1}(n) < \{\sigma^{-1}(j+1), \cdots, \sigma^{-1}(n-1)\}.$$

and if $i = j+1$ it writes

$$\sigma^{-1}(i+1) < \sigma^{-1}(i).$$

Let $1 \leq l \leq n$ be a fixed integer. Choose $\sigma \in S_n$ such that $\sigma^{-1}(i) = l$. The condition $\sigma^{-1}(i+1) < \sigma^{-1}(i) < \{\sigma^{-1}(j+1), \cdots, \sigma^{-1}(i-1)\}$ implies that the sequence $(E_1^{f^i,\sigma}, \cdots, E_n^{f^i,\sigma})$ writes $(E_1^{f,\tau}, \cdots, E_{l-1}^{f,\tau}, \{e_n\}, E_{l+1}^{f,\tau}, \cdots, E_{n-1}^{f,\tau})$ with $\tau \in S_{n-1}$ obtained as the composite $\sigma_i \sigma \delta_l$ where $\delta_l : [n-1] \to [n]$ is the map missing l and $\sigma_i : [n] \to [n-1]$ is the map repeating i. It is clear that $\epsilon(\sigma) = \epsilon(\tau)(-1)^{l+i}$. When i runs from $j+1$ to n one covers S_{n-1}. For, if $i = j+1$ then the set involving N_{j+1}^f appears before e_n and if $i > j+1$ then it appears after e_n. If $i = j+2$ then the set involving N_{j+2}^f appears before e_n and if $i > j+2$ then it appears after e_n. And so on.

It yields the computation:

$$\begin{aligned}
&\bar{\psi}\Phi(e_1 \wedge \cdots \wedge e_n \otimes p_t) \\
&= \sum_{\substack{f:\tilde{E}\to[n-1] \\ \text{order-preserving}}} \sum_{i=f(a_{p-1})+1}^{n} (-1)^{n-i}\epsilon(f) \sum_{l=1}^{n} \sum_{\substack{\sigma\in S_n, \\ \sigma^{-1}(i)=l}} \epsilon(\sigma)(E_1^{f,\sigma}, \cdots, E_n^{f,\sigma}) \otimes p_t \\
&= \sum_{\substack{f:\tilde{E}\to[n-1] \\ \text{order-preserving}}} \epsilon(f) \sum_{l=1}^{n} \sum_{\sigma\in S_{n-1}} (-1)^{n+l}\epsilon(\sigma)(E_1^{f,\sigma}, \cdots, E_{l-1}^{f,\sigma}, e_n, E_l^{f,\sigma}, \cdots, E_{n-1}^{f,\sigma}) \otimes p_t \\
&= \sum_{l=1}^{n} \sum_{\sigma\in S_{n-1}} (-1)^{n+l}\epsilon(\sigma)(e_{\sigma(1)}, \cdots, e_{\sigma(l-1)}, e_n, e_{\sigma(l+1)}, \cdots, e_{\sigma(n-1)}) \otimes p_t \\
&= \sum_{\sigma\in S_n} \epsilon(\sigma)(e_{\sigma(1)}, \cdots, e_{\sigma(n)}) \otimes p_t \\
&= \bar{\kappa}(e_1 \wedge \cdots \wedge e_n \otimes p_t). \qquad \square
\end{aligned}$$

3. Resolution of the category $\mathcal{T}_I$ and operads up to homotopy

This section is devoted to the bar and cobar constructions for differential graded categories and cocategories whose objects are the objects of $\mathcal{T}_I$. It follows closely the paper.[8] In this paper, B. Fresse works on the category of Batanin trees Epi_n. He proves that the complex we obtained with B. Richter[12] corresponds to the bar construction of the category Epi_n with coefficients in the Loday functor and the unit functor. In his paper, he proves that the category Epi_n is Koszul, yielding a minimal model $R(\mathrm{Epi}_n)$ of Epi_n.

In this paper, we work exactly in the same spirit; we have proved in Section 1 that the category $\mathcal{T}_I$ is Koszul and the purpose of the first section is to express it's minimal model $R(\mathcal{T}_I) \to \mathcal{T}_I$. The main result will be that given a species $\mathcal{M}$, then the map $R(\mathcal{T}_I) \to$ dgvs which associates $\mathcal{M}(t)$ to $t \in \mathcal{T}_I$ is a functor if and only if $\mathcal{M}$ is an operad up to homotopy.

3.1. *Bar and cobar constructions for dg categories and dg cocategories*

The bar and cobar constructions follow closely the ones for associative and coassociative algebras, and in this section we just state our notation and the theorem needed for the sequel.

From now on, we denote by $\mathrm{Ob}\mathcal{T}_I$ the set of trees in the category $\mathcal{T}_I$. A tree t has a degree $|t|$ given by the number of internal edges. A *dg graph* is a map $\Gamma : \mathrm{Ob}\mathcal{T}_I \times \mathrm{Ob}\mathcal{T}_I \to$ dgvs. We denote by $\mathcal{C}_I$ the category of differential graded connected categories whose objects are the trees $t \in \mathrm{Ob}\mathcal{T}_I$. Let $\mathcal{C}$ be such a category. Such a data is equivalent to

- A dg graph $\mathcal{C}$ which will correspond to the morphisms in the category.
- For every $a, b, c \in \mathrm{Ob}\mathcal{T}_I$, composition maps $\mathcal{C}(b,c) \otimes \mathcal{C}(a,b) \to \mathcal{C}(a,c)$ in dgvs which are associative;
- Identity elements $1_a \in \mathcal{C}(a,a)$ which are unit for the composition;
- Connectivity assumption: $\forall a \in \mathrm{Ob}\mathcal{T}_I, \mathcal{C}(a,a) = \mathbf{k}$ and $\mathcal{C}(b,a) = 0$ if $|b| < |a|$.

An example of such a category is $\mathbf{k}\mathcal{T}_I$.

For a connected dg graph Γ, we denote by $\overline{\Gamma}(s,t) = \begin{cases} 0, & \text{if } s = t, \\ \Gamma(s,t), & \text{if } s \neq t. \end{cases}$

Similarly we define $\mathcal{C}_I^c$ the category of differential graded connected cocategories whose objects are $\mathrm{Ob}\mathcal{T}_I$. A cocategory is defined the same way as a category except that the arrows go in the reverse order.

Given a dg graph $\Gamma : \mathrm{Ob}\mathcal{T}_I \times \mathrm{Ob}\mathcal{T}_I \to$ dgvs, one can form the free category generated by Γ. As a dg graph, one has

$$\mathcal{F}(\Gamma)(t,t') = \bigoplus_{t'=x_0,\cdots,x_m=t} \Gamma(x_1,x_0)\otimes\Gamma(x_2,x_1)\otimes\cdots\Gamma(x_{m-1},x_{m-2})\otimes\Gamma(x_m,x_{m-1}).$$

The compositions of maps are given by the concatenation.

Similarly the free co-category generated by Γ, denoted by $\mathcal{F}^c(\Gamma)$ is given by the same dg graph and the co-composition are given by the deconcatenation.

There is an adjunction between co-categories and categories

$$\Omega : \text{cocategories} \rightleftarrows \text{categories} : B$$

The bar construction $B(\mathcal{C})$ of the category $\mathcal{C}$ is the free cocategory $\mathcal{F}^c(s\overline{\mathcal{C}})$ with the unique coderivation lifting the composition product in $\mathcal{C}$. Namely,

$$\partial(s\alpha_1 \otimes \cdots \otimes s\alpha_p) = \sum_{i=1}^{p-1} (-)^{|s\alpha_1|+\cdots|s\alpha_i|} s\alpha_1 \otimes \cdots \otimes s(\alpha_i\alpha_{i+1}) \otimes \cdots s\alpha_p,$$

where $|s\alpha_i|$ denotes the degree of $s\alpha_i \in s\mathcal{C}(a,b)$ (see e.g.[5])

The cobar construction $\Omega(\mathcal{R})$ of a cocategory is the free category $\mathcal{F}(s^{-1}\overline{\mathcal{R}})$ with the unique derivation lifting the co-composition product in $\mathcal{R}$:

$$\begin{aligned}
&\partial(s^{-1}\alpha_1 \otimes \cdots \otimes s^{-1}\alpha_p) \\
&= \sum_{i=1}^{p} (-)^{|\alpha_1|+\cdots|\alpha_i|+i} s^{-1}\alpha_1 \otimes \cdots \otimes s^{-1}\alpha_{i,(1)} \otimes s^{-1}\alpha_{i,(2)} \otimes \cdots s^{-1}\alpha_p,
\end{aligned}$$

where we use the Sweedler's notation for the co-composition.

Lemma 3.1. *Let $\mathcal{C}$ be a category in $\mathcal{C}_I$. The counit of the adjunction is a quasi-isomorphism:*

$$\Omega B(\mathcal{C}) \to \mathcal{C}.$$

3.2. *Minimal model of the category $\mathcal{T}_I$*

In this section, we use the Koszul complex in order to build the minimal resolution of $\mathcal{T}_I$.

Lemma 3.2. *The bar construction $B(\mathbf{k}\mathcal{T}_I)(t,s)$ corresponds to the normalized bar construction $N(b_s, \mathcal{T}_I, b_t)$.*

Proof. The two definitions coincide, and we just have to check that the degrees and differentials coincide. For $t, s \in \mathrm{Ob}\mathcal{T}_I$ with $s = t/E$, an element in $N_n(b_s, \mathcal{T}_I, b_t)$ writes $(E_1, \cdots, E_n)$, with $E_1 \sqcup E_2 \sqcup \cdots \sqcup E_n = E$ and E_i is non empty for every i. It has degree n and its differential is given by $d(E_1, \cdots, E_n) = \sum_{i=1}^{n-1}(-1)^i(E_1, \cdots, E_i \sqcup E_{i+1}, \cdots, E_n)$. □

Consequently the dg graph $(t, s) \mapsto N(b_s, \mathcal{T}_I, b_t)$ is endowed with a structure of cocategory. Note that for $s = t$ one has $N(b_t, \mathcal{T}_I, b_t)$ is 1-dimensional concentrated in degree 0.

Recall that

$$K_*(b_s, \mathcal{T}_I, b_t) = \begin{cases} \Lambda^{|E|}(\mathbf{k}[E]), & \text{if } s = t/E \text{ and } * = |E|, \\ 0, & \text{elsewhere,} \end{cases}$$

with zero differential. Hence K determines a dg graph

$$\begin{aligned} K : \mathrm{Ob}\mathcal{T}_I \times \mathrm{Ob}\mathcal{T}_I &\to \quad \mathrm{dgvs} \\ (t, s) \quad &\mapsto K(b_s, \mathcal{T}_I, b_t) \end{aligned}$$

We define the co-composition on K by

$$\begin{aligned} &\Delta(h = e_1 \wedge \cdots \wedge e_n) \\ &= 1 \otimes h + h \otimes 1 + \sum_{p=1}^{n-1} \sum_{\sigma \in \mathrm{Sh}_{p,n-p}} \epsilon(\sigma) e_{\sigma(1)} \wedge \cdots \wedge e_{\sigma(p)} \otimes e_{\sigma(p+1)} \wedge \cdots \wedge e_{\sigma(n)}, \end{aligned}$$

where $\mathrm{Sh}_{p,n-p}$ denotes the set of $(p, n-p)$-shuffles.

Lemma 3.3. *The dg graph K is a subcocategory of $B(\mathbf{k}\mathcal{T}_I)$ via the map $\bar{\kappa} : K(b_s, \mathcal{T}_I, b_t) \to N(b_s, \mathcal{T}_I, b_t)$.*

Proof. The fact that the co-composition commutes with $\bar{\kappa}$ comes from the bijection between $(S_p \times S_{n-p})\mathrm{Sh}_{p,n-p}$ and S_n. □

Because the category $\mathcal{T}_I$ is Koszul (see Theorem 1.8) the morphism of co-categories $\bar{\kappa}$ is a quasi-isomorphism. Since Ω behaves well with respect to these quasi-isomorphisms, one has

Theorem 3.4. *The cobar construction of the co-category K is a minimal resolution of the category $\mathbf{k}\mathcal{T}_I$.*

3.3. *Operads up to homotopy*

Let $\mathcal{M}$ be a vector species, and consider the map

$$\begin{array}{rcl} \underline{\mathcal{M}}: \mathrm{Ob}\mathcal{T}_I \to & & \mathrm{dgvs} \\ t & \mapsto \mathcal{M}(t) = & \bigotimes_{v\in E_t} \mathcal{M}(\mathrm{In}(v)), \end{array}$$

defined in Section 2.2.

Theorem 3.5. *Let $\mathcal{M}$ be a vector species. The map $\underline{\mathcal{M}}$ determines a functor $\Omega(K) \to$ dgvs if and only if $\mathcal{M}$ is an operad up to homotopy.*

Proof. Recall that $\Omega(K) = \mathcal{F}(s^{-1}K)$.

Assume that $\underline{\mathcal{M}}$ is a functor. Since ΩK is the free category generated by $s^{-1}K$, one has for every $t, s = t/E$ a composition map

$$\circ_E : \mathcal{M}(t) \to \mathcal{M}(s),$$

of degree $|E| - 1$. Let us write $E = e_1 \wedge \cdots \wedge e_n$ a generator of the one dimensional vector space $K(b_t, \mathcal{T}_I, b_I)$. In $\Omega(K)$ one has

$$d(s^{-1}E) = \sum_{p=1}^{n-1} \sum_{\sigma\in \mathrm{Sh}_{p,n-p}} \epsilon(\sigma) s^{-1}(e_{\sigma(1)}\wedge\cdots\wedge e_{\sigma(p)})\otimes s^{-1}(e_{\sigma(p+1)}\wedge\cdots\wedge e_{\sigma(n)}).$$

In terms of functors, it writes

$$\partial(\circ_E) = \sum_{\substack{F\sqcup G=E,\\ F,G\neq\emptyset}} \epsilon(F,G) \circ_F \circ_G,$$

where $\epsilon(F,G) = \epsilon(\sigma)$ for the shuffle σ corresponding to the sets F and G when an order of elements in E is given. This is exactly the definition of an operad up to homotopy in [23, Formula 4.2.2] (see also[22]). □

Acknowledgments

It is a pleasure here to thank the Nankai University and the Chern Institute and the organizers of the summer school and the conference "Operads and universal algebra" held in July 2010. There, I had valuable discussions with P.-L. Curien, P. Malbos and Y. Guiraud. This work is inspired by the paper[8] of B. Fresse. I'd like to thank him for the discussion we had on this subject.

References

1. Hans-Joachim Baues, Mamuka Jibladze, and Andy Tonks, *Cohomology of monoids in monoidal categories*, Operads: Proceedings of Renaissance Conferences (Hartford, CT/Luminy, 1995) (Providence, RI), Contemp. Math., vol. 202, Amer. Math. Soc., 1997, pp. 137–165.
2. Clemens Berger and Ieke Moerdijk, *The Boardman-Vogt resolution of operads in monoidal model categories*, Topology **45** (2006), no. 5, 807–849.
3. J. Michael Boardman and Rainer M. Vogt, *Homotopy invariant algebraic structures on topological spaces*, Springer-Verlag, Berlin, 1973, Lecture Notes in Mathematics, Vol. 347.
4. Claude Cibils, *Cohomology of incidence algebras and simplicial complexes*, J. Pure Appl. Algebra **56** (1989), no. 3, 221–232.
5. Samuel Eilenberg and Saunders Mac Lane, *On the groups of* $H(\Pi, n)$. *I*, Ann. of Math. (2) **58** (1953), 55–106.
6. Benoit Fresse, *Homotopy of Operads and Grothendieck-Teichmüller Groups*, Book in progress, http://math.univ-lille1.fr/~fresse/OperadHomotopyBook/index.html.
7. ______, *Koszul duality of operads and homology of partition posets*, Homotopy theory: relations with algebraic geometry, group cohomology, and algebraic K-theory, Contemp. Math., vol. 346, Amer. Math. Soc., Providence, RI, 2004, pp. 115–215.
8. ______, *La catégorie des arbres élagués est de Koszul*, Preprint, 2009.
9. Ezra Getzler and John D. S. Jones, *Operads, homotopy algebra and iterated integrals for double loop spaces*, preprint, hep-th/9403055, 1994.
10. Ezra Getzler and Michael M. Kapranov, *Modular operads*, Compositio Math. **110** (1998), no. 1, 65–126.
11. Victor Ginzburg and Mikhail Kapranov, *Koszul duality for operads*, Duke Math. J. **76** (1994), no. 1, 203–272.
12. Muriel Livernet and Birgit. Richter, *An interpretation of* E_n*-homology as functor homology*, arXiv:0907.1283, DOI: 10.1007/s00209-010-0722-5 online first in Mathematische Zeitschrift, 2010.
13. Martin Markl, Steve Shnider, and Jim Stasheff, *Operads in algebra, topology and physics*, Mathematical Surveys and Monographs, vol. 96, American Mathematical Society, Providence, RI, 2002.
14. J. Peter May, *The geometry of iterated loop spaces*, Springer-Verlag, Berlin, 1972, Lectures Notes in Mathematics, Vol. 271.
15. J.Peter May and Robert W. Thomason, *The uniqueness of infinite loop space machines.*, Topology **17** (1978), 205–224.
16. Barry Mitchell, *Rings with several objects*, Advances in Math. **8** (1972), 1–161.
17. Victor Ostrik, *Module categories, weak Hopf algebras and modular invariants*, Transform. Groups **8** (2003), no. 2, 177–206.
18. Patrick Polo, *On Cohen-Macaulay posets, Koszul algebras and certain modules associated to Schubert varieties*, Bull. London Math. Soc. **27** (1995), no. 5, 425–434.

19. Charles W. Rezk, *Spaces of algebra structures and cohomology of operads*, MIT PhD Thesis, 1996.
20. Jean-Pierre Serre, *Algèbre locale. Multiplicités*, Cours au Collège de France, 1957–1958, rédigé par Pierre Gabriel. Seconde édition, 1965. Lecture Notes in Mathematics, vol. 11, Springer-Verlag, Berlin, 1965. MR 0201468 (34 #1352)
21. Steve Shnider and Donovan H. Van Osdol, *Operads as abstract algebras, and the Koszul property*, J. Pure Appl. Algebra **143** (1999), no. 1-3, 381–407, Special volume on the occasion of the 60th birthday of Professor Michael Barr (Montreal, QC, 1997).
22. Pepijn van der Laan, *Coloured Koszul duality and strongly homotopy operads*, math.QA/0312147, 2003.
23. ———, *Operads, Hopf algebras and coloured Koszul duality*, PhD thesis Universiteit Utrecht, 2004.
24. David Woodcock, *Cohen-Macaulay complexes and Koszul rings*, J. London Math. Soc. (2) **57** (1998), no. 2, 398–410.

Some Problems in Operad Theory

Jean-Louis Loday
Institut de Recherche Mathématique Avancée
CNRS et Université de Strasbourg
7 rue R. Descartes
67084 Strasbourg Cedex, France
E-mail: loday@math.unistra.fr

This is a list of some problems and conjectures related to various types of algebras, that is to algebraic operads. Some comments and hints are included.

Keywords: Operad; dendriform; algebra up to homotopy; Hopf algebra; octonion; Manin product.

Introduction

Since 1991 I got involved in the operad theory, namely after an enlightening lecture by Misha Kapranov in Strasbourg. During these two decades I came across many questions and problems. The following is an excerpt of this long list which might be helpful to have in mind while working in this theme. Of course this is a very personal choice.

Notation and terminology are those of [14] and [17]. Various tpes of algebras, i.e. algebraic operads, can be found in [23].

1. On the notion of group up to homotopy

The notion of associative algebra up to homotopy is well-known: it is called A_∞*-algebra* and was devised by Jim Stasheff in [21]. It has the following important property: starting with a differential graded associative algebra (A, d), if (V, d) is a deformation retract of (A, d), then (V, d) is not a dg associative algebra in general, but it is an A_∞-algebra. This is Kadeishvili's theorem [9], see [17] for a generalization and variations of it. It is called the Homotopy Transfer Theorem. Let us now start with a group G. What is the notion of a "group up to homotopy"? To make this question more precise we move, as in quantum group theory, to the group algebra $\mathbb{K}[G]$

of the group G over a field $\mathbb{K}$. It is well-known that this is not only a unital associative algebra, but it is a cocommutative Hopf algebra. So, it has a cocommutative coproduct Δ (induced by the diagonal on G), and the existence of an inverse in G translates to the existence of an antipode on the group algebra. So we can now reformulate the question as follows:

"What is the notion of cocommutative Hopf algebra up to homotopy?"

One of the criterion for the answer to be useful would be the existence of a Homotopy Transfer Theorem for cocommutative Hopf algebras. One has to be careful enough to take into account that the existence of a unit and a counit is part of the structure of a Hopf algebra. In the associative case the operad A_∞ does not take the unit into account. See [14] for the Hopf relation of a nonunital bialgebra.

The fact that the tensor product of two associative algebras is still an associative algebra plays a prominent role in the definition of a bialgebra (a fortiori a Hopf algebra). So it is clear that a first step in analyzing this problem is to check whether one can put an A_∞-structure on the tensor product of two A_∞-algebras and to unravel the properties of such a construction. A first answer has been given by Saneblidze and Umble in [20]. But this tensor product is not associative. This problem has been addressed in [15].

2. Subgroup of free group

It is well-known that a subgroup of a free group is free. The proof is topological in the sense that it consists in letting the free group act on a tree. Could one find a proof by looking at the properties of the associated group algebra (which is a Hopf algebra)? I am thinking about something similar to the theorems which claim that some algebra is free under certain condition (PBW type theorems, see [14]).

3. The octonions as an algebra over a Koszul operad

The octonions form a normed division algebra $\mathbb{O}$ of dimension 8, see for instance [3]. The product is known not to be associative contrarily to the other normed division algebras $\mathbb{R}, \mathbb{C}$ and $\mathbb{H}$. However it does satisfy some algebraic relation: it is an alternative algebra. Let us recall that an alternative algebra is a vector space equipped with a binary operation $x \cdot y$, such that the associator $(x, y, z) := (x \cdot y) \cdot z - x \cdot (y \cdot z)$ is antisymmetric:

$$(x, y, z) = -(y, x, z) = -(x, z, y).$$

It turns out that the operad of alternative algebras is not too good, because it is not a Koszul operad, cf. [7]. Whence the question: Find a (small) binary operad such that the octonions form an algebra over this operad and such that this operad is Koszul. Of course the ambiguity of the question is in the adjective "small". Because such an operad exists: it suffices to take the magmatic operad on one binary operation. But there may exists a smaller operad (i.e. a quotient of Mag), which is best. This operad need not be quadratic, that is, we may look for relations involving 4 variables, like in Jordan algebras.

4. Commutative algebras up to homotopy in positive characteristic

In positive characteristic p it is best to work with divided power algebras rather than commutative algebras. The notion of commutative algebra up to homotopy is well-known: it is the C_∞-algebras (also denoted Com_∞, see [17]). What is, explicitly, the notion of divided power algebra up to homotopy in characteristic p? Theoretically the problem can be solved as follows. One can perform the theory of Koszul duality for operads with divided powers, cf. [8]. Any such operad with divided powers $\Gamma\mathcal{P}$, which is Koszul, gives rise to a dg operad with divided powers $\Gamma\mathcal{P}_\infty$. A divided power algebra up to homotopy is an algebra over $\Gamma\mathcal{P}_\infty$. The point is to make all the steps of the theory explicit in the case $\mathcal{P} = Com$.

5. Manin black product for operads

What is the operad $Com \bullet Ass$? One is asking for a small presentation by generators and relations.

6. L-dendriform algebras and operadic black product

By definition an L-dendriform algebra is a vector space equipped with two operations $x \prec y$ and $x \succ y$ satisfying

$$(x \prec y) \prec z + y \succ (x \prec z) = x \prec (y * z) + (y \succ x) \prec z,$$
$$(x * y) \succ z + y \succ (x \succ z) = x \succ (y \succ z) + (y \prec x) \succ z,$$

where $x * y = x \prec y + x \succ y$ (cf. [1]). This is one of the numerous ways of splitting the associativity of the operation $*$.

Conjecture: the operad encoding L-dendriform algebras is a Manin black product:

$$preLie \bullet preLie = L\text{-}Dend.$$

In favor of this conjecture we have the following facts (cf. [22]):

$$preLie \bullet Com = Zinb,\ preLie \bullet Ass = Dend,\ preLie \bullet Lie = preLie,$$

and also $preLie \bullet Dend = Quad$.

Similarly it seems that $preLie \bullet Zinb = ComQuad$ and that the operads $Octo, L\text{-}Quadri, L\text{-}Octo$ introduced in [10] are also black products:

$$\begin{aligned} preLie \bullet Quadri &= Octo, \\ preLie \bullet L\text{-}Dend &= L\text{-}Quadri, \\ preLie \bullet L\text{-}Quadri &= L\text{-}Octo. \end{aligned}$$

Note. Some of these questions have been recently settled in [2].

7. Resolutions of associative algebras

Koszul duality theory for associative algebras gives a tool to construct free resolutions for some associative algebras, and even the minimal resolutions in certain cases. When the algebra is a group algebra, then there are tools to construct (at least the beginning of) a resolution by taking the free module on the set of generators, then of relations, then of relations between the relations, and so forth (syzygies), see for instance [13]. It would be very interesting to compare these various methods.

8. Hidden structure for EZ-AW maps

Given a deformation retract

$$h \circlearrowright (A, d_A) \underset{i}{\overset{p}{\rightleftarrows}} (V, d_V) \qquad ,$$

$$pi = \mathrm{id}_V, \qquad \mathrm{Id}_A - ip = d_A h + h d_A,$$

the HTT says that an algebraic structure on (A, d_A) can be transferred to some other algebraic structure (the hidden one) on (V, d_V). This principle is not special to chain complexes and can be applied to other situations as shown for crossed modules in [17]. Apply this principle to the Eilenberg-Zilber and Alexander-Whitney quasi-isomorphisms. Let us recall that, for X and Y being simplicial modules, these isomorphisms relate the chain complex $C_\bullet(X \times Y)$ to the tensor product $C_\bullet(X) \otimes C_\bullet(Y)$.

9. Interpolating between *Dend* and *Com*

The so-called E_n-operads are operads which interpolate between the homotopy class of the operad Ass, which contains A_∞, and the homotopy class of the operad Com, which contains C_∞. So, it solves the question: what is an associative algebra which is more or less commutative. The answer is: an E_n-algebra ; the larger n is, the more commutative it is.

Question: what is a dendriform algebra which is more or less commutative ? In other words we are looking for an interpolation D_n between the operads $Dend_\infty$ and $Zinb_\infty$, where $Dend$ is the operad of dendriform algebras (two generating operations $\prec$ and $\succ$ and three relations) and $Zinb$ is the operad of Zinbiel algebras (dendriform algebras such that $x \succ y = y \prec x$).

One of the motivation for finding D_2 is the following. It is known that the Grothendieck-Teichmüller group is related to the operad E_2. Knowing D_2 could lead to a dendriform version of the Grothendieck-Teichmüller group.

10. Good triples of binary quadratic operads

Let $\mathcal{P}$ be a binary quadratic operad which is Koszul (cf. for instance [17,18]). It gives a notion of $\mathcal{P}$-algebra and also a notion of $\mathcal{P}$-coalgebra. We conjecture that there is a compatibility relation which defines a notion of $\mathcal{P}^c$-$\mathcal{P}$-bialgebra such that $(\mathcal{P}, \mathcal{P}, \text{Vect})$ is a good triple of operads in the sense of [14].

`Comments`. There are many examples known:

$$\mathcal{P} = Com, As, Dend, Mag, 2\text{-}as.$$

When it holds, it gives a criterion for proving that a given $\mathcal{P}$-algebra is free.

11. On the coalgebra structure of Connes-Kreimer Hopf algebra

The Connes-Kreimer Hopf algebra is an algebra of polynomials endowed with an ad hoc coproduct, cf. [5]. It is known that the indecomposable part is not only coLie, but in fact co-pre-Lie, cf. [4]. If we linearly dualize (as graded modules), the Hopf algebra is the Grossman-Larson Hopf algebra, which is cocommutative and its primitive part is pre-Lie. I conjecture that there is some type of algebras, that is some operad $\mathcal{X}$, and some type of Com^c-$\mathcal{X}$-bialgebras, which fit into a good triple of operads

$$(Com, \mathcal{X}, preLie).$$

If so, then the Grossman-Larson algebra would be the free $\mathcal{X}$-algebra on one generator.

The solution of this problem in the noncommutative framework is given by the operad *Dend*, cf. [16].

12. Generalized bialgebras in positive characteristic

The Poincaré-Birkhoff-Witt theorem and the Cartier-Milnor-Moore theorem are structure theorems for cocommutative bialgebras in characteristic zero. They can be summarized by saying the triple of operads

$$(Com, As, Lie)$$

is a good triple. Several other good triples have been described in [14], some of them being valid in any characteristic, like the triple (As, Dup, Mag) for instance. For the classical case, it is known that, in order for the CMM theorem to be true in characteristic p, one has to replace the notion of Lie algebra by the notion of *restricted Lie algebras.* Operadically, restricted Lie algebras, divided power algebras and the like are obtained by replacing the "coinvariants" in the definition of an operad by the "invariants", cf. [8],

$$\Gamma\mathcal{P}(V) := \sum_n (\mathcal{P}(n) \otimes V^{\otimes n})^{S_n}.$$

So the PBW-CMM theorem in characteristic p can be phrased by saying that

$$(Com, As, \Gamma Lie)$$

is a good triple. Note that $As = \Gamma As$, so equivalently $(Com, \Gamma As, \Gamma Lie)$ is a good triple.

It would be very interesting to generalize the results on generalized bialgebras to positive characteristic along these lines, that is, to show that, when $(\mathcal{C}, \mathcal{A}, \mathcal{P})$ is a good triple, then so is $(\mathcal{C}, \Gamma\mathcal{A}, \Gamma\mathcal{P})$.

Similarly, $(\Gamma Com, As, Lie)$ is a good triple. One should be able to show that, when $(\mathcal{C}, \mathcal{A}, \mathcal{P})$ is a good triple, then so is $(\Gamma\mathcal{C}, \mathcal{A}, \mathcal{P})$.

13. Higher Dynkin diagrams and operads

Show that there exists some types of algebras (i.e. some operads) for which the finite dimensional simple algebras are classified by the diagrams described by Ocneanu in [19]. The toy-model is the operad *Lie* and the Dynkin diagrams.

14. Coquecigrues

It is well-known that a Lie group admits a tangent space at the unit element which is a Lie algebra. But there is also another relationship between groups and Lie algebras, more specifically between *discrete groups* and Lie algebras (over $\mathbb{Z}$). It is given by the descending central series. For G a discrete group, $G^{(1)} = [G, G]$ is its commutator subgroup, and, more generally, $G^{(n)} = [G, G^{(n-1)}]$ is the nth term of the descending central series. It is well-known that the graded abelian group $\bigoplus_n G^{(n)}/G^{(n+1)}$ is a Lie algebra whose bracket is induced by the commutator in G. The Jacobi identity is a consequence of a nice (and not so well-known) relation, valid in any group G, called the Philip Hall relation (see for instance [13] for some *drawing* of it related to the Borromean rings).

A natural question is the following. Let $\mathcal{P}$ be a variation of the operad *Lie* (we have in mind *preLie* and *Leib*). Is there some structure playing the role of groups in this realm? For Leibniz algebras the question arised naturally in my research on the periodicity properties of algebraic K-theory, cf. [11,12]. I called this conjectural object a *coquecigrue*. In fact I was more interested in the cohomology theory which should come with this new notion, to apply it further to groups. Recent progress using the notion of racks was achieved by Simon Covez in [6].

15. Homotopy groups of spheres

Let p be a prime number. Let $(\pi_\bullet^S(X))_p$ be the stable homotopy groups of the pointed connected topological space X, localized at p. Find a type of algebras such that $(\pi_\bullet^S(X))_p$ is an algebra of this type and such that $(\pi_\bullet^S(*))_p$ is the free algebra of this type over one generator (in degree $2p-3$).

`Comments`. The Toda brackets are likely to play a role in this problem.

References

1. C. Bai, L. Liu and X. Ni, Some results on L-dendriform algebras, J. Geom. Phys. **60** (2010), 940–950.
2. C. Bai, O. Bellier, L. Guo, X. Ni, Splitting of operations, Manin products and Rota-Baxter operators. Preprint arXiv:1106.6080
3. J.C. Baez, The octonions. Bull. Amer. Math. Soc. (N.S.) 39 (2002), no. 2, 145–205.
4. F. Chapoton, and M. Livernet, Pre-Lie algebras and the rooted trees operad. Internat. Math. Res. Notices 2001, no. 8, 395–408.
5. A. Connes, D. Kreimer, Hopf algebras, renormalization and noncommutative geometry. Comm. Math. Phys. 199 (1998), no. 1, 203–242.

6. S. Covez, The local integration of Leibniz algebras. Preprint: arXiv:1011.4112.
7. A. Dzhumadil'daev and P. Zusmanovich, The alternative operad is not Koszul, Experimental Mathematics 20 (2011), 138-144.
8. B. Fresse, On the homotopy of simplicial algebras over an operad. Trans. Amer. Math. Soc. 352 (2000), no. 9, 4113–4141.
9. T. V. Kadeishvili, The algebraic structure in the homology of an $A(\infty)$-algebra, Soobshch. Akad. Nauk Gruzin. SSR 108 (1982), no. 2, 249–252 (1983).
10. L. Liu, X. Ni, C. Bai, L-quadri-algebras (in Chinese), Sci. Sin. Math. 42 (2011) 105–124.
11. J.-L. Loday, Comparaison des homologies du groupe linéaire et de son algèbre de Lie. Ann. Inst. Fourier (Grenoble) 37 (1987), no. 4, 167–190.
12. J.-L. Loday, Algebraic K-theory and the conjectural Leibniz K-theory, K-theory (2003), 105–127.
13. J.-L. Loday, Homotopical syzygies, in "Une dégustation topologique: Homotopy theory in the Swiss Alps", Contemporary Mathematics no 265 (AMS) (2000), 99–127.
14. J.-L. Loday, Generalized bialgebras and triples of operads. Astérisque No. 320 (2008), x+116 pp.
15. J.-L. Loday, Geometric diagonals for the Stasheff associahedron and products of A-infinity algebras, preprint (2011).
16. J.-L. Loday and M.O. Ronco, Combinatorial Hopf algebras, in Quanta of Maths, Clay Mathematics Proceedings, vol. 11, Amer. Math. Soc., Providence, RI, 2011, pp. 347–383.
17. J.-L. Loday and B. Vallette, Algebraic Operads, (2011), submitted.
18. M. Markl, S. Shnider, J. Stasheff, Operads in algebra, topology and physics. Mathematical Surveys and Monographs, 96. American Mathematical Society, Providence, RI, 2002. x+349 pp.
19. A. Ocneanu, The classification of subgroups of quantum SU(N), Quantum symmetries in theoretical physics and mathematics (Bariloche, 2000), Contemp. Math., vol. 294, Amer. Math. Soc., Providence, RI, 2002, pp. 133–159.
20. S. Saneblidze, R. Umble, Diagonals on the permutahedra, multiplihedra and associahedra. Homology Homotopy Appl. 6 (2004), no. 1, 363–411.
21. J. D. Stasheff, Homotopy associativity of H-spaces. I, II, Trans. Amer. Math. Soc. 108 (1963), 275-292; ibid. 108 (1963), 293–312.
22. B. Vallette, Manin products, Koszul duality, Loday algebras and Deligne conjecture. J. Reine Angew. Math. 620 (2008), 105–164.
23. G.W. Zinbiel, Encyclopedia of types of algebras 2010, this volume and arXiv:1101.0267.

Hom-dendriform Algebras and Rota-Baxter Hom-algebras

A. Makhlouf*

Laboratoire de Mathématiques, Informatique et Applications
Université de Haute Alsace
4, rue des Frères Lumière F-68093 Mulhouse, France
**E-mail: Abdenacer.Makhlouf@uha.fr*

The aim of this paper is to introduce and study Rota-Baxter Hom-algebras. Moreover we introduce a generalization of dendriform algebras and tridendriform algebras by twisting the identities by mean of a linear map. Then we explore the connections between these categories of Hom-algebras.

Keywords: Hom-Lie algebra; Hom-associative algebra; Rota-Baxter operator; Rota-Baxter algebra; Hom-preLie algebra; Hom-dendriform algebra; Hom-tridendriform algebra.

Introduction

The study of nonassociative algebras was originally motivated by certain problems in physics and other branches of mathematics. The Hom-algebra structures arose first in quasi-deformation of Lie algebras of vector fields. Discrete modifications of vector fields via twisted derivations lead to Hom-Lie and quasi-Hom-Lie structures in which the Jacobi condition is twisted. The first examples of q-deformations, in which the derivations are replaced by σ-derivations, concerned the Witt and Virasoro algebras, see for example.[1,13–16,18,19,42,44,48] A general study and construction of Hom-Lie algebras are considered in[39,45,46] and a more general framework bordering color and super Lie algebras was introduced in.[39,45–47] In the subclass of Hom-Lie algebras skew-symmetry is untwisted, whereas the Jacobi identity is twisted by a single linear map and contains three terms as in Lie algebras, reducing to ordinary Lie algebras when the twisting linear map is the identity map.

The notion of Hom-associative algebras generalizing associative algebras to a situation where associativity law is twisted by a linear map was introduced in,[54] it turns out that the commutator bracket multiplication defined using the multiplication in a Hom-associative algebra leads naturally to

Hom-Lie algebras. This provided a different way of constructing Hom-Lie algebras. Also are introduced in[54] the Hom-Lie-admissible algebras and more general G-Hom-associative algebras with subclasses of Hom-Vinberg and Hom-preLie algebras, generalizing to the twisted situation Lie-admissible algebras, G-associative algebras, Vinberg and preLie algebras respectively and it is shown that for these classes of algebras the operation of taking commutator leads to Hom-Lie algebras as well. The enveloping algebras of Hom-Lie algebras were discussed in.[62] The fundamentals of the formal deformation theory and associated cohomology structures for Hom-Lie algebras have been considered initially in[56] and completed in.[4] Simultaneously, in[65] elements of homology for Hom-Lie algebras have been developed. In[55] and,[57] the theory of Hom-coalgebras and related structures are developed. Further development could be found in.[5,7,12,31,43,52,52,53,66]

Dendriform algebras were introduced by Loday in.[50] Dendriform algebras are algebras with two operations, which dichotomize the notion of associative algebra. The motivation to introduce these algebraic structures with two generating operations comes from K-theory. It turned out later that they are connected to several areas in mathematics and physics, including Hopf algebras, homotopy Gerstenhaber algebra, operads, homology, combinatorics and quantum field theory where they occur in the theory of renormalization of Connes and Kreimer. Later the notion of tridendriform algebra were introduced by Loday and Ronco in their study of polytopes and Koszul duality.[51] A tridendriform algebra is a vector space equipped with 3 binary operations satisfying seven relations.

The Rota-Baxter operator has appeared in a wide range of areas in pure and applied mathematics. The paradigmatic example of Rota-Baxter operator concerns the integration by parts formula of continuous functions. The algebraic formulation of Rota-Baxter algebra appeared first in G. Baxter's works in probability study of fluctuation theory. This algebra was intensively studied by G.C. Rota in connection with combinatorics. In A. Connes and D. Kreimer works related to their Hopf algebra approach to renormalization of quantum field theory, the Rota-Baxter identity appeared under the name "multiplicativity constraint". This seminal work gives rise to an important development including Rota-Baxter algebras and their connections to other algebraic structure.[2,3,21,22,24–30,34,35,37,38,40]

The purpose of this paper is to study Rota-Baxter Hom-algebras. We introduce Hom-dendriform and Hom-tridendriform algebras and then explore the connections between all these categories of Hom-algebras. We summarize in the first Section the basis of Hom-algebras and recall the definitions and some properties of Hom-associative, Hom-Lie and Hom-preLie algebras. In Section 2, we introduce the notions of Hom-dendriform algebras and Hom-tridendriform algebras and provide constructions of these algebras and their relationships with Hom-preLie algebras. Section 3 is dedicated to Hom-associative Rota-Baxter algebras, we extend the classical notion of associative Rota-Baxter algebra and show some constructions. In Section 4 we establish functors between the category of Hom-associative Rota-Baxter algebras and the categories of Hom-preLie, Hom-dendriform and Hom-tridendriform algebras. In Section 5 we discuss the Rota-Baxter operator in the context of Hom-nonassociative algebras mainly for Hom-Lie algebras.

1. Hom-associative, Hom-Lie and Hom-preLie algebras

In this section we summarize the definitions and some properties of Hom-associative, Hom-Lie and Hom-preLie algebraic structures[54] generalizing the well known associative, Lie and preLie algebras by twisting the identities with a linear map.

Throughout the article we let $\mathbb{K}$ be an algebraically closed field of characteristic 0. We mean by a Hom-algebra a triple (A, μ, α) consisting of a vector space A on which $\mu : A \times A \to A$ is a bilinear map (or $\mu : A \otimes A \to A$ is a linear map) and $\alpha : A \to A$ is a linear map. A Hom-algebra (A, μ, α) is said to be *multiplicative* if $\forall x, y \in A$ we have $\alpha(\mu(x, y)) = \mu(\alpha(x), \alpha(y))$.

Let (A, μ, α) and $A' = (A', \mu', \alpha')$ be two Hom-algebras of a given type. A linear map $f : A \to A'$ is a *morphism of Hom-algebras* if

$$\mu' \circ (f \otimes f) = f \circ \mu \quad \text{and} \quad f \circ \alpha = \alpha' \circ f.$$

In particular, Hom-algebras (A, μ, α) and (A, μ', α') are isomorphic if there exists a bijective linear map f such that$\mu = f^{-1} \circ \mu' \circ (f \otimes f)$ and $\alpha = f^{-1} \circ \alpha' \circ f$.

A subspace H of A is said to be a *subalgebra* if for all $x, y \in H$ we have $\mu(x, y) \in H$ and $\alpha(x) \in H$. A subspace I of A is said to be an ideal if for $x \in I$ and $y \in A$ we have $\mu(x, y) \in I$ and $\alpha(x) \in I$.

In all the examples involving the unspecified products are either given by skewsymmetry or equal to zero.

1.1. *Hom-associative algebras*

The Hom-associative algebras were introduced by the author and Silvestrov in.[54]

Definition 1.1 (Hom-associative algebra). *A* Hom-associative algebra *is a triple $(A, \cdot, \alpha)$ consisting of a vector space A on which $\cdot : A \otimes A \to A$ and $\alpha : A \to A$ are linear maps, satisfying*

$$\alpha(x) \cdot (y \cdot z) = (x \cdot y) \cdot \alpha(z). \tag{1}$$

Example 1.1. Let $\{x_1, x_2, x_3\}$ be a basis of a 3-dimensional linear space A over $\mathbb{K}$. The following multiplication $\cdot$ and linear map α on A define Hom-associative algebras over $\mathbb{K}^3$:

$$\begin{array}{ll} x_1 \cdot x_1 = a\, x_1, & x_2 \cdot x_2 = a\, x_2, \\ x_1 \cdot x_2 = x_2 \cdot x_1 = a\, x_2, & x_2 \cdot x_3 = b\, x_3, \\ x_1 \cdot x_3 = x_3 \cdot x_1 = b\, x_3, & x_3 \cdot x_2 = x_3 \cdot x_3 = 0, \end{array}$$

$$\alpha(x_1) = a\, x_1, \quad \alpha(x_2) = a\, x_2, \quad \alpha(x_3) = b\, x_3,$$

where a, b are parameters in $\mathbb{K}$. The algebras are not associative when $a \neq b$ and $b \neq 0$, since

$$(x_1 \cdot x_1) \cdot x_3 - x_1 \cdot (x_1 \cdot x_3) = (a - b)b x_3.$$

Example 1.2 (Polynomial Hom-associative algebra[63]).
Consider the polynomial algebra $\mathcal{A} = \mathbb{K}[x_1, \cdots, x_n]$ in n variables. Let α be an algebra endomorphism of $\mathcal{A}$ which is uniquely determined by the n polynomials $\alpha(x_i) = \sum \lambda_{i;r_1,\cdots,r_n} x_1^{r_1} \cdots x_n^{r_n}$ for $1 \leq i \leq n$. Define μ by

$$\mu(f, g) = f(\alpha(x_1), \cdots, \alpha(x_n)) g(\alpha(x_1), \cdots, \alpha(x_n)) \tag{2}$$

for all f, g in $\mathcal{A}$. Then, $(\mathcal{A}, \mu, \alpha)$ is a Hom-associative algebra.

Example 1.3 (Matrix Hom-associative algebra[66]).
Let $\mathcal{A} = (A, \mu, \alpha)$ be a Hom-associative algebra. Then $(\mathcal{M}_n(\mathcal{A}), \mu', \alpha')$, where $\mathcal{M}_n(\mathcal{A})$ is the vector space of $n \times n$ matrix with entries in A, is also a Hom-associative algebra in which the multiplication μ' is given by matrix multiplication, and α' is given by α in each entry.

1.2. *Hom-Lie algebras*

The notion of Hom-Lie algebra was introduced by Hartwig, Larsson and Silvestrov in[39,45,46] motivated initially by examples of deformed Lie algebras coming from twisted discretizations of vector fields. In this article, we follow notations and a slightly more general definition of Hom-Lie algebras from.[54]

Definition 1.2 (Hom-Lie algebra). *A* Hom-Lie algebra *is a triple* $(\mathfrak{g}, [\ ,\ ], \alpha)$ *consisting of a vector space* $\mathfrak{g}$ *on which* $[\ ,\ ] : \mathfrak{g} \times \mathfrak{g} \to \mathfrak{g}$ *is a bilinear map and* $\alpha : \mathfrak{g} \to \mathfrak{g}$ *a linear map satisfying*

$$[x, y] = -[y, x], \quad \textit{(skew-symmetry)} \tag{3}$$

$$\circlearrowleft_{x,y,z} [\alpha(x), [y, z]] = 0 \quad \textit{(Hom-Jacobi identity)} \tag{4}$$

for all x, y, z *in* $\mathfrak{g}$, *where* $\circlearrowleft_{x,y,z}$ *denotes summation over the cyclic permutation on* x, y, z.

We recover classical Lie algebras when $\alpha = id_{\mathfrak{g}}$ and the identity (4) is the Jacobi identity in this case.

Example 1.4. Let $\{x_1, x_2, x_3\}$ be a basis of a 3-dimensional vector space $\mathfrak{g}$ over $\mathbb{K}$. The following bracket and linear map α on $\mathfrak{g} = \mathbb{K}^3$ define a Hom-Lie algebra over $\mathbb{K}$:

$$\begin{array}{rclcrcl} [x_1, x_2] &=& ax_1 + bx_3 & & \alpha(x_1) &=& x_1 \\ [x_1, x_3] &=& cx_2 & & \alpha(x_2) &=& 2x_2 \\ [x_2, x_3] &=& dx_1 + 2ax_3, & & \alpha(x_3) &=& 2x_3 \end{array}$$

with $[x_2, x_1]$, $[x_3, x_1]$ and $[x_3, x_2]$ defined via skewsymmetry. It is not a Lie algebra if $a \neq 0$ and $c \neq 0$, since

$$[x_1, [x_2, x_3]] + [x_3, [x_1, x_2]] + [x_2, [x_3, x_1]] = acx_2.$$

Example 1.5 (Jackson $\mathfrak{sl}_2$). *The Jackson* $\mathfrak{sl}_2$ *is a* q*-deformation of the classical* $\mathfrak{sl}_2$. *This family of Hom-Lie algebras was constructed in*[47] *using a quasi-deformation scheme based on discretizing by means of Jackson* q*-derivations a representation of* $\mathfrak{sl}_2(\mathbb{K})$ *by one-dimensional vector fields (first order ordinary differential operators) and using the twisted commutator bracket defined in.*[39] *It carries a Hom-Lie algebra structure but not a Lie algebra structure. It is defined with respect to a basis* $\{x_1, x_2, x_3\}$ *by the brackets and a linear map* α *such that*

$$\begin{array}{rclcrcl} [x_1, x_2] &=& -2qx_2 & & \alpha(x_1) &=& qx_1 \\ [x_1, x_3] &=& 2x_3 & & \alpha(x_2) &=& q^2x_2 \\ [x_2, x_3] &=& -\frac{1}{2}(1+q)x_1, & & \alpha(x_3) &=& qx_3 \end{array}$$

where q *is a parameter in* $\mathbb{K}$. *if* $q = 1$ *we recover the classical* $\mathfrak{sl}_2$.

There is a functor from the category of Hom-associative algebras in the category of Hom-Lie algebras. It provides a different way for constructing Hom-Lie algebras by extending the fundamental construction of Lie algebras by associative algebras via commutator bracket.

Proposition 1.1 ([54]). *Let $(A, \cdot, \alpha)$ be a Hom-associative algebra defined on the vector space A by the multiplication $\cdot$ and a homomorphism α. Then the triple $(A, [\ ,\ ], \alpha)$, where the bracket is defined for all $x, y \in A$ by $[x, y] = x \cdot y - y \cdot x$, is a Hom-Lie algebra.*

1.3. *Hom-preLie algebras*

Hom-preLie algebras were introduced in[54] in the study of Hom-Lie admissible algebras.

Definition 1.3 (Hom-preLie algebras). *A left Hom-preLie algebra (resp. right Hom-preLie algebra) is a triple $(A, \cdot, \alpha)$ consisting of a vector space A, a bilinear map $\cdot : A \times A \to A$ and a homomorphism α satisfying*

$$\alpha(x) \cdot (y \cdot z) - (x \cdot y) \cdot \alpha(z) = \alpha(y) \cdot (x \cdot z) - (y \cdot x) \cdot \alpha(z), \tag{5}$$

resp.

$$\alpha(x) \cdot (y \cdot z) - (x \cdot y) \cdot \alpha(z) = \alpha(x) \cdot (z \cdot y) - (x \cdot z) \cdot \alpha(y). \tag{6}$$

Remark 1.1. Any Hom-associative algebra is a Hom-preLie algebras.

A left Hom-preLie algebra is the opposite algebra of the right Hom-preLie algebra. Both left and right Hom-preLie algebras are Hom-Lie-admissible algebras, that is the commutators define Hom-Lie algebras.[54]

2. Hom-dendriform algebras and Hom-Tridendriform algebras

In this Section, we introduce the notions of Hom-dendriform algebras and Hom-tridendriform algebras generalizing the classical dendriform and tridendriform algebras to Hom-algebras setting.

2.1. *Hom-dendriform algebras*

Dendriform algebras were introduced by Loday in.[50] Dendriform algebras are algebras with two operations, which dichotomize the notion of associative algebra. We generalize now this notion by twisting the identities by a linear map.

Definition 2.1 (Hom-dendriform algebra). *A* Hom-dendriform algebra *is a quadruple $(A, \prec, \succ, \alpha)$ consisting of a vector space A on which the operations $\prec, \succ : A \otimes A \to \mathfrak{g}$ and $\alpha : A \to A$ are linear maps satisfying*

$$(x \prec y) \prec \alpha(z) = \alpha(x) \prec (y \prec z + y \succ z), \tag{1}$$

$$(x \succ y) \prec \alpha(z) = \alpha(x) \succ (y \prec z), \tag{2}$$

$$\alpha(x) \succ (y \succ z) = (x \prec y + x \succ y) \succ \alpha(z). \tag{3}$$

for all x, y, z in A.

We recover classical dendriform algebra when $\alpha = id$.

Let $(A, \prec, \succ, \alpha)$ and $(A', \prec', \succ', \alpha')$ be two Hom-dendriform algebras. A linear map $f : A \to A'$ is a *Hom-dendriform algebras morphism* if

$$\prec' \circ (f \otimes f) = f \circ \prec, \quad \succ' \circ (f \otimes f) = f \circ \succ \quad \text{and} \quad f \circ \alpha = \alpha' \circ f.$$

We show now that we may construct Hom-dendriform algebras starting from a classical dendriform algebras and an algebra morphisms. We extend then the construction by composition introduced by Yau in[65] for Lie and associative algebras.

Theorem 2.1. *Let $(A, \prec, \succ)$ be a dendriform algebra and $\alpha : A \to A$ be a dendriform algebra morphism. Then $A_\alpha = (A, \prec_\alpha, \succ_\alpha, \alpha)$, where $\prec_\alpha = \alpha \circ \prec$ and $\succ_\alpha = \alpha \circ \succ$, is a Hom-dendriform algebra.*

Moreover, suppose that $(A', \prec', \succ')$ is another dendriform algebra and $\alpha' : A' \to A'$ is a dendriform algebra morphism. If $f : A \to A'$ is a dendriform algebra morphism that satisfies $f \circ \alpha = \alpha' \circ f$ then

$$f : (A, \prec_\alpha, \succ_\alpha, \alpha) \longrightarrow (A', \prec'_\alpha, \succ'_\alpha, \alpha')$$

is a morphism of Hom-dendriform algebras.

Proof. Observe that

$$(x \prec_\alpha y) \prec_\alpha \alpha(z) = \alpha^2((x \prec y) \prec z),$$
$$(x \prec_\alpha y) \succ_\alpha \alpha(z) = \alpha^2((x \prec y) \succ z),$$
$$(x \succ_\alpha y) \succ_\alpha \alpha(z) = \alpha^2((x \succ y) \succ z),$$
$$(x \succ_\alpha y) \prec_\alpha \alpha(z) = \alpha^2((x \succ y) \prec z).$$

And similarly

$$\alpha(x) \prec_\alpha (y \prec_\alpha z) = \alpha^2(x \prec (y \prec z)),$$
$$\alpha(x) \prec_\alpha (y \succ_\alpha z) = \alpha^2(x \prec (y \succ z)),$$
$$\alpha(x) \succ_\alpha (y \succ_\alpha z) = \alpha^2(x \succ (y \succ z)),$$
$$\alpha(x) \succ_\alpha (y \prec_\alpha z) = \alpha^2(x \succ (y \prec z)).$$

Therefore the identities (1),(2),(3) follow obviously from the identities satisfied by $(A, \prec, \succ)$. The second assertion is proved similarly. $\square$

In the classical case the commutative dendriform algebras are also called Zinbiel algebras.[49,50] The left and right operations are further required to identify, $x \prec y = y \succ x$. We call commutative Hom-dendriform algebras Hom-Zinbiel algebras.

Definition 2.2 (Hom-Zinbiel algebra). *A* Hom-Zinbiel algebra *is a triple $(A, \circ, \alpha)$ consisting of a vector space A on which $\circ : A \otimes A \to A$ and $\alpha : A \to A$ are linear maps satisfying*

$$(x \circ y) \circ \alpha(z) = \alpha(x) \circ (y \circ z) + \alpha(x) \circ (z \circ y), \tag{4}$$

for all x, y, z in A.

Remark 2.1. One may construct Hom-Zinbiel algebra by composition method starting from a classical Zinbiel algebra $(A, \circ)$ and an algebra endomorphism α by considering $(A, \circ_\alpha, \alpha)$, where $x \circ_\alpha y = \alpha(x \circ y)$.

We show now that Hom-dendriform algebra structure dichotomize the Hom-associative structure and provide a connection to Hom-preLie algebras.

Proposition 2.1. *Let $(A, \prec, \succ, \alpha)$ be a Hom-dendriform algebra. Let $\star : A \otimes A \to A$ be a linear map defined for all $x, y \in A$ by*

$$x \star y = x \prec y + x \succ y. \tag{5}$$

Then $(A, \star, \alpha)$ is a Hom-associative algebra.

Proof. For all $x, y, z \in A$ we have

$$\begin{aligned}
\alpha(x) \star (y \star z) &= \alpha(x) \star (y \prec z + y \succ z) \\
&= \alpha(x) \prec (y \prec z + y \succ z) + \alpha(x) \succ (y \prec z + y \succ z) \\
&= (x \prec y) \prec \alpha(z) + \alpha(x) \succ (y \prec z) + \alpha(x) \succ (y \succ z) \\
&= (x \prec y) \prec \alpha(z) + (x \succ y) \prec \alpha(z) + (x \prec y + x \succ y) \succ \alpha(z) \\
&= (x \prec y + x \succ y) \prec \alpha(z) + (x \prec y + x \succ y) \succ \alpha(z) \\
&= (x \star y) \prec \alpha(z) + (x \star y) \succ \alpha(z) \\
&= (x \star y) \star \alpha(z).
\end{aligned}$$

$\square$

Proposition 2.2. *Let $(A, \prec, \succ, \alpha)$ be a Hom-dendriform algebra. Let $\lhd : A \otimes A \to A$ and $\rhd : A \otimes A \to A$ be linear maps defined for all $x, y \in A$ by*

$$x \rhd y = x \succ y - y \prec x \quad \textit{and} \quad x \lhd y = x \prec y - y \succ x. \tag{6}$$

Then $(A, \triangleright, \alpha)$ is a left Hom-preLie algebra and $(A, \triangleleft, \alpha)$ is a right Hom-preLie algebra.

Proof. For all $x, y, z \in A$ we have

$$\alpha(x) \triangleright (y \triangleright z) = \alpha(x) \triangleright (y \succ z - z \prec y)$$
$$= \alpha(x) \succ (y \succ z) - \alpha(x) \succ (z \prec y) - (y \succ z) \prec \alpha(x) + (z \prec y) \prec \alpha(x).$$

and

$$(x \triangleright y) \triangleright \alpha(z) = (x \succ y - y \prec x) \triangleright \alpha(z)$$
$$= (x \succ y) \succ \alpha(z) - (y \prec x) \succ \alpha(z) - \alpha(z) \prec (x \succ y) + \alpha(z) \prec (y \prec x).$$

Using (1) and (3), we may write

$$\alpha(x) \triangleright (y \triangleright z) = (x \prec y) \succ \alpha(z) + (x \succ y) \succ \alpha(z) - \alpha(x) \succ (z \prec y)$$
$$-(y \succ z) \prec \alpha(x) + \alpha(z) \prec (y \prec x) + \alpha(z) \prec (y \succ x).$$

Direct simplification and identity (1) lead to

$$\alpha(x) \triangleright (y \triangleright z) - (x \triangleright y) \triangleright \alpha(z) - \alpha(y) \triangleright (x \triangleright z + (y \triangleright x) \triangleright \alpha(z) = 0.$$

Similar proof shows the right Hom-preLie structure. □

Remark 2.2. If $(A, \prec, \succ, \alpha)$ is a commutative Hom-Dendriform algebra then the corresponding left and right Hom-preLie algebras vanish.

2.2. *Hom-tridendriform algebras*

The notion of tridendriform algebra were introduced by Loday and Ronco in.[51] A tridendriform algebra is a vector space equipped with 3 binary operations $\prec, \succ, \cdot$ satisfying seven relations. We extend this notion to Hom situation as follows:

Definition 2.3 (Hom-Tridendriform algebra). *A* Hom-tridendriform algebra *is a quintuple $(A, \prec, \succ, \cdot, \alpha)$ consisting of a vector space A on which the operations $\prec, \succ, \cdot : A \otimes A \to A$ and $\alpha : A \to A$ are linear maps satisfying*

$$(x \prec y) \prec \alpha(z) = \alpha(x) \prec (y \prec z + y \succ z + y \cdot z), \tag{7}$$
$$(x \succ y) \prec \alpha(z) = \alpha(x) \succ (y \prec z), \tag{8}$$
$$\alpha(x) \succ (y \succ z) = (x \prec y + x \succ y + x \cdot y) \succ \alpha(z), \tag{9}$$
$$(x \prec y) \cdot \alpha(z) = \alpha(x) \cdot (y \succ z), \tag{10}$$
$$(x \succ y) \cdot \alpha(z) = \alpha(x) \succ (y \cdot z), \tag{11}$$

$$(x \cdot y) \prec \alpha(z) = \alpha(x) \cdot (y \prec z), \tag{12}$$

$$(x \cdot y) \cdot \alpha(z) = \alpha(x) \cdot (y \cdot z), \tag{13}$$

for all x, y, z in A.

We recover classical tridendriform algebra when $\alpha = id$.

Remark 2.3. Any Hom-tridendriform algebra gives a Hom-dendriform algebra by setting $x \cdot y = 0$ for all $x, y \in A$.

As in Theorem 2.1, Given a classical tridendriform algebra and an algebra morphism we may construct by composition a Hom-tridendriform algebra.

Proposition 2.3. *Let $(A, \prec, \succ, \cdot)$ be a tridendriform algebra and $\alpha : A \to A$ be a tridendriform algebra morphism. Then $A_\alpha = (A, \prec_\alpha, \succ_\alpha, \cdot_\alpha, \alpha)$, where $\prec_\alpha = \alpha \circ \prec$, $\succ_\alpha = \alpha \circ \succ$ and $\cdot_\alpha = \alpha \circ \cdot$, is a Hom-tridendriform algebra.*

Moreover, suppose that $(A', \prec', \succ', \cdot')$ is another tridendriform algebra and $\alpha' : A' \to A'$ is a tridendriform algebra morphism. If $f : A \to A'$ is a tridendriform algebra morphism that satisfies $f \circ \alpha = \alpha' \circ f$ then

$$f : (A, \prec_\alpha, \succ_\alpha, \cdot_\alpha, \alpha) \longrightarrow (A', \prec'_\alpha, \succ'_\alpha, \cdot'_\alpha \alpha')$$

is a morphism of Hom-tridendriform algebras.

Similarly as in,[21] we obtain in the Hom-algebras setting the following new operation:

Proposition 2.4. *Let $(A, \prec, \succ, \cdot, \alpha)$ be a Hom-tridendriform algebra and $* : A \otimes A \to A$ be an operation defined by $x * y = x \prec y + x \succ y + x \cdot y$. Then $(A, *, \alpha)$ is a Hom-associative algebra.*

Proof. Using the axioms of Hom-tridendriform algebras we have for all $x, y, z \in A$

$$\begin{aligned}
\alpha(x) * (y * z) &= \alpha(x) * (y \prec z + y \succ z + y \cdot z) \\
&= \alpha(x) \prec (y \prec z + y \succ z + y \cdot z) + \alpha(x) \succ (y \prec z + y \succ z + y \cdot z) \\
&\quad + \alpha(x) \cdot (y \prec z + y \succ z + y \cdot z) \\
&= (x \prec y) \prec \alpha(z) + (x \succ y) \prec \alpha(z) + (x \prec y + x \succ y + x \cdot y) \succ \alpha(z) \\
&\quad + (x \succ y) \cdot \alpha(z) + (x \cdot y) \prec \alpha(z) + (x \prec y) \cdot \alpha(z) + (x \cdot y) \cdot \alpha(z) \\
&= (x \prec y + x \succ y + x \cdot y) \prec \alpha(z) + (x \prec y + x \succ y + x \cdot y) \succ \alpha(z)
\end{aligned}$$

$$+(x \prec y + x \succ y + x \cdot y) \cdot \alpha(z)$$
$$= (x * y) * \alpha(z). \qquad \square$$

3. Rota-Baxter operators and Hom-associative algebras

We extend in this section the notion of Rota-Baxter algebra to Hom-associative algebras.

Definition 3.1. A Hom-associative Rota-Baxter algebra is a Hom-associative algebra $(A, \cdot, \alpha)$ endowed with a linear map $R : A \to A$ subject to the relation

$$R(x) \cdot R(y) = R(R(x) \cdot y + x \cdot R(y) + \lambda x \cdot y), \tag{1}$$

where $\lambda \in \mathbb{K}$.

The map R is called *Rota-Baxter operator* of weight λ and the identity (1) is called *Rota-Baxter identity.* We denote the Hom-associative Rota-Baxter algebra by a quadruple $(A, \cdot, \alpha, R)$. We recover classical Rota-Baxter associative algebras when $\alpha = id$ and we denote them by triples $(A, \cdot, R)$.

Remark 3.1. Let $(A, \cdot, \alpha, R)$ be a Hom-associative Rota-Baxter algebra, where R is a Rota-Baxter operator of weight λ. Then $(A, \cdot, \alpha, \lambda id - R)$ is a Hom-associative Rota-Baxter algebra. Indeed, the proof is straightforward and does not use the Hom-associativity of the algebra.

In the following we provide some constructions of Rota-Baxter Hom-algebras starting from classical Rota-Baxter algebra. Also we construct new Rota-Baxter Hom-algebras from a given Rota-Baxter Hom-algebra. These constructions extend to Rota-Baxter Hom-algebras the composition method, nth derived Hom-algebra construction and a construction involving elements of the centroid.

Theorem 3.1. *Let $(A, \cdot, R)$ be an associative Rota-Baxter algebra and $\alpha : A \to A$ be an algebra morphism commuting with R. Then $(A, \cdot_\alpha, \alpha, R)$, where $x \cdot_\alpha y = \alpha(x \cdot y)$, is a Hom-associative Rota-Baxter algebra.*

Proof. The Hom-associative structure of the algebra follows from Yau's Theorem in.[65]

Now we check that R is still a Rota-Baxter operator for the Hom-associative algebra.

$$\begin{aligned} R(x) \cdot_\alpha R(y) &= \alpha(R(x) \cdot R(y)) \\ &= \alpha(R(R(x) \cdot y + x \cdot R(y) + \lambda x \cdot y)) \\ &= \alpha(R(R(x) \cdot y)) + \alpha(R(x \cdot R(y))) + \alpha(R(\lambda x \cdot y))). \end{aligned}$$

Since α and R commute then

$$\begin{aligned} R(x)\cdot_\alpha R(y) &= R(\alpha(R(x)\cdot y)) + R(\alpha(x\cdot R(y))) + R(\alpha(\lambda x\cdot y))) \\ &= R(R(x)\cdot_\alpha y + x\cdot_\alpha R(y) + \lambda x\cdot_\alpha y)). \end{aligned}$$

□

More generally, given a Hom-associative Rota-Baxter algebra (A,μ,α,R), one may ask whether this Hom-associative Rota-Baxter algebra is induced by an ordinary associative Rota-Baxter algebra $(A,\widetilde{\mu},R)$, that is α is an algebra morphism with respect to $\widetilde{\mu}$ and $\mu=\alpha\circ\widetilde{\mu}$.

Let (A,μ,α) be a multiplicative Hom-associative algebra. It was observed in[33] that in case α is invertible, the composition method using α^{-1} leads to an associative algebra. If α is an algebra morphism with respect to $\widetilde{\mu}$ then α is also an algebra morphism with respect to μ. Indeed,

$$\mu(\alpha(x),\alpha(y)) = \alpha\circ\widetilde{\mu}(\alpha(x),\alpha(y)) = \alpha\circ\alpha\circ\widetilde{\mu}(x,y) = \alpha\circ\mu(x,y).$$

If α is bijective then α^{-1} is also an algebra automorphism. Therefore one may use an untwist operation on the Hom-associative algebra in order to recover the associative algebra ($\widetilde{\mu}=\alpha^{-1}\circ\mu$).

Proposition 3.1. *Let (A,μ,α,R) be a multiplicative Hom-associative Rota-Baxter algebra where α is invertible and such that α and R commute. Then $(A,\mu_{\alpha^{-1}}=\alpha^{-1}\circ\mu,R)$ is a Hom-associative Rota-Baxter algebra.*

Proof. The associativity condition follows from

$$\begin{aligned} 0 &= \alpha^{-2}\mu(\alpha(x),\mu(y,z)) - \mu(\mu(x,y),\alpha(z)) \\ &= \alpha^{-1}\mu(x,\alpha^{-1}\mu(y,z)) - \alpha^{-1}\mu(\alpha^{-1}\mu(x,y),z) \\ &= \mu_{\alpha^{-1}}(x,\mu_{\alpha^{-1}}(y,z)) - \mu_{\alpha^{-1}}(\mu_{\alpha^{-1}}(x,y),z). \end{aligned}$$

Since α and R commute then α^{-1} and R commute as well. Hence R is a Rota-Baxter operator for the new multiplication. □

We may also derive new Hom-associative algebras from a given multiplicative Hom-associative algebra using the following procedure. We split the definition given in[69] into two types of nth derived Hom-algebras.

Definition 3.2 ([69]). *Let (A,μ,α) be a multiplicative Hom-algebra and $n\geq 0$. The type 1 nth derived Hom-algebra of A is defined by*

$$A^n = \left(A,\mu^{(n)}=\alpha^n\circ\mu,\alpha^{n+1}\right), \tag{2}$$

and the type 2 *nth derived Hom-algebra of* A *is defined by*

$$A^n = \left(A, \mu^{(n)} = \alpha^{2^n-1} \circ \mu, \alpha^{2^n}\right). \tag{3}$$

Note that in both cases $A^0 = A$ *and* $A^1 = \left(A, \mu^{(1)} = \alpha \circ \mu, \alpha^2\right)$.

Observe that for $n \geq 1$ and $x, y, z \in A$ we have

$$\begin{aligned}\mu^{(n)}(\mu^{(n)}(x,y), \alpha^{n+1}(z)) &= \alpha^n \circ \mu(\alpha^n \circ \mu(x,y), \alpha^{n+1}(z)) \\ &= \alpha^{2n} \circ \mu(\mu(x,y), \alpha(z)).\end{aligned}$$

Therefore, following,[69] one obtains the following result.

Theorem 3.2. *Let* (A, μ, α, R) *be a multiplicative Hom-associative Rota-Baxter algebra such that* α *and* R *commute.Then the nth derived Hom-algebra of type* 1 *is also a Hom-associative Rota-Baxter algebra.*

Proof. The operator R is a Rota-Baxter operator for the new multiplication since

$$\alpha^n(\mu(R(x), R(y))) = \alpha^n(R(\mu(x, R(y))) + R(\mu(R(x), y)) + \lambda R(\mu(x,y))). \square$$

In the following we construct Hom-associative Rota-Baxter algebras involving elements of the centroid of associative Rota-Baxter algebras. The construction of Hom-algebras using elements of the centroid was initiated in[10] for Lie algebras.

Let $(A, \cdot)$ be an associative algebra. An endomorphism $\alpha \in End(A)$ is said to be an element of the centroid if $\alpha(x \cdot y) = \alpha(x) \cdot y = x \cdot \alpha(y)$ for all $x, y \in A$. The centroid of A is defined by

$$Cent(A) = \{\alpha \in End(A) : \alpha(x \cdot y) = \alpha(x) \cdot y = x \cdot \alpha(y),\ \forall x, y \in A\}.$$

The same definition of the centroid is assumed for Hom-associative algebras.

Proposition 3.2. *Let* (A, μ, R) *be an associative Rota-Baxter algebra where* R *is a Rota-Baxter operator of weight* λ. *Let* $\alpha \in Cent(A)$ *and set for all* $x, y \in A$

$$\mu^1_\alpha(x,y) = \mu(\alpha(x), y) \quad \text{and} \quad \mu^2_\alpha(x,y) = \mu(\alpha(x), \alpha(y)).$$

Assume that α *and* R *commute. Then* $(A, \mu^1_\alpha, \alpha, R)$ *and* $(A, \mu^2_\alpha, \alpha, R)$ *are Hom-associative Rota-Baxter algebras.*

Proof. Observe that for all $x, y, z \in A$

$$\begin{aligned}\mu_\alpha^1(\alpha(x), \mu_\alpha^1(y,z)) &= \mu(\alpha^2(x), \mu(\alpha(y), z)) = \mu(\alpha^2(x), \alpha\mu(y,z)) \\ &= \alpha(\mu(\alpha(x), \mu(y,z))) = \alpha^2(\mu(x, \mu(y,z))).\end{aligned}$$

Similarly

$$\begin{aligned}\mu_\alpha^2(\alpha(x), \mu_\alpha^2(y,z)) &= \mu(\alpha^2(x), \alpha\mu(\alpha(y), \alpha(z))) \\ &= \mu(\alpha^2(x), \alpha^2\mu(y,z)) = \alpha^2(\mu(x, \mu(y,z))).\end{aligned}$$

The triple $(A, \mu_\alpha^1, \alpha)$ and $(A, \mu_\alpha^2, \alpha)$ are Hom-associative algebras. They define also Rota-Baxter algebras since

$$\begin{aligned}\mu_\alpha^1(R(x), R(y)) &= \mu(\alpha(R(x)), R(y)) = \alpha(\mu(R(x), R(y))) \\ &= \alpha(R(\mu(x, R(y)) + \mu(R(x), y) + \lambda\mu(x,y))) \\ &= R(\mu(\alpha(x), R(y)) + \mu(\alpha(R(x)), y) + \lambda\mu(\alpha(x), y)) \\ &= R(\mu_\alpha^1(x, R(y)) + \mu_\alpha^1(R(x), y) + \lambda\mu_\alpha^1(x,y)).\end{aligned}$$

and

$$\begin{aligned}\mu_\alpha^2(R(x), R(y)) &= \mu(\alpha(R(x)), \alpha(R(y))) = \alpha(\mu(R(x), \alpha(R(y)))) \\ &= \alpha^2(\mu(R(x), R(y))) \\ &= \alpha^2(R(\mu(x, R(y)) + \mu(R(x), y) + \lambda\mu(x,y))) \\ &= R(\alpha(\mu(\alpha(x), R(y)))) + \alpha(\mu(\alpha(R(x)), y)) + \lambda\alpha(\mu(\alpha(x), y))) \\ &= R(\mu(\alpha(R(y)), \alpha(x)) + \mu(\alpha(y), \alpha(R(x))) + \lambda\mu(\alpha(y), \alpha(x))) \\ &= R(\mu_\alpha^2(x, R(y) + \mu_\alpha^2(R(x), y) + \lambda\mu_\alpha^2(x,y)).\end{aligned}$$

$\square$

4. Some functors

We show in this section that there is a functor from the category of Hom-associative Rota-Baxter algebras to the category of Hom-preLie algebras and then a functor to the categories of Hom-dendriform algebras and Hom-tridendriform algebras.

Proposition 4.1. *Let $(A, \cdot, \alpha, R)$ be a Hom-associative Rota-Baxter algebra where R is a Rota-Baxter operator of weight $\lambda = 0$ or -1. Assume that α and R commute. We define the operation $*$ on A by*

$$x * y = R(x) \cdot y - y \cdot R(x) + \lambda x \cdot y. \tag{1}$$

Then $(A, *, \alpha)$ is a Hom-preLie algebra.

Proof. In case $\lambda = 0$, we have for all $x, y, z \in A$

$$\begin{aligned}
&\alpha(x) * (y * z) \\
&= \alpha(x) * (R(y) \cdot z - z \cdot R(y)) \\
&= R(\alpha(x)) \cdot (R(y) \cdot z - z \cdot R(y)) - (R(y) \cdot z - z \cdot R(y)) \cdot R(\alpha(x)) \\
&= R(\alpha(x)) \cdot (R(y) \cdot z) - R(\alpha(x)) \cdot (z \cdot R(y)) \\
&\quad -(R(y) \cdot z) \cdot R(\alpha(x)) + (z \cdot R(y)) \cdot R(\alpha(x)) \\
&= \alpha(R(x)) \cdot (R(y) \cdot z) - \alpha(R(x)) \cdot (z \cdot R(y)) \\
&\quad -(R(y) \cdot z) \cdot \alpha(R(x)) + (z \cdot R(y)) \cdot \alpha(R(x)),
\end{aligned}$$

and

$$\begin{aligned}
&(x * y) * \alpha(z) \\
&= (R(x) \cdot y - y \cdot R(x)) * \alpha(z) \\
&= R(R(x) \cdot y - y \cdot R(x)) \cdot \alpha(z) - \alpha(z) \cdot R(R(x) \cdot y - y \cdot R(x)) \\
&= R(R(x) \cdot y) \cdot \alpha(z) - R(y \cdot R(x)) \cdot \alpha(z) \\
&\quad -\alpha(z) \cdot R(R(x) \cdot y) + \alpha(z) \cdot R(y \cdot R(x)).
\end{aligned}$$

Then

$$\begin{aligned}
&\alpha(x) * (y * z) - (x * y) * \alpha(z) - \alpha(y) * (x * z) + (y * x) * \alpha(z) = \\
&\alpha(R(x)) \cdot (R(y) \cdot z) - \alpha(R(x)) \cdot (z \cdot R(y)) - (R(y) \cdot z) \cdot \alpha(R(x)) \\
&+(z \cdot R(y)) \cdot \alpha(R(x)) - R(R(x) \cdot y) \cdot \alpha(z) + R(y \cdot R(x)) \cdot \alpha(z) \\
&+\alpha(z) \cdot R(R(x) \cdot y) - \alpha(z) \cdot R(y \cdot R(x)) - \alpha(R(y)) \cdot (R(x) \cdot z) \\
&+\alpha(R(y)) \cdot (z \cdot R(x)) + (R(x) \cdot z) \cdot \alpha(R(y)) - (z \cdot R(x)) \cdot \alpha(R(y)) \\
&+R(R(y) \cdot x) \cdot \alpha(z) - R(x \cdot R(y)) \cdot \alpha(z) - \alpha(z) \cdot R(R(y) \cdot x) \\
&+\alpha(z) \cdot R(x \cdot R(y)).
\end{aligned}$$

Using the Rota-Baxter identity 1 we gather the 5th and 14th, 6th and 13th, 7th and 16th, 8th and 15th terms. Therefore we obtain

$$\begin{aligned}
&\alpha(x) * (y * z) - (x * y) * \alpha(z) - \alpha(y) * (x * z) + (y * x) * \alpha(z) = \\
&\alpha(R(x)) \cdot (R(y) \cdot z) - \alpha(R(x)) \cdot (z \cdot R(y)) - (R(y) \cdot z) \cdot \alpha(R(x)) \\
&+(z \cdot R(y)) \cdot \alpha(R(x)) - (R(x) \cdot R(y)) \cdot \alpha(z) + (R(y) \cdot R(x)) \cdot \alpha(z) \\
&+\alpha(z) \cdot (R(x) \cdot R(y)) - \alpha(z) \cdot (R(y) \cdot R(x)) - \alpha(R(y)) \cdot (R(x) \cdot z) \\
&+\alpha(R(y)) \cdot (z \cdot R(x)) + (R(x) \cdot z) \cdot \alpha(R(y)) - (z \cdot R(x)) \cdot \alpha(R(y)).
\end{aligned}$$

Then Hom-associativity leads to

$$\alpha(x) * (y * z) - (x * y) * \alpha(z) - \alpha(y) * (x * z) + (y * x) * \alpha(z) = 0.$$

In case $\lambda = -1$, we have for all $x, y, z \in A$

$$\begin{aligned} \alpha(x) * (y * z) \qquad &= R(\alpha(x)) \cdot (R(y) \cdot z - z \cdot R(y) \\ &-y \cdot z) - (R(y) \cdot z - z \cdot R(y) - y \cdot z) \cdot R(\alpha(x)) \\ &-\alpha(x) \cdot (R(y) \cdot z - z \cdot R(y) - y \cdot z), \end{aligned}$$

and

$$\begin{aligned} (x * y) * \alpha(z) &= R(R(x) \cdot y - y \cdot R(x) - x \cdot y) \cdot \alpha(z) \\ &-\alpha(z) \cdot R(R(x) \cdot y - y \cdot R(x) - x \cdot y) \\ &-(R(x) \cdot y - y \cdot R(x) - x \cdot y) \cdot \alpha(z). \end{aligned}$$

Then using the fact that α and R commute, and the Hom-associativity we obtain

$$\begin{aligned} &\alpha(x) * (y * z) - (x * y) * \alpha(z) - \alpha(y) * (x * z) + (y * x) * \alpha(z) = \\ &\alpha(R(x)) \cdot (R(y) \cdot z) + (z \cdot R(y)) \cdot \alpha(R(x)) - R(R(x) \cdot y) \cdot \alpha(z) \\ &+R(y \cdot R(x)) \cdot \alpha(z) + \alpha(z) \cdot R(R(x) \cdot y) - \alpha(z) \cdot R(y \cdot R(x)) \\ &-\alpha(R(y)) \cdot (R(x) \cdot z) - (z \cdot R(x)) \cdot \alpha(R(y)) + R(R(y) \cdot x) \cdot \alpha(z) \\ &-R(x \cdot R(y)) \cdot \alpha(z) - \alpha(z) \cdot R(R(y) \cdot x) + \alpha(z) \cdot R(x \cdot R(y)) \\ &+R(x \cdot y) \cdot \alpha(z) + \alpha(z) \cdot R(y \cdot x) - R(y \cdot x) \cdot \alpha(z) - \alpha(z) \cdot R(x \cdot y). \end{aligned}$$

Then it vanishes using the Rota-Baxter identity (1). □

Now we connect Hom-associative Rota-Baxter algebras to Hom-dendriform algebras. We generalize to Hom-algebras setting, the result given by Aguiar for weight 0 Rota-Baxter algebras in[3] and extended by Ebrahimi-Fard in[21] to any Rota-Baxter algebras.

Proposition 4.2. *Let $(A, \cdot, \alpha, R)$ be a Hom-associative Rota-Baxter algebra where R is a Rota-Baxter operator of weight λ. Assume that α and R commute. We define the operation $\prec$ and $\succ$ on A by*

$$x \prec y = x \cdot R(y) + \lambda x \cdot y, \quad \textit{and} \quad x \succ y = R(x) \cdot y. \tag{2}$$

Then $(A, \prec, \succ, \alpha)$ is a Hom-dendriform algebra.

Proof. For simplicity we provide the proof for $\lambda = 0$. Let $x, y, z \in A$, we have by using Hom-associativity, identity (1) and the fact that α and R commute:

$$\begin{aligned}&(x \prec y) \prec \alpha(z) - \alpha(x) \prec (y \prec z + y \succ z)\\ &= (x \cdot R(y)) \cdot R(\alpha(z)) - \alpha(x) \cdot R(y \cdot R(z) + R(y) \cdot z)\\ &= (x \cdot R(y)) \cdot \alpha(R(z)) - \alpha(x) \cdot (R(y) \cdot R(z)) = 0.\end{aligned}$$

$$\begin{aligned}&(x \succ y) \prec \alpha(z) - \alpha(x) \succ (y \prec z)\\ &= (R(x) \cdot y) \cdot R(\alpha(z)) - R(\alpha(x)) \cdot (y \cdot R(z))\\ &= (R(x) \cdot y) \cdot \alpha(R(z)) - \alpha(R(x)) \cdot (y \cdot R(z)) = 0.\end{aligned}$$

$$\begin{aligned}&\alpha(x) \succ (y \succ z) - (x \prec y + x \succ y) \succ \alpha(z)\\ &= R(\alpha(x)) \cdot (R(y) \cdot z) - R(x \cdot R(y) + R(x) \cdot y) \cdot \alpha(z)\\ &= \alpha(R(x)) \cdot (R(y) \cdot z) - (R(x) \cdot R(y)) \cdot \alpha(z) = 0.\end{aligned}$$

The general case follows from similar and straightforward calculations. For example

$$\begin{aligned}&(x \succ y) \prec \alpha(z) - \alpha(x) \succ (y \prec z)\\ &= (R(x) \cdot y) \prec \alpha(z) - \alpha(x) \succ (y \cdot R(z) + \lambda y \cdot z)\\ &= (R(x) \cdot y) \cdot R(\alpha(z)) + \lambda(R(x) \cdot y) \cdot \alpha(z) - R(\alpha(x)) \cdot (y \cdot R(z) + \lambda y \cdot z)\\ &= 0.\end{aligned}$$

□

Remark 4.1. Proposition 4.1 could be obtained as a corollary of Proposition 4.2 and Proposition 2.2 which leads to a construction of right Hom-preLie algebra with the following multiplication

$$x \lhd y = x \cdot R(y) - R(y) \cdot x.$$

Considering the associated categories and denoting by $HRBass_\lambda$ the category of Hom-associative Rota-Baxter algebras, $HpreLie$ the category of Hom-preLie algebras and $Hdend$ the category of Hom-dendriform algebras, we summarize the previous results in the following proposition

Proposition 4.3. *The following diagram is commutative*

$$\begin{array}{ccc} Hdend & \rightarrow & HpreLie \\ \downarrow & & \downarrow \\ HRBass_0 & \rightarrow & HpreLie \end{array}$$

We show now a connection between Hom-associative Rota-Baxter algebras and Hom-tridendriform algebras. The classical case was stated in.[21]

Proposition 4.4. *Let $(A, \cdot, \alpha, R)$ be a Hom-associative Rota-Baxter algebra where R is a Rota-Baxter operator of weight λ. Assume that α and R commute. We define the operation $\prec$, $\succ$ and $\bullet$ on A by*

$$x \prec y = x \cdot R(y), \qquad x \succ y = R(x) \cdot y \quad \text{and} \quad x \bullet y = \lambda x \cdot y. \tag{3}$$

Then $(A, \prec, \succ, \bullet, \alpha)$ is a Hom-tridendriform algebra.

Proof. The first three axioms follow from Proposition 4.2 and the last four use Hom-associativity and the commutation between R and α. For example

$$\begin{aligned}(x \prec y) \bullet \alpha(z) - \alpha(x) \bullet (y \succ z) &= \lambda(x \cdot R(y)) \cdot \alpha(z) - \lambda\alpha(x) \cdot (R(y) \cdot z) \\ &= 0.\end{aligned}$$ □

Observe that the category of Hom-dendriform algebras can be identified with a subcategory of objects in the category of Hom-tridendriform algebras.

Following Proposition 2.4, we derive a new Hom-associative multiplication defined by

$$x * y = x \cdot R(y) + R(x) \cdot y + \lambda x \cdot y.$$

As in the classical case it satisfies

$$R(x * y) = R(x) \cdot R(y) \quad \text{and} \quad \widetilde{R}(x * y) = -\widetilde{R}(x) \cdot \widetilde{R}(y)$$

where $\widetilde{R}(x) = -\lambda x - R(x)$.

5. Rota-Baxter operators and Hom-Nonassociative algebras

Rota-Baxter operator in the context of Lie algebras were introduced independently by Belavin and Drinfeld and Semenov-Tian-Shansky[9,60] in the 1980th and were related to solutions of the (modified) classical Yang-Baxter equation. The theory were developed later by Ebrahimi-Fard.[21]

We may extend the theory of Hom-associative Rota-Baxter algebras developed above to any Hom-Nonassociative algebra. We set the following definition

Definition 5.1. A Hom-Nonassociative Rota-Baxter algebra is a Hom-Nonassociative algebra $(A, [\ ,\ ], \alpha)$ endowed with a linear map $R : A \to A$

subject to the relation

$$[R(x), R(y)] = R([R(x), y] + [x, R(y)] + \lambda[x, y]), \tag{1}$$

where $\lambda \in \mathbb{K}$.

The map R is called *Rota-Baxter operator* of weight λ.

We obtain the following construction by composition of Hom-Lie Rota-Baxter algebras, extending the construction of Hom-Lie algebras given by Yau in[63] to Rota-Baxter algebras.

Theorem 5.1. *Let $(\mathfrak{g}, [\ ,\ ], R)$ be a Lie Rota-Baxter algebra and $\alpha : \mathfrak{g} \to \mathfrak{g}$ be a Lie algebra endomorphism commuting with R. Then $(\mathfrak{g}, [\ ,\ ]_\alpha, \alpha, R)$, where $[\ ,\ ]_\alpha = \alpha \circ [\ ,\ ]$, is a Hom-Lie Rota-Baxter algebra.*

Proof. Observe that $[\alpha(x), [y, z]_\alpha]_\alpha = \alpha[\alpha(x), \alpha[y, z]] = \alpha^2[x, [y, z]]$. Therefore the Hom-Jacobi identity for $\mathfrak{g}_\alpha = (\mathfrak{g}, [\ ,\ ]_\alpha, \alpha)$ follows obviously from the Jacobi identity of $(\mathfrak{g}, [\ ,\ ])$. The skew-symmetry is proved similarly.

Now we check that R is still a Rota-Baxter operator for the Hom-Lie algebra.

$$\begin{aligned}
[R(x), R(y)]_\alpha &= \alpha[R(x), R(y)] \\
&= \alpha(R([R(x), y] + [x, R(y)] + \lambda[x, y])) \\
&= \alpha(R([R(x), y])) + \alpha(R([x, R(y)])) + \alpha(R(\lambda[x, y])).
\end{aligned}$$

Since α and R commute then

$$\begin{aligned}
[R(x), R(y)]_\alpha &= R(\alpha([R(x), y])) + R(\alpha([x, R(y)])) + R(\alpha(\lambda[x, y])) \\
&= R([R(x), y]_\alpha + [x, R(y)]_\alpha + \lambda[x, y]_\alpha).
\end{aligned}$$

□

Remark 5.1. In particular the proposition is valid when α is an involution.

Let $(\mathfrak{g}, [\ ,\ ], \alpha)$ be a Hom-Lie algebra. It was observed in[33] that in case α is invertible, the composition method using α^{-1} leads to a Lie algebra.

Proposition 5.1. *Let $(\mathfrak{g}, [\ ,\ ], \alpha, R)$ be a Hom-Lie Rota-Baxter algebra such that α and R commute. Then $(\mathfrak{g}, [\ ,\ ]_{\alpha^{-1}} = \alpha^{-1} \circ [\ ,\ ], R)$ is a Lie Rota-Baxter algebra.*

Proof. The Jacobi identity follows from

$$\begin{aligned}
\circlearrowleft_{x,y,z} [x, [y, z]_{\alpha^{-1}}]_{\alpha^{-1}} = \circlearrowleft_{x,y,z} \alpha^{-1}([x, \alpha^{-1}([y, z])]) = \circlearrowleft_{x,y,z} \alpha^{-2}[\alpha(x), [y, z]] \\
= 0.
\end{aligned}$$

Since α and R commute then α^{-1} and R commute as well. Hence R is a Rota-Baxter operator for the new multiplication. □

We construct now new Hom-Lie Rota Baxter algebras from a given multiplicative Hom-Lie Rota-Baxter algebra using nth derived Hom-Lie algebras.

Definition 5.2 ([69]). *Let $(\mathfrak{g}, [\ ,\ ], \alpha)$ be a multiplicative Hom-Lie algebra and $n \geq 0$. The nth derived Hom-algebra of $\mathfrak{g}$ is defined by*

$$\mathfrak{g}_{(n)} = \left(\mathfrak{g}, [\ ,\ ]^{(n)} = \alpha^n \circ [\ ,\ ], \alpha^{n+1}\right), \tag{2}$$

Note that $\mathfrak{g}_{(0)} = \mathfrak{g}$ and $\mathfrak{g}_{(1)} = \left(\mathfrak{g}, [\ ,\ ]^{(1)} = \alpha \circ [\ ,\ ], \alpha^2\right)$.

Observe that for $n \geq 1$ and $x, y, z \in \mathfrak{g}$ we have

$$\begin{aligned}
[[x,y]^{(n)}, \alpha^{n+1}(z)]^{(n)} &= \alpha^n([\alpha^n([x,y]), \alpha^{n+1}(z)]) \\
&= \alpha^{2n}([[x,y], \alpha(z)]).
\end{aligned}$$

Theorem 5.2. *Let $(\mathfrak{g}, [\ ,\ ], \alpha, R)$ be a multiplicative Hom-Lie Rota-Baxter algebra and assume that α and R commute. Then its nth derived Hom-algebra is a Hom-Lie Rota-Baxter algebra.*

Proof. The nth derived Hom-algebra is a Hom-Lie algebra according to.[69] It is also a Rota-Baxter algebra since

$$\alpha^n([R(x), R(y)]) = \alpha^n(R([x, R(y)]) + R([R(x), y]) + \lambda R([x,y])). \qquad \square$$

In the following we construct Hom-Lie Rota-Baxter algebras involving elements of the centroid of Lie Rota-Baxter algebras. Let $(\mathfrak{g}, [\cdot,\cdot], R)$ be a Lie Rota-Baxter algebra. An endomorphism $\alpha \in End(\mathfrak{g})$ is said to be an element of the centroid if $\alpha[x,y] = [\alpha(x), y]$ for all $x, y \in \mathfrak{g}$. The centroid is defined by $Cent(\mathfrak{g}) = \{\alpha \in End(\mathfrak{g}) : \alpha[x,y] = [\alpha(x), y],\ \forall x, y \in \mathfrak{g}\}$. The same definition of the centroid is assumed for Hom-Lie Rota-Baxter algebra.

Proposition 5.2. *Let $(\mathfrak{g}, [\cdot,\cdot], R)$ be a Lie Rota-Baxter algebra where R is a Rota-Baxter operator of weight λ. Let $\alpha \in Cent(\mathfrak{g})$ and set for all $x, y \in \mathfrak{g}$*

$$[x,y]^1_\alpha = [\alpha(x), y] \quad \text{and} \quad [x,y]^2_\alpha = [\alpha(x), \alpha(y)].$$

Assume that α and R commute. Then $(\mathfrak{g}, [\cdot,\cdot]^1_\alpha, \alpha, R)$ and $(\mathfrak{g}, [\cdot,\cdot]^2_\alpha, \alpha, R)$ are Hom-Lie Rota-Baxter algebras.

Proof. The triple $(\mathfrak{g}, [\cdot,\cdot]^1_\alpha, \alpha)$ and $(\mathfrak{g}, [\cdot,\cdot]^2_\alpha, \alpha)$ are Hom-Lie algebras according to [10, Proposition 1.12].

They define also Rota-Baxter algebras since

$$\begin{aligned}
[R(x), R(y)]^1_\alpha &= [\alpha(R(x)), R(y)] = \alpha([R(x), R(y)]) \\
&= \alpha(R([x, R(y)] + [R(x), y] + \lambda[x, y])) \\
&= R([\alpha(x), R(y)] + [\alpha(R(x)), y] + \lambda[\alpha(x), y]) \\
&= R([x, R(y)]^1_\alpha + [R(x), y]^1_\alpha + \lambda[x, y]^1_\alpha).
\end{aligned}$$

and similarly

$$\begin{aligned}
[R(x), R(y)]^2_\alpha &= [\alpha(R(x)), \alpha(R(y))] = \alpha([R(x), \alpha(R(y))]) \\
&= -\alpha^2([R(y), R(x)]) = \alpha^2([R(x), R(y)]) \\
&= \alpha^2(R([x, R(y)] + [R(x), y] + \lambda[x, y])) \\
&= R(\alpha([\alpha(x), R(y)]) + \alpha([\alpha(R(x)), y]) + \lambda\alpha([\alpha(x), y])) \\
&= -R(\alpha([R(y), \alpha(x)]) + \alpha([y, \alpha(R(x))]) + \lambda\alpha([y, \alpha(x)])) \\
&= -R([\alpha(R(y)), \alpha(x)] + [\alpha(y), \alpha(R(x))] + \lambda[\alpha(y), \alpha(x)]) \\
&= R([x, R(y)]^2_\alpha + [R(x), y]^2_\alpha + \lambda[x, y]^2_\alpha).
\end{aligned}$$

□

We may obtain similar connections to Hom-preLie and Hom-dendriform algebras as for Hom-associative algebras. For example we have

Proposition 5.3. *Let $(A, [\ ,\ ], \alpha, R)$ be a Hom-Lie Rota-Baxter algebra where R is a Rota-Baxter operator of weight 0. Assume that α and R commute. We define the operation $*$ on A by*

$$x * y = [R(x), y]. \tag{3}$$

Then $(A, *, \alpha)$ is a Hom-preLie algebra.

Remark 5.2. The connection between Rota-Baxter Hom-algebras and Yang-Baxter equation will be developed in a forthcoming paper, as well as free Hom-associative Rota-Baxter algebra.

Acknowledgments

The author is grateful to K. Ebrahimi-Fard and the referee for their valuable remarks.

References

1. N. Aizawa, H. Sato : *q-Deformation of the Virasoro algebra with central extension,* Physics Letters B, Phys. Lett. B **256**, no. 1, 185–190 (1991). Hiroshima University pre-print HUPD-9012 (1990).
2. M. Aguiar, *Pre-Poisson algebras*, Lett. Math. Phys., **54** (4), (2000) 263–277.
3. M. Aguiar, *Infinitesimal bialgebras, preLie and dendriform algebras*, in: Hopf algebras in: Lect. Notes Pure Appl. Math., vol **237**, Marcel Dekker, New york, 2004, 1–33.
4. F. Ammar, Z. Ejbehi and A. Makhlouf, *Cohomology and Deformations of Hom-algebras,* Journal of Lie Theory **21** No. 4, (2011) 813–836.
5. F. Ammar and A. Makhlouf, *Hom-Lie algebras and Hom-Lie admissible superalgebras*, J. of Algebra, Vol. **324** (7), (2010) 1513–1528.
6. H. An and C. Bai, *From Rota-Baxter algebras to preLie algebras*, e-print arXiv:0711.1389v1.
7. H. Ataguema, A. Makhlouf and S. Silvestrov, *Generalization of n-ary Nambu algebras and beyond*, Journal of Mathematical Physics **50**, 1 (2009).
8. G. Baxter, *An analytic problem whose solution folllows from a simple algebraic identity*, Pacific J. Math. **10** (1960) 731–742.
9. A. A. Belavin and V. G. Drinfeld, *Solutions of the classical Yang-Baxter equation for simple Lie algebras,* Funct. Anal. Appl., **16**, (1982) 159–180.
10. S. Benayadi and A. Makhlouf, *Hom-Lie algebras with symmetric invariant nondegenerate bilinear form*, e-print arXiv:1009.4226v1 (2010).
11. P. Cartier, *On the structure of free Baxter algebras*, Advances in Math., **9**, (1972) 253–265.
12. S. Caenepeel and I. Goyvaerts, *Monoidal Hom-Hopf algebras*, e-print arXiv:0907.0187v1 (2009).
13. M. Chaichian, D. Ellinas and Z. Popowicz, *Quantum conformal algebra with central extension,* Phys. Lett. B **248**, no. 1-2, (1990) 95–99.
14. M. Chaichian, A. P. Isaev, J. Lukierski, Z. Popowicz and P. Prešnajder, *q-Deformations of Virasoro algebra and conformal dimensions,* Phys. Lett. B **262** (1), (1991) 32–38.
15. M. Chaichian, P. Kulish and J. Lukierski, *q-Deformed Jacobi identity, q-oscillators and q-deformed infinite-dimensional algebras,* Phys. Lett. B **237**, no. 3-4, (1990) 401–406.
16. M. Chaichian, Z. Popowicz and P. Prešnajder, *q-Virasoro algebra and its relation to the q-deformed KdV system,* Phys. Lett. B **249**, no. 1, (1990) 63–65.
17. A. Connes and D. Kreimer, *Hopf algebras, Renormalization and Noncommutative Geometry,* Comm. in Math. Phys., **199** (203) (1998).
18. T. L. Curtright and C. K. Zachos, *Deforming maps for quantum algebras,* Phys. Lett. B **243**, no. 3, (1990) 237–244.
19. C. Daskaloyannis, *Generalized deformed Virasoro algebras*, Modern Phys. Lett. A **7** no. 9, (1992) 809–816.
20. J. Dixmier, *Enveloping algebras*, Graduate studies in Math, **11**, AMS, 1996.
21. K. Ebrahimi-Fard, *Loday-type Algebras and the Rota-Baxter Relation*, Lett. Math. Phys. **61** (2002) 139–147.

22. K. Ebrahimi-Fard and L. Guo, *Free Rota-Baxter algebras and rooted trees*, J. Algebra App. **7** (2) (2008) 1–28.
23. ______ *Rota-Baxter algebras and dendriform algebras*, J. Pure Appl. Algebra **212** (2008) 320–339.
24. ______ *Rota-Baxter Algebras in Renormalization of Perturbative Quantum Field Theory*, Fields Institute Communications, **50**, (2007) 47–105.
25. K. Ebrahimi-Fard, L. Guo and D. Kreimer, *Spitzer's Identity and the Algebraic Birkhoff Decomposition in pQFT*, J. Phys. A: Math. Gen., **37**, (2004) 11037–11052.
26. K. Ebrahimi-Fard, L. Guo and D. Manchon, *Birkhoff type decompositions and the Baker-Campbell-Hausdorff recursion*, Comm. in Math. Phys., **267** no.3, (2006) 821–845.
27. K. Ebrahimi-Fard and D. Manchon *Dendriform equations*, J. of Algebra **322** (2009) 4053–4079.
28. ______ *Twisted Dendriform algebras and the preLie Magnus expansion*, e-print arXiv:0910.2166 (2009).
29. K. Ebrahimi-Fard, J.M. Gracia-Bondia and F. Patras, *Rota-Baxter algebras and new combinatorial identities*, Lett. Math. Phys. **81** (2007) 61–75.
30. K. Ebrahimi-Fard, D. Manchon and F. Patras, *New identities in Dendriform algebras*, J. of Algebra **320** (2008) 708–727.
31. M. Elhamadadi and A. Makhlouf, *Cohomology and Deformations of Hom-Alternative,* J. Gen. Lie Theory Appl., Special issue in Deformation Theory, (2011) (to appear).
32. Y. Fregier, A. Gohr and S. Silvestrov, *Unital algebras of Hom-associative type and surjective or injective twistings*, J. Gen. Lie Theory Appl. Vol. **3** (4), (2009) 285-295.
33. A. Gohr, *On Hom-algebras with surjective twisting*, e-print arXiv:0906.3270v3 (2009).
34. L. Guo, *Properties of free Baxter algebras*, Adv. Math. **151** (2000) 346–375.
35. ______ Baxter algebras and differential algebras, e-print arXiv:0407180 (2004), in Differential Algebra and Related Topics, World Scientific Publishing Company, 2002, 281–305.
36. ______ *WHAT IS a Rota-Baxter algebra*, Notice of Amer. Math. Soc. **56** (2009) 1436–1437.
37. L. Guo and W. Keigher, *Baxter algebras and shuffle products*, Adv. Math. **150** (2000) 117–149.
38. ______ *On free Baxter algebras: completions and the internal construction*, Adv. Math. **151** (2000) 101–127.
39. J. T. Hartwig, D. Larsson and S. Silvestrov, *Deformations of Lie algebras using σ-derivations*, J. of Algebra **295**, (2006) 314-361.
40. X. Li, D. Hou and C. Bai, *Rota-Baxter operators on preLie algebras*, Journal of Nonlinear Mathematical Physics, **14** (2) (2007) 269-289.
41. L. Hellström and S. Silvestrov, *Commuting elements in q-deformed Heisenberg algebras*, World Scientific (2000).
42. N. Hu, *q-Witt algebras, q-Lie algebras, q-holomorph structure and representations,* Algebra Colloq. **6**, no. 1, (1999) 51–70.

43. Q. Jin and X. Li, *Hom-Lie algebra structures on semi-simple Lie algebras*, J. of Algebra, Volume **319**, Issue 4, (2008) 1398–1408
44. C. Kassel, *Cyclic homology of differential operators, the Virasoro algebra and a q-analogue*, Commun. Math. Phys. **146** (1992) 343–351.
45. D. Larsson and S. Silvestrov, *Quasi-Hom-Lie algebras, Central Extensions and 2-cocycle-like identities,* J. of Algebra **288**, (2005) 321–344.
46. ——— *Quasi-Lie algebras,* in *Noncommutative Geometry and Representation Theory in Mathematical Physics,* Contemp. Math., **391**, Amer. Math. Soc., Providence, RI, (2005) 241–248.
47. ——— *Quasi-deformations of $sl_2(\mathbb{F})$ using twisted derivations*, Comm. in Algebra **35**, (2007) 4303 - 4318.
48. Liu, Ke Qin, *Characterizations of the quantum Witt algebra*, Lett. Math. Phys. **24** , no. 4, (1992) 257–265.
49. J-L. Loday, *Cup-product for Leibniz cohomology and dual Leibniz algebras*, Math . Scand. **77** (2) (1995) 189–196.
50. ———*Dialgebras*, in Lecture Notes in Math., vol. **1763**, Springer, Berlin, 2001, 7–66.
51. J.L. Loday and M. Ronco, *Trialgebras and families of polytopes*, in Homotopy Theory: relations with Algebraic Geometry , Group Cohomology, and Algebraic K-Theory, in Contemp. Math. vol **346**, Amer. Math. Soc., Providence, RI, 2004, 369–673.
52. A. Makhlouf, *Hom-alternative algebras and Hom-Jordan algebras*, International Electronic Journal of Algebra, Volume **8** (2010) 177–190.
53. ——— *Paradigm of Nonassociative Hom-algebras and Hom-superalgebras,* Proceedings of Jordan Structures in Algebra and Analysis Meeting, Eds: J. Carmona Tapia, A. Morales Campoy, A. M. Peralta Pereira, M. I. Ram?rez ?lvarez, Publishing house: Circulo Rojo (2010) 145–177.
54. A. Makhlouf and S. Silvestrov, *Hom-algebra structures*, J. Gen. Lie Theory Appl. **2** (2) , (2008) 51–64.
55. ———*Hom-Lie admissible Hom-coalgebras and Hom-Hopf algebras*, Published as Chapter 17, pp 189-206, S. Silvestrov, E. Paal, V. Abramov, A. Stolin, (Eds.), Generalized Lie theory in Mathematics, Physics and Beyond, Springer-Verlag, Berlin, Heidelberg, (2008).
56. ———*Notes on Formal deformations of Hom-Associative and Hom-Lie algebras*, Forum Mathematicum, vol. **22** (4) (2010) 715–759.
57. ———*Hom-Algebras and Hom-Coalgebras*, J. of Algebra and its Applications, Vol. **9**, (2010).
58. G.C. Rota, *Baxter algebras and combinatorial identities*, I, II, Bull. Amer. Math. Soc. **75** (1969) 325-329; Bull. Amer. Math. Soc. **75** (1969) 330-334.
59. G.C. Rota, *Baxter operators, an introduction*, In Gian-Carlo Rota on combinatorics, Contemp. Mathematicians, pages 504?512. Birkh?auser Boston, Boston, MA, 1995.
60. M.A. Semenov-Tian-Shansky, *What is a classical r-matrix?*, Funct. Ana. Appl., **17**, no.4., (1983) 259–272.
61. Y. Sheng, *Representations of hom-Lie algebras*, e-print arXiv:1005.0140v1 (2010).

62. D. Yau, *Enveloping algebra of Hom-Lie algebras*, J. Gen. Lie Theory Appl. **2** (2) (2008) 95–108.
63. ——— *Hom-algebras and homology,* J. Lie Theory **19** (2009) 409–421.
64. ——— *Hom-Yang-Baxter equation, Hom-Lie algebras, and quasi-triangular bialgebras,* J. Phys. A **42**, (2009) 165–202.
65. ——— *Hom-algebras as deformations and homology,* e-print arXiv:0712.3515v1 (2007).
66. ——— *Hom-bialgebras and comodule algebras*, e-print arXiv:0810.4866v1 (2008).
67. ——— *The Hom-Yang-Baxter equation and Hom-Lie algebras,* e-print arXiv:0905.1887v2 (2009).
68. ——— *The classical Hom-Yang-Baxter equation and Hom-Lie bialgebras,* e-print arXiv:0905.1890v1 (2009).
69. ——— *Hom-Malsev, Hom-alternative, and Hom-Jordan algebras,* e-print arXiv:1002.3944 (2010).

Free Field Realizations of the Current Algebras Associated with (Super) Lie Algebras

Wen-Li Yang*

Institute of Modern Physics, Northwest University,
Xian, 710069, P. R. China
**E-mail: wlyang@nwu.edu.cn*

Based on the particular orderings introduced for the positive roots of a finite dimensional simple basic Lie superalgebra $\mathcal{G}$, we construct the explicit free field realization of the corresponding current (or affine) superalgebra $\widehat{\mathcal{G}}_k$ at an arbitrary level k.

Keywords: Conformal field theory; current algebra; free field realization.

1. Introduction

The interest in two-dimensional non-linear σ-models with supergroups or their cosets as target spaces has grown drastically over the last ten years because of their applications ranging from string theory[5,9] and logarithmic conformal field theories (CFTs)[28,42] (for a review, see e.g.,[20,25] and references therein) to modern condensed matter physics.[3,6,10,17,29,37,38,40] The Wess-Zumino-Novikov-Witten(WZNW) models associated with supergroups stand out as an important class of such σ-models. This is due to the fact that, besides their own importance, the WZNW models are also the "building blocks" for other coset models which can be obtained by gauging or coset constructions.[2,4,33,39] In these models, current or affine (super)algebras[21,32] are the underlying symmetry algebras and are relevant to integrability of the model.

The Wakimoto free field realizations of current algebras[43] have been proved very powerful in the study of the WZNW models on bosonic groups.[1,7,16,18,24,27] Since the work of Wakimoto on the $sl(2)$ current algebra, much effort has been made to obtain similar results for the general case.[8,11,12,19,26,30,41] In these constructions, the explicit differential operator realizations of the corresponding finite dimensional (super)algebras play a key role. However, explicit differential operator expressions heavily depend on the choice of local coordinate systems in the so-called big cell $\mathcal{U}$.[23] Thus it is at least very involved, if not impossible, to obtain explicit differential operator expressions for higher-rank (super)algebras in the usual coordinate systems.[11,13–15,26,30,41] Recently it was shown in[44,45,47,48] that there exists a certain coordinate system in $\mathcal{U}$, which drastically simplifies the computation involved in the construction of explicit differential operator expressions for higher-rank (super)algebras. We call such a coordinate system the "good coordinate system".

In this paper we show how to establish such a "good coordinate system" of the big cell $\mathcal{U}$ for an arbitrary finite-dimensional basic Lie (super)algebra.[22,31] It has been shown[46] that the "good coordinate system" *indeed* exits and is related to a particular ordering for the positive roots of the (super)algebra. Based on such an ordering of the positive roots, we construct the "good coordinate system" for the three infinite series superalgebras $gl(r|n)$, $osp(2r|2n)$ and $osp(2r+1|2n)$, and derive their explicit differential operator representations. We then apply these differential operators to construct explicit free field representations of the associated current algebras. These free field realizations of the $gl(r|n)$, $osp(2r|2n)$ and $osp(2r+1|2n)$ current algebras give rise to the Fock representations of the current algebras. They provide explicit realizations of the vertex operator construction[36] of representations for affine superalgebras $gl(r|n)_k$, $osp(2r|2n)_k$ and $osp(2r+1|2n)_k$. These representations are in general not irreducible. To obtain irreducible representations, one needs the associated screening charges and to perform the cohomology analysis as in.[8,12,18,19]

This paper is organized as follows. In section 2, we briefly review finite-dimensional simple basic Lie superalgebras and their corresponding current algebras, which also introduces our notation and some basic ingredients. In section 3, we introduce the particular orderings for the positive roots of the superalgebras $gl(r|n)$, $osp(2r|2n)$ and $osp(2r+1|2n)$. Based on the orderings, we construct the explicit differential operator representations of the superalgebras. In section 4 we apply these differential operator expressions to construct the explicit free field realizations of the $gl(r|n)$, $osp(2r|2n)$ and $osp(2r+1|2n)$ currents.

2. Notation and preliminaries

Let $\mathcal{G} = \mathcal{G}_{\bar{0}} + \mathcal{G}_{\bar{1}}$ be a finite dimensional simple basic Lie superalgebra[22,31] with a $\mathbb{Z}_2$-grading:

$$[a] = \begin{cases} 0 \text{ if } a \in \mathcal{G}_{\bar{0}}, \\ 1 \text{ if } a \in \mathcal{G}_{\bar{1}}. \end{cases}$$

The superdimension of $\mathcal{G}$, denoted by sdim, is defined by

$$\mathrm{sdim}\,(\mathcal{G}) = \dim\,(\mathcal{G}_{\bar{0}}) - \dim\,(\mathcal{G}_{\bar{1}})\,. \tag{1}$$

For any two homogenous elements (i.e. elements with definite $\mathbb{Z}_2$-gradings) $a, b \in \mathcal{G}$, the Lie bracket is defined by

$$[a, b] = a\,b - (-1)^{[a][b]} b\,a.$$

This (anti)commutator extends to inhomogenous elements through linearity. Let $\{E_i | i = 1, \ldots, d\}$, where $d = \dim(\mathcal{G})$, be the basis of $\mathcal{G}$, which satisfy (anti)commutation relations,

$$[E_i,\, E_j] = \sum_{l=1}^{d} f_{ij}^{l}\, E_l. \tag{2}$$

The coefficients f_{ij}^{l} are the structure constants of $\mathcal{G}$. Alternatively, one can use the associated root system[22] to label the generators of $\mathcal{G}$ as follows. Let H be the Cartan subalgebra of $\mathcal{G}$. A root α of $\mathcal{G}$ ($\alpha \neq 0$) will be an element in H^*, the dual of H, such that:

$$\mathcal{G}_\alpha = \{a \in \mathcal{G} |\, [h, a] = \alpha(h)\, a, \quad \forall h \in H\} \neq 0. \tag{3}$$

The set of roots is denoted by Δ. Let $\Pi\!:= \{\alpha_i | i = 1, \ldots, r\}$ be the simple roots of $\mathcal{G}$, where the rank of $\mathcal{G}$ is equal to $r = \dim(H)$. With respect to Π, the set of positive roots is denoted by Δ_+, and we write $\alpha > 0$ if $\alpha \in \Delta_+$. A root α is called even or bosonic (odd or fermionic) if $\mathcal{G}_\alpha \in \mathcal{G}_{\bar{0}}$ ($\mathcal{G}_\alpha \in \mathcal{G}_{\bar{1}}$). The set of even roots is denoted by $\Delta_{\bar{0}}$, while the set of odd roots is denoted by $\Delta_{\bar{1}}$. Associated with each positive root α, there is a raising operator E_α, a lowering operator F_α and a Cartan generator H_α. These operators have definite $\mathbb{Z}_2$-gradings:

$$[H_\alpha] = 0, \quad [E_\alpha] = [F_\alpha] = \begin{cases} 0, \, \alpha \in \Delta_{\bar{0}} \bigcap \Delta_+, \\ 1, \, \alpha \in \Delta_{\bar{1}} \bigcap \Delta_+. \end{cases}$$

Moreover, one has the Cartan-Weyl decomposition of $\mathcal{G}$

$$\mathcal{G} = \mathcal{G}_- \oplus H \oplus \mathcal{G}_+, \tag{4}$$

where $\mathcal{G}_-$ is a span of lowering operators $\{F_\alpha\}$ and $\mathcal{G}_+$ is a span of raising operators $\{E_\alpha\}$, and $\mathcal{G}_\pm$ respectively generates an nilpotent subalgebra of $\mathcal{G}$.

One can introduce a nondegenerate and invariant supersymmetric metric or bilinear form for $\mathcal{G}$, which is denoted by (E_i, E_j). Then the affine Lie superalgebra $\mathcal{G}_k$ (or $\mathcal{G}$ current algebra) associated to $\mathcal{G}$ is generated by $\{E_i^n | i = 1, \ldots, d;\ n \in \mathbb{Z}\}$ satisfying (anti)commutation relation:

$$[E_i^n, E_j^m] = \sum_{l=1}^{d} f_{ij}^l E_l^{n+m} + nk(E_i, E_j)\delta_{n+m,0}. \tag{5}$$

Introduce currents

$$E_i(z) = \sum_{n\in\mathbb{Z}} E_i^n\, z^{-n-1}, \quad i = 1, \ldots, d.$$

Then the (anti)commutation relations (5) can be re-expressed in terms of the OPEs[21] of the currents,

$$E_i(z)E_j(w) = k\frac{(E_i, E_j)}{(z-w)^2} + \frac{\sum_{m=1}^{d} f_{ij}^l E_l(w)}{(z-w)}, \qquad i, j = 1, \ldots, d, \tag{6}$$

where f_{ij}^l are the structure constants (2). The aim of this paper is to construct *explicit* free field realizations of the current algebras associated with the unitary series $gl(r|n)$ and the orthosymplectic series $osp(2r|2n)$ and $osp(2r+1|2n)$ at an arbitrary level k.

3. Differential operator realizations of superalgebras

Let G be a Lie supergroup with $\mathcal{G}$ being its Lie superalgebra, and X be the flag manifold G/B_-, where B_- is the Borel subgroup corresponding to the subalgebra $\mathcal{G}_- \oplus H$. The differential operator realization of $\mathcal{G}$ can be obtained from the infinitesimal action of the corresponding group element on sections of a line bundle over X[35] or an η-invariant lifting of the vector fields on X which form a representation of $\mathcal{G}$.[23] As an open set of X, we will take the big cell $\mathcal{U}$, which is the orbit of the unit coset under the action of subgroup N_+ with Lie superalgebra $\mathcal{G}_+$. After choosing some local coordinates of $\mathcal{U}$, all the generators of $\mathcal{G}$ in principle can be realized by first-order differential operators of the coordinates. In this section we show that there are "good coordinate systems" which enable us to obtain the explicit differential operator realizations of all basic Lie superalgebras. We shall construct such coordinate systems for the three infinite series of basic superalgebras $sl(r|n)$, $osp(2r|2n)$ and $osp(2r+1|2n)$ with generic r and n. Our coordinate system in $\mathcal{U}$ is based on a particular ordering introduced for positive roots Δ_+ of the corresponding superalgebra. We call this ordering the normal ordering[34] of Δ_+.

Definition 3.1. The roots of Δ_+ are in normal ordering if all roots are ordered in such a way that: (i) for any pairwise non-colinear roots $\alpha, \beta, \gamma \in \Delta_+$ such that $\gamma = \alpha + \beta$, γ is between α and β; (ii) for α, $2\alpha \in \Delta_+$, 2α is located on the nearest right of α.

Such an ordering was constructed explicitly for all (super)algebras with rank less than 3 in.[34] In the following, we shall give the normal ordering of positive roots for each of the three infinite series superalgebras $gl(r|n)$, and $osp(2r|2n)$ and $osp(2r+1|2n)$.

3.1. *Differential operator realization of* $gl(r|n)$

Hereafter, let us fix two non-negative integers n and r such that $2 \le n + r$. Let us introduce $n + r$ linear-independent vectors: $\{\delta_i | i = 1, \ldots, n\}$ and $\{\epsilon_i | i = 1, \ldots r\}$. These vectors are endowed with a symmetric inner product such that

$$(\delta_m, \delta_l) = \delta_{ml}, \ (\delta_m, \epsilon_i) = 0, \ (\epsilon_i, \epsilon_j) = -\delta_{ij}. \tag{1}$$

The root system Δ of $gl(r|n)$ (or $A(r-1, n-1)$) can be expressed in terms of the vectors:

$$\Delta = \{\epsilon_i - \epsilon_j, \delta_m - \delta_l, \delta_m - \epsilon_i, \epsilon_i - \delta_m\}, \quad 1 \le i \ne j \le r, \ 1 \le m \ne l \le n,$$

while the even roots $\Delta_{\bar{0}}$ and the odd roots $\Delta_{\bar{1}}$ are given respectively by

$$\Delta_{\bar{0}} = \{\epsilon_i - \epsilon_j, \delta_m - \delta_l\}, \ \Delta_{\bar{1}} = \{\pm(\delta_m - \epsilon_i)\}, \quad 1 \le i \ne j \le r, \ 1 \le m \ne l \le n.$$

The distinguished simple roots are

$$\begin{aligned} &\alpha_1 = \delta_1 - \delta_2, \ldots, \alpha_{n-1} = \delta_{n-1} - \delta_n, \alpha_n = \delta_n - \epsilon_1, \\ &\alpha_{n+1} = \epsilon_1 - \epsilon_2, \ldots, \alpha_{n+r-1} = \epsilon_{r-1} - \epsilon_r. \end{aligned} \tag{2}$$

With regard to the simple roots, the corresponding positive roots Δ_+ are

$$\delta_m - \delta_l, \quad \epsilon_i - \epsilon_j, \quad 1 \le i < j \le r, \ 1 \le m < l \le n, \tag{3}$$

$$\delta_m - \epsilon_i, \quad 1 \le m \le n, \ 1 \le i \le r. \tag{4}$$

Among these positive roots, $\{\delta_m - \epsilon_i | i = 1, \ldots, r, \ m = 1, \ldots, n\}$ are odd and the others are even. Then we construct the normal ordering of the corresponding positive roots.

Proposition 3.1. *A normal ordering of* Δ_+ *for* $gl(r|n)$ *is given by*

$$\begin{aligned}&\epsilon_{r-1}-\epsilon_r;\ldots;\epsilon_1-\epsilon_r,\ldots,\epsilon_1-\epsilon_2;\ \delta_n-\epsilon_r,\ldots,\delta_n-\epsilon_1;\\&\ldots;\ \delta_1-\epsilon_r,\ldots,\delta_1-\epsilon_1,\delta_1-\delta_n,\ldots,\delta_1-\delta_2.\end{aligned}\tag{5}$$

Proof. One can directly verify that the above ordering of the positive roots (3)-(4) of $gl(r|n)$ fulfills all requirements of Definition 3.1. □

Let us introduce a bosonic coordinate ($x_{m,l}$, $y_{i,j}$ for $m<l$ and $i<j$) with a $\mathbb{Z}_2$-grading zero: $[x]=[y]=0$ associated with each positive even root (resp. $\delta_m-\delta_l$, $\epsilon_i-\epsilon_j$ for $m<l$ and $i<j$), and a fermionic coordinate ($\theta_{l,i}$) with a $\mathbb{Z}_2$-grading one: $[\theta]=1$ associated with each positive odd root (resp. $\delta_l-\epsilon_i$). These coordinates satisfy the following (anti)commutation relations:

$$[x_{i,j},x_{m,l}]=0,\ [\partial_{x_{i,j}},\partial_{x_{m,l}}]=0,\ [\partial_{x_{i,j}},x_{m,l}]=\delta_{im}\delta_{jl},\tag{6}$$

$$[y_{i,j},y_{m,l}]=0,\ [\partial_{y_{i,j}},\partial_{y_{m,l}}]=0,\ [\partial_{y_{i,j}},y_{m,l}]=\delta_{im}\delta_{jl},\tag{7}$$

$$[\theta_{i,j},\theta_{m,l}]=0,\ [\partial_{\theta_{i,j}},\partial_{\theta_{m,l}}]=0,\ [\partial_{\theta_{i,j}},\theta_{m,l}]=\delta_{im}\delta_{jl},\tag{8}$$

and the other (anti)commutation relations vanish.

It is well-known that the big cell $\mathcal{U}$ is isomorphic to the subgroup N_+ and hence to the subalgebra $\mathcal{G}_+$ via the exponential map. Therefore we can choose the following coordinate system $G_+(x,y,\theta)$ for the associated big cell $\mathcal{U}$:

$$G_+(x;y;\theta)=(G_{n+r-1,n+r})\ldots(G_{j,n+r}\ldots G_{j,j+1})\,(G_{1,n+r}\ldots G_{1,2}).\tag{9}$$

Here, for $i<j$, $G_{i,j}$ is given by

$$G_{i,j}=\begin{cases}e^{y_{i,j}E_{\epsilon_i-\epsilon_j}}, & \text{if } 1\le i<j\le r,\\ e^{\theta_{l,i}E_{\delta_l-\epsilon_i}}, & \text{if } 1\le l\le n,\ 1\le i\le r,\\ e^{x_{m,l}E_{\delta_m-\delta_l}}, & \text{if } 1\le m<l\le n.\end{cases}\tag{10}$$

Hereafter, let us adopt the convention that

$$E_i\equiv E_{\alpha_i},\quad F_i\equiv F_{\alpha_i},\quad i=1,\ldots,n+r.\tag{11}$$

Let $\langle\Lambda|$ be the highest weight vector of the representation of $gl(r|n)$ with highest weights $\{\lambda_i\}$, satisfying the following conditions:

$$\langle\Lambda|F_i=0,\qquad 1\le i\le n+r,\tag{12}$$

$$\langle\Lambda|H_i=\lambda_i\,\langle\Lambda|,\qquad 1\le i\le n+r.\tag{13}$$

Here the generators H_i are expressed in terms of some linear combinations of H_α.[47] An arbitrary vector in the corresponding Verma module is parametrized by $\langle\Lambda|$ and the corresponding bosonic and fermionic coordinates as

$$\langle\Lambda; x; y; \theta| = \langle\Lambda|G_+(x; y; \theta). \tag{14}$$

One can define a differential operator realization $\rho^{(d)}$ of the generators of $gl(r|n)$ by

$$\rho^{(d)}(g)\,\langle\Lambda; x; y; \theta| \equiv \langle\Lambda; x; y; \theta|\, g, \qquad \forall g \in gl(r|n). \tag{15}$$

Here $\rho^{(d)}(g)$ is a differential operator of the coordinates $\{x; y; \theta\}$ associated with the generator g, which can be obtained from the defining relation (15). The defining relation also assures that the differential operator realization is actually a representation of $gl(r|n)$. Therefore it is sufficient to give the differential operators related to the simple roots, as the others can be constructed through the simple ones by the (anti)commutation relations. The coordinate system (9)-(10) enabled us[47] to obtain the explicit differential operator realization of $gl(r|n)$.

Proposition 3.2. *The differential operator representations of the generators associated with the simple roots of $gl(r|n)$ are given by*

$$\begin{aligned}
\rho^{(d)}(E_j) &= \sum_{k\leq j-1} x_{k,j}\,\partial_{x_{k,j+1}} + \partial_{x_{j,j+1}},\ 1 \leq j \leq r-1, \qquad (16)\\
\rho^{(d)}(E_r) &= \sum_{k\leq r-1} x_{k,r}\,\partial_{\theta_{k,1}} + \partial_{\theta_{r,1}},\\
\rho^{(d)}(E_{r+j}) &= \sum_{k\leq r} \theta_{k,j}\,\partial_{\theta_{k,1+j}} + \sum_{k\leq j-1} y_{k,j}\,\partial_{y_{k,1+j}} + \partial_{y_{j,1+j}},\\
&\qquad 1 \leq j \leq n-1,\\
\rho^{(d)}(E_{j,j}) &= \sum_{k\leq j-1} x_{k,j}\,\partial_{x_{k,j}} - \sum_{j+1\leq k\leq r} x_{j,k}\,\partial_{x_{j,k}} - \sum_{k\leq n} \theta_{j,k}\,\partial_{\theta_{j,k}} + \lambda_j,\\
&\qquad 1 \leq j \leq r-1,\\
\rho^{(d)}(E_{r,r}) &= \sum_{k\leq r-1} x_{k,r}\,\partial_{x_{k,r}} - \sum_{k\leq n} \theta_{r,k}\,\partial_{\theta_{r,k}} + \lambda_r,\\
\rho^{(d)}(E_{r+j,r+j}) &= \sum_{k\leq r} \theta_{k,j}\,\partial_{\theta_{k,j}} + \sum_{k\leq j-1} y_{k,j}\,\partial_{y_{k,j}}\\
&\quad - \sum_{j+1\leq k\leq n} y_{j,k}\,\partial_{y_{j,k}} + \lambda_{r+j},\ 1 \leq j \leq n,
\end{aligned}$$

$$\rho^{(d)}(F_j) = \sum_{k\leq j-1} x_{k,j+1}\,\partial_{x_{k,j}} - \sum_{j+2\leq k\leq r} x_{j,k}\,\partial_{x_{j+1,k}} - \sum_{k\leq n}\theta_{j,k}\,\partial_{\theta_{j+1,k}}$$
$$-x_{j,j+1}\left(\sum_{j+1\leq k\leq r} x_{j,k}\,\partial_{x_{j,k}} + \sum_{k\leq n}\theta_{j,k}\,\partial_{\theta_{j,k}}\right)$$
$$+x_{j,j+1}\left(\sum_{j+2\leq k\leq r} x_{j+1,k}\,\partial_{x_{j+1,k}} + \sum_{k\leq n}\theta_{j+1,k}\,\partial_{\theta_{j+1,k}}\right)$$
$$+x_{j,j+1}\,(\lambda_j - \lambda_{j+1})\,,\ 1\leq j\leq r-1,$$
$$\rho^{(d)}(F_r) = \sum_{k\leq r-1}\theta_{k,1}\,\partial_{x_{k,r}} + \sum_{2\leq k\leq n}\theta_{r,k}\partial_{y_{1,k}} + \theta_{r,1}\,(\lambda_r + \lambda_{r+1})$$
$$-\theta_{r,1}\left(\sum_{2\leq k\leq n}(\theta_{r,k}\,\partial_{\theta_{r,k}} + y_{1,k}\,\partial_{y_{1,k}})\right)$$
$$\rho^{(d)}(F_{r+j}) = \sum_{k\leq r}\theta_{k,1+j}\,\partial_{\theta_{k,j}} + \sum_{k\leq j-1} y_{k,1+j}\,\partial_{y_{k,j}} - \sum_{j+2\leq k\leq n} y_{j,k}\,\partial_{y_{1+j,k}}$$
$$-y_{j,1+j}\sum_{j+1\leq k\leq n} y_{j,k}\,\partial_{y_{j,k}} + y_{j,1+j}\sum_{j+2\leq k\leq n} y_{1+j,k}\,\partial_{y_{1+j,k}}$$
$$+y_{j,1+j}\,(\lambda_{r+j} - \lambda_{r+1+j})\,,\ 1\leq j\leq n-1. \tag{17}$$

A direct computation shows that these differential operators (17)-(16) satisfy the $gl(r|n)$ (anti)commutation relations corresponding to the simple roots and the associated Serre relations. This implies that the differential representation of non-simple generators can be consistently constructed from the simple ones. Hence, we have obtained an explicit differential realization of $gl(r|n)$.

3.2. *Differential operator realization of* $osp(2r|2n)$

The root system Δ of $osp(2r|2n)$ (or $D(r,n)$) can be expressed in terms of the vectors $\{\delta_l\}$ and $\{\epsilon_i\}$ (1) as follows:

$$\Delta = \{\pm\epsilon_i \pm \epsilon_j,\ \pm\delta_m \pm \delta_l,\ \pm 2\delta_l,\ \pm\delta_l \pm \epsilon_i\}\,,\quad 1\leq i\neq j\leq r,\ 1\leq m\neq l\leq n,$$

while the even roots $\Delta_{\bar{0}}$ and the odd roots $\Delta_{\bar{1}}$ are given by

$$\Delta_{\bar{0}} = \{\pm\epsilon_i \pm \epsilon_j,\ \pm\delta_m \pm \delta_l,\ \pm 2\delta_l\}\,,\quad \Delta_{\bar{1}} = \{\pm\delta_l \pm \epsilon_i\}\,,$$
$$1\leq i\neq j\leq r,\ 1\leq m\neq l\leq n.$$

The distinguished simple roots are

$$\alpha_1 = \delta_1 - \delta_2, \ldots, \alpha_{n-1} = \delta_{n-1} - \delta_n,\ \alpha_n = \delta_n - \epsilon_1, \tag{18}$$
$$\alpha_{n+1} = \epsilon_1 - \epsilon_2, \ldots, \alpha_{n+r-1} = \epsilon_{r-1} - \epsilon_r,\ \alpha_{n+r} = \epsilon_{r-1} + \epsilon_r. \tag{19}$$

With regard to the simple roots, the corresponding positive roots Δ_+ are

$$\delta_m - \delta_l, \quad 2\delta_l, \quad \delta_m + \delta_l, \quad 1 \le m < l \le n, \tag{20}$$

$$\delta_l - \epsilon_i, \quad \delta_l + \epsilon_i, \quad 1 \le i \le r, 1 \le l \le n, \tag{21}$$

$$\epsilon_i - \epsilon_j, \quad \epsilon_i + \epsilon_j, \quad 1 \le i < j \le r. \tag{22}$$

In order to obtain an explicit differential operator realization of $osp(2r|2n)$, let us introduce the normal ordering of its positive roots.

Proposition 3.3. *A normal ordering of Δ_+ for $osp(2r|2n)$ is given by*

$$\begin{aligned} &\epsilon_{r-1}+\epsilon_r,\ \epsilon_{r-1}-\epsilon_r;\ldots;\ \epsilon_1+\epsilon_2,\ldots,\epsilon_1+\epsilon_r,\ \epsilon_1-\epsilon_r,\ldots,\epsilon_1-\epsilon_2;\\ &\quad \delta_n+\epsilon_1,\ldots,\delta_n+\epsilon_r,\ 2\delta_n,\ \delta_n-\epsilon_r,\ldots,\delta_n-\epsilon_1;\ldots;\\ &\quad \delta_1+\delta_2,\ldots,\delta_1+\delta_n,\ \delta_1+\epsilon_1,\ldots,\delta_1+\epsilon_r,\ 2\delta_1,\\ &\qquad \delta_1-\epsilon_r,\ldots,\delta_1-\epsilon_1,\delta_1-\delta_n,\ldots,\delta_1-\delta_2. \end{aligned} \tag{23}$$

Proof. One can directly verify that the above ordering of the positive roots (20)-(22) of $osp(2r|2n)$ obeys all requirements of Definition 1. □

For the case $r = 0$, the ordering (23) gives rise to the normal ordering of the positive roots of $sp(2n)$, while for the case $n = 0$ it yields the normal ordering of the positive roots of $so(2r)$. Based on these orderings, a "good coordinate system" in each of the associated big cells for $so(2n)$ and $sp(2n)$ was constructed in.[44] Here we use the ordering (23) to construct the "good coordinate system" in the associated big cell $\mathcal{U}$ and the explicit differential operator realization of $osp(2r|2n)$.

In addition to the coordinates $\{x_{m,l}; y_{i,j}; \theta_{l,i}\}$, which are associated with the positive roots $\{\delta_m - \delta_l; \epsilon_i - \epsilon_j; \delta_l - \epsilon_i\}$, we also need to introduce extra coordinates$\{\bar{x}_{m,l}; \bar{y}_{i,j}; x_l; \bar{\theta}_{l,i}\}$ associated with the positive roots $\{\delta_m+\delta_l; \epsilon_i+\epsilon_j; 2\delta_l; \delta_l+\epsilon_i\}$ respectively. The coordinates $\{x_{m,l}; y_{i,j}; \theta_{l,i}\}$ and their differentials satisfy the same (anti)commutation relations as (6)-(8). The other non-trivial relations are

$$[\bar{x}_{i,j}, \bar{x}_{m,l}] = 0, \ [\partial_{\bar{x}_{i,j}}, \partial_{\bar{x}_{m,l}}] = 0, \ [\partial_{\bar{x}_{i,j}}, \bar{x}_{m,l}] = \delta_{im}\delta_{jl}, \tag{24}$$

$$[x_m, x_l] = 0, \ [\partial_{x_m}, \partial_{x_l}] = 0, \ [\partial_{x_m}, x_l] = \delta_{ml}, \tag{25}$$

$$[\bar{y}_{i,j}, \bar{y}_{m,l}] = 0, \ [\partial_{\bar{y}_{i,j}}, \partial_{\bar{y}_{m,l}}] = 0, \ [\partial_{\bar{y}_{i,j}}, \bar{y}_{m,l}] = \delta_{im}\delta_{jl}, \tag{26}$$

$$[\bar{\theta}_{i,j}, \bar{\theta}_{m,l}] = 0, \ [\partial_{\bar{\theta}_{i,j}}, \partial_{\bar{\theta}_{m,l}}] = 0, \ [\partial_{\bar{\theta}_{i,j}}, \bar{\theta}_{m,l}] = \delta_{im}\delta_{jl}, \tag{27}$$

and the other (anti)commutation relations vanish.

Based on the very ordering (23) of the positive roots of $osp(2r|2n)$, we may introduce the following coordinate system $G_+(x, \bar{x}; y, \bar{y}; \theta, \bar{\theta})$ for the associated big cell $\mathcal{U}$:

$$\begin{aligned}G_+(x,\bar{x};y,\bar{y};\theta,\bar{\theta}) = &\left(\bar{G}_{n+r-1,n+r}\,G_{n+r-1,n+r}\right)\cdots\\ &\times\left(\bar{G}_{n+1,n+2}\ldots\bar{G}_{n+1,n+r}\,G_{n+1,n+r}\ldots G_{n+1,n+2}\right)\\ &\times\left(\bar{G}_{n,n+1}\ldots\bar{G}_{n,n+r}\,G_n\,G_{n,n+r}\ldots G_{n,n+1}\right)\cdots\\ &\times\left(\bar{G}_{1,2}\ldots\bar{G}_{1,n+r}\,G_1\,G_{1,n+r}\ldots G_{1,2}\right). \qquad (28)\end{aligned}$$

Here $G_{i,j}$, $\bar{G}_{i,j}$ and G_i are given by

$$G_{m,l}=e^{x_{m,l}E_{\delta_m-\delta_l}},\quad \bar{G}_{m,l}=e^{\bar{x}_{m,l}E_{\delta_m+\delta_l}},\qquad 1\le m<l\le n, \qquad (29)$$

$$G_l=e^{x_lE_{2\delta_l}},\ G_{l,n+i}=e^{\theta_{l,i}E_{\delta_l-\epsilon_i}},\ \bar{G}_{l,n+i}=e^{\bar{\theta}_{l,i}E_{\delta_l+\epsilon_i}},\quad 1\le l\le n,\ 1\le i\le r, \qquad (30)$$

$$G_{n+i,n+j}=e^{y_{i,j}E_{\epsilon_i-\epsilon_j}},\quad \bar{G}_{n+i,n+j}=e^{\bar{y}_{i,j}E_{\epsilon_i+\epsilon_j}},\quad 1\le i<j\le r. \qquad (31)$$

Thus all generators of $osp(2r|2n)$ can be realized in terms of the first order differential operators of the coordinates $\{x,\bar{x};y,\bar{y};\theta,\bar{\theta}\}$ as the similar procedure as that of the $gl(r|n)$ case. Here we present the result of the generators corresponding to simple roots.[46]

Proposition 3.4. *The differential operator representations of the generators associated with the simple roots of* $osp(2r|2n)$ *are given by*

$$\rho^{(d)}(E_l)=\sum_{m=1}^{l-1}\left(x_{m,l}\partial_{x_{m,l+1}}-\bar{x}_{m,l+1}\partial_{\bar{x}_{m,l}}\right)+\partial_{x_{l,l+1}},\ 1\le l\le n-1, \qquad (32)$$

$$\rho^{(d)}(E_n)=\sum_{m=1}^{n-1}\left(x_{m,n}\partial_{\theta_{m,1}}+\bar{\theta}_{m,1}\partial_{\bar{x}_{m,n}}\right)+\partial_{\theta_{n,1}},$$

$$\begin{aligned}\rho^{(d)}(E_{n+i})=&\sum_{m=1}^{n}\left(\theta_{m,i}\partial_{\theta_{m,i+1}}-\bar{\theta}_{m,i+1}\partial_{\bar{\theta}_{m,i}}\right)\\ &+\sum_{m=1}^{i-1}\left(y_{m,i}\partial_{y_{m,i+1}}-\bar{y}_{m,i+1}\partial_{\bar{y}_{m,i}}\right)+\partial_{y_{i,i+1}},\quad 1\le i\le r-1,\end{aligned}$$

$$\begin{aligned}\rho^{(d)}(E_{n+r})=&\sum_{m=1}^{n}\left(2\theta_{m,r-1}\theta_{m,r}\partial_{x_m}+\theta_{m,r-1}\partial_{\bar{\theta}_{m,r}}-\theta_{m,r}\partial_{\bar{\theta}_{m,r-1}}\right)\\ &+\sum_{m=1}^{r-2}\left(y_{m,r-1}\partial_{\bar{y}_{m,r}}-y_{m,r}\partial_{\bar{y}_{m,r-1}}\right)+\partial_{\bar{y}_{r-1,r}},\end{aligned}$$

$$\rho^{(d)}(H_l) = \sum_{m=1}^{l-1} \left(x_{m,l}\partial_{x_{m,l}} - \bar{x}_{m,l}\partial_{\bar{x}_{m,l}}\right) - \sum_{m=l+1}^{n} \left(x_{l,m}\partial_{x_{l,m}} + \bar{x}_{l,m}\partial_{\bar{x}_{l,m}}\right)$$
$$- \sum_{m=1}^{r} \left(\theta_{l,m}\partial_{\theta_{l,m}} + \bar{\theta}_{l,m}\partial_{\bar{\theta}_{l,m}}\right) - 2x_l\partial_{x_l} + \lambda_l, \qquad 1 \leq l \leq n,$$
$$\rho^{(d)}(H_{n+i}) = \sum_{m=1}^{n} \left(\theta_{m,i}\partial_{\theta_{m,i}} - \bar{\theta}_{m,i}\partial_{\bar{\theta}_{m,i}}\right) + \sum_{m=1}^{i-1} \left(y_{m,i}\partial_{y_{m,i}} - \bar{y}_{m,i}\partial_{\bar{y}_{m,i}}\right)$$
$$- \sum_{m=i+1}^{r} \left(y_{i,m}\partial_{y_{i,m}} + \bar{y}_{i,m}\partial_{\bar{y}_{i,m}}\right) + \lambda_{n+i}, \qquad 1 \leq i \leq r,$$

$$\rho^{(d)}(F_l) = \sum_{m=1}^{l-1} \left(x_{m,l+1}\partial_{x_{m,l}} - \bar{x}_{m,l}\partial_{\bar{x}_{m,l+1}}\right) - x_l\partial_{\bar{x}_{l,l+1}} - 2\bar{x}_{l,l+1}\partial_{x_{l+1}}$$
$$+ \sum_{m=l+2}^{n} \left(x_{l,m}\bar{x}_{l,m}\partial_{\bar{x}_{l,l+1}} - x_{l,m}\partial_{x_{l+1,m}} - 2\bar{x}_{l,m}x_{l+1,m}\partial_{x_{l+1}} - \bar{x}_{l,m}\partial_{\bar{x}_{l+1,m}}\right)$$
$$- \sum_{m=1}^{r} \left(\theta_{l,m}\bar{\theta}_{l,m}\partial_{\bar{x}_{l,l+1}} + \theta_{l,m}\partial_{\theta_{l+1,m}} + 2\bar{\theta}_{l,m}\theta_{l+1,m}\partial_{x_{l+1}} + \bar{\theta}_{l,m}\partial_{\bar{\theta}_{l+1,m}}\right)$$
$$-x_{l,l+1}^2\partial_{x_{l,l+1}} + 2x_{l,l+1}x_{l+1}\partial_{x_{l+1}} - 2x_{l,l+1}x_l\partial_{x_l}$$
$$+x_{l,l+1}\left[\sum_{m=l+2}^{n}\left(x_{l+1,m}\partial_{x_{l+1,m}} + \bar{x}_{l+1,m}\partial_{\bar{x}_{l+1,m}} - x_{l,m}\partial_{x_{l,m}} - \bar{x}_{l,m}\partial_{\bar{x}_{l,m}}\right)\right]$$
$$+x_{l,l+1}\left[\sum_{m=1}^{r}\left(\theta_{l+1,m}\partial_{\theta_{l+1,m}} + \bar{\theta}_{l+1,m}\partial_{\bar{\theta}_{l+1,m}} - \theta_{l,m}\partial_{\theta_{l,m}} - \bar{\theta}_{l,m}\partial_{\bar{\theta}_{l,m}}\right)\right]$$
$$+x_{l,l+1}(\lambda_l - \lambda_{l+1}), \qquad 1 \leq l \leq n-1,$$
$$\rho^{(d)}(F_n) = \sum_{m=1}^{n-1} \left(\theta_{m,1}\partial_{x_{m,n}} - \bar{x}_{m,n}\partial_{\bar{\theta}_{m,1}}\right) - x_n\partial_{\bar{\theta}_{n,1}}$$
$$+ \sum_{m=2}^{r} \left(\theta_{n,m}\partial_{y_{1,m}} - \theta_{n,m}\bar{\theta}_{n,m}\partial_{\bar{\theta}_{n,1}} + \bar{\theta}_{n,m}\partial_{\bar{y}_{1,m}}\right)$$
$$-\theta_{n,1}\sum_{m=2}^{r} \left(\theta_{n,m}\partial_{\theta_{n,m}} + \bar{\theta}_{n,m}\partial_{\bar{\theta}_{n,m}} + y_{1,m}\partial_{y_{1,m}} + \bar{y}_{1,m}\partial_{\bar{y}_{1,m}}\right)$$
$$-2\theta_{n,1}x_n\partial_{x_n} - 2\theta_{n,1}\bar{\theta}_{n,1}\partial_{\bar{\theta}_{n,1}} + \theta_{n,1}(\lambda_n + \lambda_{n+1}),$$

$$\rho^{(d)}(F_{n+i}) = \sum_{m=1}^{n} (\theta_{m,i+1}\partial_{\theta_{m,i}} - \bar{\theta}_{m,i}\partial_{\bar{\theta}_{m,i+1}}) + \sum_{m=1}^{i-1} (y_{m,i+1}\partial_{y_{m,i}} - \bar{y}_{m,i}\partial_{\bar{y}_{m,i+1}})$$
$$+ \sum_{m=i+2}^{r} \left(y_{i,m}\bar{y}_{i,m}\partial_{\bar{y}_{i,i+1}} - y_{i,m}\partial_{y_{i+1,m}} - \bar{y}_{i,m}\partial_{\bar{y}_{i+1,m}}\right)$$
$$+y_{i,i+1}\sum_{m=i+2}^{r} \left(y_{i+1,m}\partial_{y_{i+1,m}} + \bar{y}_{i+1,m}\partial_{\bar{y}_{i+1,m}} - y_{i,m}\partial_{y_{i,m}} - \bar{y}_{i,m}\partial_{\bar{y}_{i,m}}\right)$$
$$-y_{i,i+1}^2\partial_{y_{i,i+1}} + y_{i,i+1}(\lambda_{n+i} - \lambda_{n+i+1}), \qquad 1 \leq i \leq r-1,$$

$$\rho^{(d)}(F_{n+r}) = \sum_{m=1}^{n} \left(\bar{\theta}_{m,r}\partial_{\theta_{m,r-1}} + 2\bar{\theta}_{m,r-1}\bar{\theta}_{m,r}\partial_{x_m} - \bar{\theta}_{m,r-1}\partial_{\theta_{m,r}}\right)$$
$$+ \sum_{m=1}^{r-2} \left(\bar{y}_{m,r}\partial_{y_{m,r-1}} - \bar{y}_{m,r-1}\partial_{y_{m,r}}\right) - \bar{y}_{r-1,r}^2\partial_{\bar{y}_{r-1,r}}$$
$$+\bar{y}_{r-1,r}(\lambda_{n+r-1} + \lambda_{n+r}). \tag{33}$$

A direct computation shows that these differential operators (32)-(33) satisfy the $osp(2r|2n)$ (anti)commutation relations corresponding to the simple roots and the associated Serre relations. This implies that the differential representation of non-simple generators can be consistently constructed from the simple ones. Hence, we have obtained an explicit differential realization of $osp(2r|2n)$.

3.3. *Differential operator realization of* $osp(2r+1|2n)$

The root system Δ of $osp(2r+1|2n)$ (or $B(r,n)$) can be expressed in terms of the vectors $\{\delta_l\}$ and $\{\epsilon_i\}$ (1) as follows:

$$\Delta = \{\pm\epsilon_i \pm \epsilon_j,\ \pm\epsilon_i,\ \pm\delta_m \pm \delta_l,\ \pm\delta_l,\ \pm 2\delta_l,\ \pm\delta_l \pm \epsilon_i\},$$
$$1 \le i \ne j \le r,\ 1 \le m \ne l \le n,$$

while the even roots $\Delta_{\bar{0}}$ and the odd roots $\Delta_{\bar{1}}$ are given respectively by

$$\Delta_{\bar{0}} = \{\pm\epsilon_i \pm \epsilon_j,\ \pm\epsilon_i,\ \pm\delta_m \pm \delta_l,\ \pm 2\delta_l\}, \quad \Delta_{\bar{1}} = \{\pm\delta_l \pm \epsilon_i,\ \pm\delta_l\},$$
$$1 \le i \ne j \le r,\ 1 \le m \ne l \le n.$$

The distinguished simple roots are

$$\alpha_1 = \delta_1 - \delta_2, \ldots, \alpha_{n-1} = \delta_{n-1} - \delta_n,\ \alpha_n = \delta_n - \epsilon_1, \tag{34}$$

$$\alpha_{n+1} = \epsilon_1 - \epsilon_2, \ldots, \alpha_{n+r-1} = \epsilon_{r-1} - \epsilon_r,\ \alpha_{n+r} = \epsilon_r. \tag{35}$$

With regard to the simple roots, the corresponding positive roots Δ_+ are

$$\delta_m - \delta_l, \quad 2\delta_l, \quad \delta_m + \delta_l, \quad 1 \le m < l \le n, \tag{36}$$

$$\delta_l - \epsilon_i, \quad \delta_l + \epsilon_i, \quad \delta_l, \quad 1 \le i \le r, 1 \le l \le n, \tag{37}$$

$$\epsilon_i - \epsilon_j, \quad \epsilon_i + \epsilon_j, \quad \epsilon_i, \quad 1 \le i < j \le r. \tag{38}$$

Among these positive roots, $\{\delta_l,\ \delta_l \pm \epsilon_i | \, i = 1, \ldots, r,\ l = 1 \ldots, n\}$ are odd and the others are even.To obtain an explicit expression of the differential operator realization of $osp(2r+1|2n)$, let us introduce the normal ordering of its positive roots.

Proposition 3.5. *A normal ordering of* Δ_+ *for* $osp(2r+1|2n)$ *is given by*

$$\begin{aligned}
&\epsilon_r;\ \epsilon_{r-1} + \epsilon_r,\ \epsilon_{r-1},\ \epsilon_{r-1} - \epsilon_r; \ldots;\\
&\epsilon_1 + \epsilon_2, \ldots, \epsilon_1 + \epsilon_r,\ \epsilon_1,\ \epsilon_1 - \epsilon_r, \ldots, \epsilon_1 - \epsilon_2;\\
&\quad \delta_n + \epsilon_1, \ldots, \delta_n + \epsilon_r,\ 2\delta_n,\ \delta_n,\ \delta_n - \epsilon_r, \ldots, \delta_n - \epsilon_1; \ldots;\\
&\quad\quad \delta_1 + \delta_2, \ldots, \delta_1 + \delta_n,\ \delta_1 + \epsilon_1, \ldots, \delta_1 + \epsilon_r,\ 2\delta_1,\ \delta_1,\\
&\quad\quad\quad \delta_1 - \epsilon_r, \ldots, \delta_1 - \epsilon_1, \delta_1 - \delta_n, \ldots, \delta_1 - \delta_2.
\end{aligned} \tag{39}$$

Proof. One can directly verify that the above ordering of the positive roots (36)-(38) of $osp(2r+1|2n)$ satisfies all requirements of Definition 1. □

For the case $n = 0$, the ordering (39) gives rise to the normal ordering of the positive roots of $so(2r+1)$. Based on this ordering a "good coordinate system" in the associated big cell of $so(2r+1)$ was constructed in.[44] Here we use the ordering (39) to construct a "good coordinate system" in the associated big cell $\mathcal{U}$ and the explicit differential operator realization of $osp(2r+1|2n)$.

In addition to the coordinates $\{x_{m,l}, \bar{x}_{m,l}; x_m; y_{i,j}, \bar{y}_{i,j}; \theta_{l,i}, \bar{\theta}_{l,i}\}$, which are associated with the positive roots $\{\delta_m - \delta_l, \delta_m + \delta_l; 2\delta_l; \epsilon_i - \epsilon_j, \epsilon_i + \epsilon_j; \delta_l - \epsilon_i, \delta_l + \epsilon_i\}$, we also need to introduce $n + r$ extra coordinates $\{\theta_l | l = 1, \ldots, n\}$ and $\{y_i | i = 1, \ldots, r\}$ associated with the positive roots $\{\delta_l | l = 1, \ldots, n\}$ and $\{\epsilon_i | i = 1, \ldots, r\}$ respectively. The coordinates $\{x_{m,l}, \bar{x}_{m,l}; x_m; y_{i,j}, \bar{y}_{i,j}; \theta_{l,i}, \bar{\theta}_{l,i}\}$ and their differentials satisfy the same (anti)commutation relations as (6)-(8) and (24)-(27). The other non-trivial relations are

$$[y_i, y_j] = [\partial_{y_i}, \partial_{y_j}] = 0, \quad [\partial_{y_i}, y_j] = \delta_{ij}, \quad i, j = 1, \ldots, r. \tag{40}$$

$$[\theta_m, \theta_l] = [\partial_{\theta_m}, \partial_{\theta_l}] = 0, \quad [\partial_{\theta_m}, \theta_l] = \delta_{ml}, \quad m, l = 1, \ldots, n. \tag{41}$$

Based on the very ordering (39) of the positive roots of $osp(2r+1|2n)$, we introduce the following coordinate system $G_+(x, \bar{x}; y, \bar{y}; \theta, \bar{\theta})$ for the associated big cell $\mathcal{U}$:

$$\begin{aligned} &G_+(x, \bar{x}; y, \bar{y}; \theta, \bar{\theta}) \qquad (42)\\ &= (G_{n+r}) \left(\bar{G}_{n+r-1,n+r}\, G_{n+r-1}\, G_{n+r-1,n+r}\right) \cdots \\ &\quad \times \left(\bar{G}_{n+1,n+2} \ldots \bar{G}_{n+1,n+r}\, G_{n+1}\, G_{n+1,n+r} \ldots G_{n+1,n+2}\right) \\ &\quad \times \left(\bar{G}_{n,n+1} \ldots \bar{G}_{n,n+r}\, \bar{G}_n G_n\, G_{n,n+r} \ldots G_{n,n+1}\right) \cdots \\ &\quad \times \left(\bar{G}_{1,2} \ldots \bar{G}_{1,n+r}\, \bar{G}_1 G_1\, G_{1,n+r} \ldots G_{1,2}\right). \end{aligned} \tag{43}$$

Here $G_{i,j}$,$\bar{G}_{i,j}$,G_i and $\bar{G}_i$ are given by

$$G_{m,l} = e^{x_{m,l} E_{\delta_m - \delta_l}}, \quad \bar{G}_{m,l} = e^{\bar{x}_{m,l} E_{\delta_m + \delta_l}}, \quad 1 \le m < l \le n, \tag{44}$$

$$\bar{G}_l = e^{x_l E_{2\delta_l}}, \; G_l = e^{\theta_l E_{\delta_l}}, \; G_{n+i} = e^{y_i E_{\epsilon_i}}, \; 1 \le l \le n, \, 1 \le i \le r, \tag{45}$$

$$G_{l,n+i} = e^{\theta_{l,i} E_{\delta_l - \epsilon_i}}, \quad \bar{G}_{l,n+i} = e^{\bar{\theta}_{l,i} E_{\delta_l + \epsilon_i}}, \quad 1 \le l \le n, \, 1 \le i \le r, \tag{46}$$

$$G_{n+i,n+j} = e^{y_{i,j} E_{\epsilon_i - \epsilon_j}}, \quad \bar{G}_{n+i,n+j} = e^{\bar{y}_{i,j} E_{\epsilon_i + \epsilon_j}}, \quad 1 \le i < j \le r. \tag{47}$$

Then we can obtain the first order differential operator realization of the generators of $osp(2r+1|2n)$.[46]

Proposition 3.6. *The differential operator representation of the generators associated with the simple roots of osp*$(2r+1|2n)$ *are given by*

$$\begin{aligned}
\rho^{(d)}(E_l) &= \sum_{m=1}^{l-1}\left(x_{m,l}\partial_{x_{m,l+1}} - \bar{x}_{m,l+1}\partial_{\bar{x}_{m,l}}\right) + \partial_{x_{l,l+1}}, \\
&\quad 1 \le l \le n-1, \qquad\qquad (48) \\
\rho^{(d)}(E_n) &= \sum_{m=1}^{n-1}\left(x_{m,n}\partial_{\theta_{m,1}} + \bar{\theta}_{m,1}\partial_{\bar{x}_{m,n}}\right) + \partial_{\theta_{n,1}}, \\
\rho^{(d)}(E_{n+i}) &= \sum_{m=1}^{n}\left(\theta_{m,i}\partial_{\theta_{m,i+1}} - \bar{\theta}_{m,i+1}\partial_{\bar{\theta}_{m,i}}\right) \\
&\quad + \sum_{m=1}^{i-1}\left(y_{m,i}\partial_{y_{m,i+1}} - \bar{y}_{m,i+1}\partial_{\bar{y}_{m,i}}\right) + \partial_{y_{i,i+1}}, \qquad 1 \le i \le r-1, \\
\rho^{(d)}(E_{n+r}) &= \sum_{m=1}^{n}\left(\theta_m\partial_{\bar{\theta}_{m,r}} - \theta_{m,r}\partial_{\theta_m} - \theta_{m,r}\theta_m\partial_{x_m}\right) \\
&\quad + \sum_{m=1}^{r-1}\left(y_{m,r}\partial_{y_m} - y_m\partial_{\bar{y}_{m,r}}\right) + \partial_{y_r},
\end{aligned}$$

$$\begin{aligned}
\rho^{(d)}(F_l) = {} & \sum_{m=1}^{l-1}\left(x_{m,l+1}\partial_{x_{m,l}} - \bar{x}_{m,l}\partial_{\bar{x}_{m,l+1}}\right) - x_l\partial_{\bar{x}_{l,l+1}} - 2\bar{x}_{l,l+1}\partial_{x_{l+1}} \\
& + \sum_{m=l+2}^{n}\left(x_{l,m}\bar{x}_{l,m}\partial_{\bar{x}_{l,l+1}} - x_{l,m}\partial_{x_{l+1,m}} - 2\bar{x}_{l,m}x_{l+1,m}\partial_{x_{l+1}} - \bar{x}_{l,m}\partial_{\bar{x}_{l+1,m}}\right) \\
& - \sum_{m=1}^{r}\left(\theta_{l,m}\bar{\theta}_{l,m}\partial_{\bar{x}_{l,l+1}} + \theta_{l,m}\partial_{\theta_{l+1,m}} + 2\bar{\theta}_{l,m}\theta_{l+1,m}\partial_{x_{l+1}} + \bar{\theta}_{l,m}\partial_{\bar{\theta}_{l+1,m}}\right) \\
& - \theta_l\partial_{\theta_{l+1}} - \theta_l\theta_{l+1}\partial_{x_{l+1}} + x_{l,l+1}\theta_{l+1}\partial_{\theta_{l+1}} - x_{l,l+1}\theta_l\partial_{\theta_l} \\
& - x_{l,l+1}^2\partial_{x_{l,l+1}} + 2x_{l,l+1}x_{l+1}\partial_{x_{l+1}} - 2x_{l,l+1}x_l\partial_{x_l} \\
& + x_{l,l+1}\left[\sum_{m=l+2}^{n}\left(x_{l+1,m}\partial_{x_{l+1,m}} + \bar{x}_{l+1,m}\partial_{\bar{x}_{l+1,m}} - x_{l,m}\partial_{x_{l,m}} - \bar{x}_{l,m}\partial_{\bar{x}_{l,m}}\right)\right] \\
& + x_{l,l+1}\left[\sum_{m=1}^{r}\left(\theta_{l+1,m}\partial_{\theta_{l+1,m}} + \bar{\theta}_{l+1,m}\partial_{\bar{\theta}_{l+1,m}} - \theta_{l,m}\partial_{\theta_{l,m}} - \bar{\theta}_{l,m}\partial_{\bar{\theta}_{l,m}}\right)\right] \\
& + x_{l,l+1}(\lambda_l - \lambda_{l+1}), \qquad 1 \le l \le n-1,
\end{aligned}$$

$$\begin{aligned}
\rho^{(d)}(F_n) = {} & \sum_{m=1}^{n-1}\left(\theta_{m,1}\partial_{x_{m,n}} - \bar{x}_{m,n}\partial_{\bar{\theta}_{m,1}}\right) - x_n\partial_{\bar{\theta}_{n,1}} \\
& + \sum_{m=2}^{r}\left(\theta_{n,m}\partial_{y_{1,m}} - \theta_{n,m}\bar{\theta}_{n,m}\partial_{\bar{\theta}_{n,1}} + \bar{\theta}_{n,m}\partial_{\bar{y}_{1,m}}\right) - \theta_n\partial_{y_1} \\
& - \theta_{n,1}\sum_{m=2}^{r}\left(\theta_{n,m}\partial_{\theta_{n,m}} + \bar{\theta}_{n,m}\partial_{\bar{\theta}_{n,m}} + y_{1,m}\partial_{y_{1,m}} + \bar{y}_{1,m}\partial_{\bar{y}_{1,m}}\right) \\
& - 2\theta_{n,1}x_n\partial_{x_n} - 2\theta_{n,1}\bar{\theta}_{n,1}\partial_{\bar{\theta}_{n,1}} - \theta_{n,1}\theta_n\partial_{\theta_n} - \theta_{n,1}y_1\partial_{y_1} \\
& + \theta_{n,1}(\lambda_n + \lambda_{n+1}),
\end{aligned}$$

$$\begin{aligned}
\rho^{(d)}(F_{n+i}) = & \sum_{m=1}^{n}(\theta_{m,i+1}\partial_{\theta_{m,i}} - \bar{\theta}_{m,i}\partial_{\bar{\theta}_{m,i+1}}) + \sum_{m=1}^{i-1}(y_{m,i+1}\partial_{y_{m,i}} - \bar{y}_{m,i}\partial_{\bar{y}_{m,i+1}}) \\
& + \sum_{m=i+2}^{r}\left(y_{i,m}\bar{y}_{i,m}\partial_{\bar{y}_{i,i+1}} - y_{i,m}\partial_{y_{i+1,m}} - \bar{y}_{i,m}\partial_{\bar{y}_{i+1,m}}\right) \\
& - y_i\partial_{y_{i+1}} + \frac{y_i^2}{2}\partial_{\bar{y}_{i,i+1}} + y_{i,i+1}y_{i+1}\partial_{y_{i+1}} - y_{i,i+1}y_i\partial_{y_i} \\
& + y_{i,i+1}\sum_{m=i+2}^{r}\left(y_{i+1,m}\partial_{y_{i+1,m}} + \bar{y}_{i+1,m}\partial_{\bar{y}_{i+1,m}} - y_{i,m}\partial_{y_{i,m}} - \bar{y}_{i,m}\partial_{\bar{y}_{i,m}}\right) \\
& - y_{i,i+1}^2\partial_{y_{i,i+1}} + y_{i,i+1}(\lambda_{n+i} - \lambda_{n+i+1}), \qquad 1 \le i \le r-1, \\
\rho^{(d)}(F_{n+r}) = & \sum_{m=1}^{n}\left(\bar{\theta}_{m,r}\partial_{\theta_m} - \bar{\theta}_{m,r}\theta_m\partial_{x_m} - \theta_m\partial_{\theta_{m,r}}\right) \\
& + \sum_{m=1}^{r-1}\left(y_m\partial_{y_{m,r}} - \bar{y}_{m,r}\partial_{y_m}\right) - \frac{y_r^2}{2}\partial_{y_r} + y_r\lambda_{n+r},
\end{aligned}$$

$$\begin{aligned}
\rho^{(d)}(H_l) = & \sum_{m=1}^{l-1}\left(x_{m,l}\partial_{x_{m,l}} - \bar{x}_{m,l}\partial_{\bar{x}_{m,l}}\right) - \sum_{m=l+1}^{n}\left(x_{l,m}\partial_{x_{l,m}} + \bar{x}_{l,m}\partial_{\bar{x}_{l,m}}\right) \\
& - \sum_{m=1}^{r}\left(\theta_{l,m}\partial_{\theta_{l,m}} + \bar{\theta}_{l,m}\partial_{\bar{\theta}_{l,m}}\right) - 2x_l\partial_{x_l} - \theta_l\partial_{\theta_l} + \lambda_l, \quad 1 \le l \le n, \\
\rho^{(d)}(H_{n+i}) = & \sum_{m=1}^{n}\left(\theta_{m,i}\partial_{\theta_{m,i}} - \bar{\theta}_{m,i}\partial_{\bar{\theta}_{m,i}}\right) + \sum_{m=1}^{i-1}\left(y_{m,i}\partial_{y_{m,i}} - \bar{y}_{m,i}\partial_{\bar{y}_{m,i}}\right) \\
& - \sum_{m=i+1}^{r}\left(y_{i,m}\partial_{y_{i,m}} + \bar{y}_{i,m}\partial_{\bar{y}_{i,m}}\right) - y_i\partial_{y_i} + \lambda_{n+i}, \quad 1 \le i \le r.
\end{aligned} \tag{49}$$

A direct computation shows that these differential operators (48)-(49) satisfy the $osp(2r+1|2n)$ (anti)commutation relations corresponding to the simple roots and the associated Serre relations. This implies that the differential representation of non-simple generators can be consistently constructed from the simple ones. Hence, we have obtained an explicit differential realization of $osp(2r+1|2n)$.

4. Free field realization of current superalgebras

4.1. *Current superalgebra* $gl(r|n)_k$

With the help of the explicit differential operator expressions of $gl(r|n)$ given by (16)-(17) we can construct the explicit free field representation of the $gl(r|n)$ current algebra at arbitrary level k in terms of $\frac{n(n-1)}{2} + \frac{r(r-1)}{2}$ bosonic β-γ pairs $\{(\beta_{i,j}, \gamma_{i,j}),\ (\beta'_{i',j'}, \gamma'_{i',j'})\, 1 \le i < j \le n,\ 1 \le i' < j' \le r\}$, nr fermionic $b-c$ pairs $\{(\Psi^{+}_{i,j}, \Psi_{i,j})\, 1 \le i \le n,\ 1 \le j \le r\}$ and $n+r$ free scalar fields ϕ_i, $i = 1, \ldots, n+r$. These free fields obey the following OPEs:

$$\beta_{i,j}(z)\,\gamma_{m,l}(w) = -\gamma_{m,l}(z)\,\beta_{i,j}(w) = \frac{\delta_{im}\delta_{jl}}{(z-w)}, \tag{1}$$

$$\begin{aligned}
&\beta'_{i,j}(z)\,\gamma'_{m,l}(w) = -\gamma'_{m,l}(z)\,\beta'_{i,j}(w) = \frac{\delta_{im}\delta_{jl}}{(z-w)},\\
&\Psi^{+}_{m,i}(z)\,\Psi_{l,j}(w) = \Psi_{l,j}(z)\,\Psi^{+}_{m,i}(w) = \frac{\delta_{ml}\delta_{ij}}{(z-w)},\\
&\phi_m(z)\phi_l(w) = \delta_{ml}\,\ln(z-w),\ 1 \le m,l \le n,\\
&\phi_{n+i}(z)\phi_{n+j}(w) = -\delta_{ij}\,\ln(z-w),\ 1 \le i,j \le r,
\end{aligned} \tag{2}$$

and the other OPEs are trivial. Then we have[47]

Theorem 4.1. *The currents associated with the simple roots of the $gl(r|n)$ current algebra at a generic level k are given in terms of the free fields (1)-(2) as*

$$\begin{aligned}
E_j(z) &= \sum_{l\le j-1}\gamma_{l,j}(z)\beta_{l,j+1}(z) + \beta_{j,j+1}(z),\ 1\le j\le r-1,\\
E_r(z) &= \sum_{l\le r-1}\gamma_{l,m}(z)\Psi_{l,1}(z) + \Psi_{r,1}(z),\\
E_{r+j}(z) &= \sum_{l\le r}\Psi^{\dagger}_{l,j}(z)\Psi_{l,j+1}(z) + \sum_{l\le j-1}\gamma'_{l,j}(z)\beta'_{l,j+1}(z) + \beta'_{j,j+1}(z),\\
&\qquad 1\le j\le n-1,\\
E_{j,j}(z) &= \sum_{l\le j-1}\gamma_{l,j}(z)\beta_{l,j}(z) - \sum_{j+1\le l\le r}\gamma_{j,l}(z)\beta_{j,l}(z) - \sum_{l\le n}\Psi^{\dagger}_{j,l}(z)\Psi_{j,l}(z)\\
&\quad + \sqrt{k+r-n}\,\partial\phi_j(z) - \frac{1+\alpha}{2\sqrt{k+r-n}}\sum_{l=1}^{r+n}\partial\phi_l(z),\\
&\qquad 1\le j\le r,\\
E_{r+j,r+j}(z) &= \sum_{l\le r}\Psi^{\dagger}_{l,j}(z)\Psi_{l,j}(z) + \sum_{l\le j-1}\gamma'_{l,j}(z)\beta'_{l,j}(z) - \sum_{j+1\le l\le n}\gamma'_{j,l}(z)\beta'_{j,l}(z)\\
&\quad + \sqrt{k+r-n}\,\partial\phi_{r+j}(z) + \frac{1+\alpha}{2\sqrt{k+r-n}}\sum_{l=1}^{r+n}\partial\phi_l(z),\\
&\qquad 1\le j\le n,
\end{aligned}$$

$$\begin{aligned}
F_j(z) &= \sum_{l\le j-1}\gamma_{l,j+1}(z)\beta_{l,j}(z) - \sum_{j+2\le l\le r}\gamma_{j,l}(z)\beta_{j+1,l}(z) - \sum_{l\le n}\Psi^{\dagger}_{j,l}(z)\Psi_{j+1,l}(z)\\
&\quad -\gamma_{j,j+1}(z)\left(\sum_{j+1\le l\le r}\gamma_{j,l}(z)\beta_{j,l}(z) + \sum_{l\le n}\Psi^{\dagger}_{j,l}(z)\Psi_{j,l}(z)\right)\\
&\quad +\gamma_{j,j+1}(z)\left(\sum_{j+2\le l\le r}\gamma_{j+1,l}(z)\beta_{j+1,l}(z) + \sum_{l\le n}\Psi^{\dagger}_{j+1,l}(z)\Psi_{j+1,l}(z)\right)\\
&\quad +\sqrt{k+r-n}\,\gamma_{j,j+1}(z)\left(\partial\phi_j(z) - \partial\phi_{j+1}(z)\right) + (k+j-1)\partial\gamma_{j,j+1}(z),\\
&\qquad 1\le j\le r-1,
\end{aligned}$$

$$F_r(z) = \sum_{l\le r-1} \Psi^\dagger_{l,1}(z)\beta_{l,m}(z) + \sum_{2\le l\le n} \Psi^\dagger_{r,l}(z)\beta'_{1,l}(z)$$
$$-\Psi^\dagger_{r,1}(z)\left(\sum_{2\le l\le n}\left(\Psi^\dagger_{r,l}(z)\,\Psi_{r,l}(z) + \gamma'_{1,l}(z)\beta'_{1,l}(z)\right)\right)$$
$$+\sqrt{k+r-n}\Psi^\dagger_{r,1}(z)\left(\partial\phi_r(z) + \partial\phi_{r+1}(z)\right)$$
$$+(k+r-1)\partial\Psi^\dagger_{r,1}(z),$$

$$F_{r+j}(z) = \sum_{l\le r} \Psi^\dagger_{l,j+1}(z)\Psi_{l,j}(z) + \sum_{l\le j-1} \gamma'_{l,j+1}(z)\beta'_{l,j}(z)$$
$$-\sum_{j+2\le l\le n} \gamma'_{j,l}(z)\beta'_{j+1,l}(z)$$
$$-\gamma'_{j,j+1}(z)\left(\sum_{j+1\le l\le n}\gamma'_{j,l}(z)\beta'_{j,l}(z) - \sum_{j+2\le l\le n}\gamma'_{j+1,l}(z)\beta'_{j+1,l}(z)\right)$$
$$+\sqrt{k+r-n}\gamma'_{j,j+1}\left(\partial\phi_{r+j}(z) - \partial\phi_{r+j+1}(z)\right)$$
$$-(k+r+1-j)\partial\gamma'_{j,j+1}(z),\ 1\le j\le n-1,$$

where $\alpha = 1 + \frac{2k}{r-n} - \frac{2\sqrt{k(k+r-n)}}{r-n}$. *Here and throughout normal ordering of free fields is implied whenever necessary.*

It is remarked that for $m = n$, α can be chosen as $\alpha = \lim_{m\to n}(1 + \frac{2k}{m-n} - \frac{2\sqrt{k(k+m-n)}}{m-n}) = 0$. For the case of $m = n = 4$, it recovers the result of $gl(4|4)$.[47]

4.2. *Current superalgebra* $osp(2r|2n)_k$

With the help of the explicit differential operator expressions of $osp(2r|2n)$ given by (32)-(33) we can construct the explicit free field representation of the $osp(2r|2n)$ current algebra at arbitrary level k in terms of $n^2 + r^2 - r$ bosonic β-γ pairs $\{(\beta_{i,j},\gamma_{i,j}),\ (\bar\beta_{i,j},\bar\gamma_{i,j}),\ (\beta_i,\gamma_i),\ (\beta'_{i',j'},\gamma'_{i',j'}),\ (\bar\beta'_{i',j'},\bar\gamma'_{i',j'}),\ 1\le i<j\le n,\ 1\le i'<j'\le r\}$, $2nr$ fermionic $b-c$ pairs $\{(\Psi^+_{i,j},\Psi_{i,j}),\ (\bar\Psi^+_{i,j},\bar\Psi_{i,j}),\ 1\le i\le n,\ 1\le j\le r\}$ and $n+r$ free scalar fields ϕ_i, $i = 1,\ldots,n+r$. The free fields $\{(\beta_{i,j},\gamma_{i,j}),\ (\beta'_{i',j'},\gamma'_{i',j'})\}$, $\{(\Psi^+_{i,j},\Psi_{i,j})\}$ and $\{\phi_i\}$ obey the same OPEs as (1)-(2). The other non-trivial

OPEs are

$$\bar{\beta}_{i,j}(z)\,\bar{\gamma}_{m,l}(w) = -\bar{\gamma}_{m,l}(z)\,\bar{\beta}_{i,j}(w) = \frac{\delta_{im}\delta_{jl}}{(z-w)}, \tag{3}$$

$$\beta_m(z)\,\gamma_l(w) = -\gamma_m(z)\,\beta_l(w) = \frac{\delta_{ml}}{(z-w)},\ 1 \le m, l \le n,$$

$$\bar{\beta}'_{i,j}(z)\,\bar{\gamma}'_{m,l}(w) = -\bar{\gamma}'_{m,l}(z)\,\bar{\beta}'_{i,j}(w) = \frac{\delta_{im}\delta_{jl}}{(z-w)},$$

$$\bar{\Psi}^+_{m,i}(z)\,\bar{\Psi}_{l,j}(w) = \bar{\Psi}_{l,j}(z)\,\bar{\Psi}^+_{m,i}(w) = \frac{\delta_{ml}\delta_{ij}}{(z-w)}, \tag{4}$$

and the other OPEs are trivial. Then we have[46]

Theorem 4.2. *The currents associated with the simple roots of the $osp(2r|2n)$ current algebra at a generic level k are given in terms of the free fields (1)-(2) and (3)-(4) as*

$$E_l(z) = \sum_{m=1}^{l-1}\left(\gamma_{m,l}(z)\beta_{m,l+1}(z) - \bar{\gamma}_{m,l+1}(z)\bar{\beta}_{m,l}(z)\right) + \beta_{l,l+1}(z),$$
$$1 \le l \le n-1, \tag{5}$$

$$E_n(z) = \sum_{m=1}^{n-1}\left(\gamma_{m,n}(z)\Psi_{m,1}(z) + \bar{\Psi}^+_{m,1}(z)\bar{\beta}_{m,n}(z)\right) + \Psi_{n,1}(z), \tag{6}$$

$$\begin{aligned}
E_{n+i}(z) &= \sum_{m=1}^{n}\left(\Psi^+_{m,i}(z)\Psi_{m,i+1}(z) - \bar{\Psi}^+_{m,i+1}(z)\bar{\Psi}_{m,i}(z)\right) \\
&\quad + \sum_{m=1}^{i-1}\left(\gamma'_{m,i}(z)\beta'_{m,i+1}(z) - \bar{\gamma}'_{m,i+1}(z)\bar{\beta}'_{m,i}(z)\right) + \beta'_{i,i+1}(z), \quad 1 \le i \le r-1, \\
E_{n+r}(z) &= \sum_{m=1}^{n}\left(2\Psi^+_{m,r-1}(z)\Psi^+_{m,r}(z)\beta_m(z) + \Psi^+_{m,r-1}(z)\bar{\Psi}_{m,r}(z) - \Psi^+_{m,r}(z)\bar{\Psi}_{m,r-1}(z)\right) \\
&\quad + \sum_{m=1}^{r-2}\left(\gamma'_{m,r-1}(z)\bar{\beta}'_{m,r}(z) - \gamma'_{m,r}(z)\bar{\beta}'_{m,r-1}(z)\right) + \bar{\beta}'_{r-1,r}(z), \\
F_l(z) &= \sum_{m=1}^{l-1}\left(\gamma_{m,l+1}(z)\beta_{m,l}(z) - \bar{\gamma}_{m,l}(z)\bar{\beta}_{m,l+1}(z)\right) \\
&\quad - \gamma_l(z)\bar{\beta}_{l,l+1}(z) - 2\bar{\gamma}_{l,l+1}(z)\beta_{l+1}(z) \\
&\quad + \sum_{m=l+2}^{n}\left(\gamma_{l,m}(z)\bar{\gamma}_{l,m}(z)\bar{\beta}_{l,l+1}(z) - \gamma_{l,m}(z)\beta_{l+1,m}(z)\right) \\
&\quad - \sum_{m=l+2}^{n}\left(2\bar{\gamma}_{l,m}(z)\gamma_{l+1,m}(z)\beta_{l+1}(z) + \bar{\gamma}_{l,m}(z)\bar{\beta}_{l+1,m}(z)\right) \\
&\quad - \sum_{m=1}^{r}\left(\Psi^+_{l,m}(z)\bar{\Psi}^+_{l,m}(z)\bar{\beta}_{l,l+1}(z) + \Psi^+_{l,m}(z)\Psi_{l+1,m}(z)\right) \\
&\quad - \sum_{m=1}^{r}\left(2\bar{\Psi}^+_{l,m}(z)\Psi^+_{l+1,m}(z)\beta_{l+1}(z) + \bar{\Psi}^+_{l,m}(z)\bar{\Psi}_{l+1,m}(z)\right) \\
&\quad - \gamma^2_{l,l+1}(z)\beta_{l,l+1}(z) - \gamma_{l,l+1}(z)\sum_{m=l+2}^{n}\left(\gamma_{l,m}(z)\beta_{l,m}(z) + \bar{\gamma}_{l,m}(z)\bar{\beta}_{l,m}(z)\right)
\end{aligned}$$

$$
\begin{aligned}
&+\gamma_{l,l+1}(z)\sum_{m=l+2}^{n}\left(\gamma_{l+1,m}(z)\beta_{l+1,m}(z)+\bar{\gamma}_{l+1,m}(z)\bar{\beta}_{l+1,m}(z)\right)\\
&-\gamma_{l,l+1}(z)\sum_{m=1}^{r}\left(\Psi^{+}_{l,m}(z)\Psi_{l,m}(z)+\bar{\Psi}^{+}_{l,m}(z)\bar{\Psi}_{l,m}(z)\right)\\
&+\gamma_{l,l+1}(z)\sum_{m=1}^{r}\left(\Psi^{+}_{l+1,m}(z)\Psi_{l+1,m}(z)+\bar{\Psi}^{+}_{l+1,m}(z)\bar{\Psi}_{l+1,m}(z)\right)\\
&+2\gamma_{l,l+1}(z)\gamma_{l+1}(z)\beta_{l+1}(z)-2\gamma_{l,l+1}(z)\gamma_{l}(z)\beta_{l}(z)\\
&+\sqrt{k+2(r-n-1)}\gamma_{l,l+1}(z)\left(\partial\phi_{l}(z)-\partial\phi_{l+1}(z)\right)\\
&+(-k+2(l-1))\partial\gamma_{l,l+1}(z),\quad 1\le l\le n-1,
\end{aligned}
$$

$$
\begin{aligned}
F_{n}(z)=&\sum_{m=1}^{n-1}\left(\Psi^{+}_{m,1}(z)\beta_{m,n}(z)-\bar{\gamma}_{m,n}(z)\bar{\Psi}_{m,1}(z)\right)-\gamma_{n}(z)\bar{\Psi}_{n,1}(z)\\
&+\sum_{m=2}^{r}\left(\Psi^{+}_{n,m}(z)\beta'_{1,m}(z)-\Psi^{+}_{n,m}(z)\bar{\Psi}^{+}_{n,m}(z)\bar{\Psi}_{n,1}(z)+\bar{\Psi}^{+}_{n,m}(z)\bar{\beta}'_{1,m}(z)\right)\\
&-\Psi^{+}_{n,1}(z)\sum_{m=2}^{r}\left(\Psi^{+}_{n,m}(z)\Psi_{n,m}(z)+\bar{\Psi}^{+}_{n,m}(z)\bar{\Psi}_{n,m}(z)\right)-2\Psi^{+}_{n,1}(z)\bar{\Psi}^{+}_{n,1}(z)\bar{\Psi}_{n,1}(z)\\
&-\Psi^{+}_{n,1}(z)\sum_{m=2}^{r}\left(\gamma'_{1,m}(z)\beta'_{1,m}(z)+\bar{\gamma}'_{1,m}(z)\bar{\beta}'_{1,m}(z)\right)-2\Psi^{+}_{n,1}(z)\gamma_{n}(z)\beta_{n}(z)\\
&+\sqrt{k+2(r-n-1)}\Psi^{+}_{n,1}(z)\left(\partial\phi_{n}(z)+\partial\phi_{n+1}(z)\right)\\
&+(-k+2(n-1))\partial\,\Psi^{+}_{n,1}(z),\\
F_{n+i}(z)=&\sum_{m=1}^{n}\left(\Psi^{+}_{m,i+1}(z)\Psi_{m,i}(z)-\bar{\Psi}^{+}_{m,i}(z)\bar{\Psi}_{m,i+1}(z)\right)\\
&+\sum_{m=1}^{i-1}\left(\gamma'_{m,i+1}(z)\beta'_{m,i}(z)-\bar{\gamma}'_{m,i}(z)\bar{\beta}'_{m,i+1}(z)\right)\\
&+\sum_{m=i+2}^{r}\left(\gamma'_{i,m}(z)\bar{\gamma}'_{i,m}(z)\bar{\beta}'_{i,i+1}(z)-\gamma'_{i,m}(z)\beta'_{i+1,m}(z)-\bar{\gamma}'_{i,m}(z)\bar{\beta}'_{i+1,m}(z)\right)\\
&+\gamma'_{i,i+1}(z)\sum_{m=i+2}^{r}\left(\gamma'_{i+1,m}(z)\beta'_{i+1,m}(z)+\bar{\gamma}'_{i+1,m}(z)\bar{\beta}'_{i+1,m}(z)\right)\\
&-\gamma'_{i,i+1}(z)\sum_{m=i+2}^{r}\left(\gamma'_{i,m}(z)\beta'_{i,m}(z)+\bar{\gamma}'_{i,m}(z)\bar{\beta}'_{i,m}(z)\right)\\
&-\gamma'_{i,i+1}(z)\gamma'_{i,i+1}(z)\beta'_{i,i+1}(z)\\
&+\sqrt{k+2(r-n-1)}\gamma'_{i,i+1}(z)\left(\partial\phi_{n+i}(z)-\partial\phi_{n+i+1}(z)\right)\\
&+(k+2(i-n-1))\,\partial\gamma'_{i,i+1}(z),\qquad 1\le i\le r-1,\\
F_{n+r}(z)=&\sum_{m=1}^{n}\left(\bar{\Psi}^{+}_{m,r}(z)\Psi_{m,r-1}(z)+2\bar{\Psi}^{+}_{m,r-1}(z)\bar{\Psi}^{+}_{m,r}(z)\beta_{m}(z)-\bar{\Psi}^{+}_{m,r-1}(z)\Psi_{m,r}(z)\right)\\
&+\sum_{m=1}^{r-2}\left(\bar{\gamma}'_{m,r}(z)\beta'_{m,r-1}(z)-\bar{\gamma}'_{m,r-1}(z)\beta'_{m,r}(z)\right)\\
&-\bar{\gamma}'_{r-1,r}(z)\bar{\gamma}'_{r-1,r}(z)\bar{\beta}'_{r-1,r}(z)\\
&+\sqrt{k+2(r-n-1)}\bar{\gamma}'_{r-1,r}(z)\left(\partial\phi_{n+r-1}(z)+\partial\phi_{n+r}(z)\right)\\
&+(k+2(r-n-2))\,\partial\bar{\gamma}'_{r-1,r}(z),
\end{aligned}
$$

$$
\begin{aligned}
H_{l}(z)=&\sum_{m=1}^{l-1}\left(\gamma_{m,l}(z)\beta_{m,l}(z)-\bar{\gamma}_{m,l}(z)\bar{\beta}_{m,l}(z)\right)-\sum_{m=l+1}^{n}\left(\gamma_{l,m}(z)\beta_{l,m}(z)+\bar{\gamma}_{l,m}(z)\bar{\beta}_{l,m}(z)\right)\\
&-2\gamma_{l}(z)\beta_{l}(z)-\sum_{m=1}^{r}\left(\Psi^{+}_{l,m}(z)\Psi_{l,m}(z)+\bar{\Psi}^{+}_{l,m}(z)\bar{\Psi}_{l,m}(z)\right)\\
&+\sqrt{k+2(r-n-1)}\partial\phi_{l}(z),\qquad 1\le l\le n,
\end{aligned}
$$

$$\begin{aligned} H_{n+i}(z) =& \sum_{m=1}^{n}\left(\Psi^{+}_{m,i}(z)\Psi_{m,i}(z)-\bar{\Psi}^{+}_{m,i}(z)\bar{\Psi}_{m,i}(z)\right)+\sum_{m=1}^{i-1}\left(\gamma'_{m,i}(z)\beta'_{m,i}(z)-\bar{\gamma}'_{m,i}(z)\bar{\beta}'_{m,i}(z)\right)\\ &-\sum_{m=i+1}^{r}\left(\gamma'_{i,m}(z)\beta'_{i,m}(z)+\bar{\gamma}'_{i,m}(z)\bar{\beta}'_{i,m}(z)\right)\\ &+\sqrt{k+2(r-n-1)}\partial\phi_{n+i}(z), \qquad 1\le i\le r. \end{aligned} \tag{7}$$

Here normal ordering of free fields is implied.

Some remarks are in order. The free field realization of the currents associated with the non-simple roots can be obtained from the OPEs of the simple ones. For $n = r$, our result reduces to the free field realization of the $osp(2n|2n)$ current algebra.[45] When $n = 0$ (or $r = 0$), our result recovers the free field realization of $so(2r)$ (or $sp(2n)$) current algebra proposed in.[44]

4.3. *Current superalgebra* $osp(2r+1|2n)_k$

With the help of the explicit differential operator expressions of $osp(2r+1|2n)$ given by (48)-(49) we can construct the explicit free field representation of the $osp(2r+1|2n)$ current algebra at an arbitrary level k in terms of n^2+r^2 bosonic β-γ pairs $\{(\beta_{i,j},\gamma_{i,j}),\ (\bar{\beta}_{i,j},\bar{\gamma}_{i,j}),\ (\beta_i,\gamma_i),\ (\beta'_{i',j'},\gamma'_{i',j'}),\ (\bar{\beta}'_{i',j'},\bar{\gamma}'_{i',j'}),\ (\beta'_{i'},\gamma'_{i'}),\ 1\le i<j\le n,\ 1\le i'<j'\le r\}$, $n(2r+1)$ fermionic $b-c$ pairs $\{(\Psi^{+}_{i,j},\Psi_{i,j}),\ (\bar{\Psi}^{+}_{i,j},\bar{\Psi}_{i,j}),\ (\Psi^{+}_{i},\Psi_{i}),\ 1\le i\le n,\ 1\le j\le r\}$ and $n+r$ free scalar fields ϕ_i, $i=1,\ldots,n+r$. The free fields $\{(\beta_{i,j},\gamma_{i,j}),\ (\bar{\beta}_{i,j},\bar{\gamma}_{i,j}),\ (\beta_i,\gamma_i),\ (\beta'_{i',j'},\gamma'_{i',j'}),\ (\bar{\beta}'_{i',j'},\bar{\gamma}'_{i',j'})\}$, $\{(\Psi^{+}_{i,j},\Psi_{i,j}),\ (\bar{\Psi}^{+}_{i,j},\bar{\Psi}_{i,j})\}$ and $\{\phi_i\}$ obey the same OPEs as (1)-(2) and (3)-(4). The other non-trivial OPEs are

$$\beta'_i(z)\,\gamma'_j(w) = -\gamma'_j(z)\,\beta'_i(w) = \frac{\delta_{ij}}{(z-w)}, \qquad 1\le i,j\le r, \tag{8}$$

$$\Psi^{+}_{m}(z)\,\Psi_l(w) = \Psi_l(z)\,\Psi^{+}_{m}(w) = \frac{\delta_{ml}}{(z-w)}, \qquad 1\le m,l\le n. \tag{9}$$

Theorem 4.3. *The currents associated with the simple roots of the* $osp(2r+1|2n)$ *current algebra at a generic level* k *are given in terms of the free fields (3)-(4) and (8)-(9) as*

$$\begin{aligned} E_l(z) =& \sum_{m=1}^{l-1}\left(\gamma_{m,l}(z)\beta_{m,l+1}(z)-\bar{\gamma}_{m,l+1}(z)\bar{\beta}_{m,l}(z)\right)+\beta_{l,l+1}(z),\\ & 1\le l\le n-1, \end{aligned} \tag{10}$$

$$E_n(z) = \sum_{m=1}^{n-1}\left(\gamma_{m,n}(z)\Psi_{m,1}(z)+\bar{\Psi}^{+}_{m,1}(z)\bar{\beta}_{m,n}(z)\right)+\Psi_{n,1}(z),$$

$$E_{n+i}(z) = \sum_{m=1}^{n} \left(\Psi^+_{m,i}(z)\Psi_{m,i+1}(z) - \bar{\Psi}^+_{m,i+1}(z)\bar{\Psi}_{m,i}(z)\right)$$
$$+\sum_{m=1}^{i-1} \left(\gamma'_{m,i}(z)\beta'_{m,i+1}(z) - \bar{\gamma}'_{m,i+1}(z)\bar{\beta}'_{m,i}(z)\right) + \beta'_{i,i+1}(z),$$
$$1 \le i \le r-1,$$
$$E_{n+r}(z) = \sum_{m=1}^{n}\left(\Psi^+_{m}(z)\bar{\Psi}_{m,r}(z) - \Psi^+_{m,r}(z)\Psi_m(z) - \Psi^+_{m,r}(z)\Psi^+_m(z)\beta_m(z)\right)$$
$$+\sum_{m=1}^{r-1}\left(\gamma'_{m,r}(z)\beta'_m(z) - \gamma'_m(z)\bar{\beta}'_{m,r}(z)\right) + \beta'_r(z),$$
$$H_l(z) = \sum_{m=1}^{l-1}\left(\gamma_{m,l}(z)\beta_{m,l}(z) - \bar{\gamma}_{m,l}(z)\bar{\beta}_{m,l}(z)\right)$$
$$-\sum_{m=l+1}^{n}\left(\gamma_{l,m}(z)\beta_{l,m}(z) + \bar{\gamma}_{l,m}(z)\bar{\beta}_{l,m}(z)\right)$$
$$-2\gamma_l(z)\beta_l(z) - \sum_{m=1}^{r}\left(\Psi^+_{l,m}(z)\Psi_{l,m}(z) + \bar{\Psi}^+_{l,m}(z)\bar{\Psi}_{l,m}(z)\right)$$
$$-\Psi^+_l(z)\Psi_l(z) + \sqrt{k+2r-2n-1}\,\partial\phi_l(z), \qquad 1 \le l \le n,$$
$$H_{n+i}(z) = \sum_{m=1}^{n}\left(\Psi^+_{m,i}(z)\Psi_{m,i}(z) - \bar{\Psi}^+_{m,i}(z)\bar{\Psi}_{m,i}(z)\right)$$
$$+\sum_{m=1}^{i-1}\left(\gamma'_{m,i}(z)\beta'_{m,i}(z) - \bar{\gamma}'_{m,i}(z)\bar{\beta}'_{m,i}(z)\right)$$
$$-\sum_{m=i+1}^{r}\left(\gamma'_{i,m}(z)\beta'_{i,m}(z) + \bar{\gamma}'_{i,m}(z)\bar{\beta}'_{i,m}(z)\right) - \gamma'_i(z)\beta'_i(z)$$
$$+\sqrt{k+2r-2n-1}\,\partial\phi_{n+i}(z), \qquad 1 \le i \le r,$$
$$F_{n+r}(z) = \sum_{m=1}^{n}\left(\bar{\Psi}^+_{m,r}(z)\Psi_m(z) - \bar{\Psi}^+_{m,r}(z)\Psi^+_m(z)\beta_m(z) - \Psi^+_m(z)\Psi_{m,r}(z)\right)$$
$$+\sum_{m=1}^{r-1}\left(\gamma'_m(z)\beta'_{m,r}(z) - \bar{\gamma}'_{m,r}(z)\beta'_m(z)\right) - \frac{1}{2}\gamma'_r(z)\gamma'_r(z)\beta'_r(z)$$
$$+\sqrt{k+2r-2n-1}\,\gamma'_r(z)\partial\phi_{n+r}(z) + \left(k+2(r-n-1)\right)\partial\gamma'_r(z),$$

$$F_{n+i}(z) = \sum_{m=1}^{n}\left(\Psi^+_{m,i+1}(z)\Psi_{m,i}(z) - \bar{\Psi}^+_{m,i}(z)\bar{\Psi}_{m,i+1}(z)\right)$$

$$
\begin{aligned}
&+\sum_{m=1}^{i-1}\left(\gamma'_{m,i+1}(z)\beta'_{m,i}(z)-\bar{\gamma}'_{m,i}(z)\bar{\beta}'_{m,i+1}(z)\right)\\
&+\sum_{m=i+2}^{r}\left(\gamma'_{i,m}(z)\bar{\gamma}'_{i,m}(z)\bar{\beta}'_{i,i+1}(z)-\gamma'_{i,m}(z)\beta'_{i+1,m}(z)\right.\\
&\qquad\qquad\left.-\bar{\gamma}'_{i,m}(z)\bar{\beta}'_{i+1,m}(z)\right)\\
&-\gamma'_i(z)\beta'_{i+1}(z)+\frac{1}{2}\gamma'_i(z)\gamma'_i(z)\bar{\beta}'_{i,i+1}(z)\\
&+\gamma'_{i,i+1}(z)\sum_{m=i+2}^{r}\left(\gamma'_{i+1,m}(z)\beta'_{i+1,m}(z)+\bar{\gamma}'_{i+1,m}(z)\bar{\beta}'_{i+1,m}(z)\right)\\
&-\gamma'_{i,i+1}(z)\sum_{m=i+2}^{r}\left(\gamma'_{i,m}(z)\beta'_{i,m}(z)+\bar{\gamma}'_{i,m}(z)\bar{\beta}'_{i,m}(z)\right)\\
&-\gamma'_{i,i+1}(z)\left(\gamma'_{i,i+1}(z)\beta'_{i,i+1}(z)-\gamma'_{i+1}(z)\beta'_{i+1}(z)+\gamma'_i(z)\beta'_i(z)\right)\\
&+\sqrt{k+2r-2n-1}\,\gamma'_{i,i+1}(z)\left(\partial\phi_{n+i}(z)-\partial\phi_{n+i+1}(z)\right)\\
&+\left(k+2(i-n-1)\right)\partial\gamma'_{i,i+1}(z),\qquad 1\le i\le r-1,
\end{aligned}
$$

$$
\begin{aligned}
F_n(z)=&\sum_{m=1}^{n-1}\left(\Psi^+_{m,1}(z)\beta_{m,n}(z)-\bar{\gamma}_{m,n}(z)\bar{\Psi}_{m,1}(z)\right)-\gamma_n(z)\bar{\Psi}_{n,1}(z)\\
&+\sum_{m=2}^{r}\left(\Psi^+_{n,m}(z)\beta'_{1,m}(z)-\Psi^+_{n,m}(z)\bar{\Psi}^+_{n,m}(z)\bar{\Psi}_{n,1}(z)\right.\\
&\qquad\qquad\left.+\bar{\Psi}^+_{n,m}(z)\bar{\beta}'_{1,m}(z)\right)\\
&-\Psi^+_{n,1}(z)\sum_{m=2}^{r}\left(\Psi^+_{n,m}(z)\Psi_{n,m}(z)+\bar{\Psi}^+_{n,m}(z)\bar{\Psi}_{n,m}(z)\right)\\
&-2\Psi^+_{n,1}(z)\bar{\Psi}^+_{n,1}(z)\bar{\Psi}_{n,1}(z)\\
&-\Psi^+_{n,1}(z)\sum_{m=2}^{r}\left(\gamma'_{1,m}(z)\beta'_{1,m}(z)+\bar{\gamma}'_{1,m}(z)\bar{\beta}'_{1,m}(z)\right)\\
&-2\Psi^+_{n,1}(z)\gamma_n(z)\beta_n(z)\\
&-\Psi^+_n(z)\beta'_1(z)-\Psi^+_{n,1}(z)\Psi^+_n(z)\Psi_n(z)-\Psi^+_{n,1}(z)\gamma'_1(z)\beta'_1(z)\\
&+\sqrt{k+2r-2n-1}\,\Psi^+_{n,1}(z)\left(\partial\phi_n(z)+\partial\phi_{n+1}(z)\right)\\
&+(-k+2(n-1))\partial\,\Psi^+_{n,1}(z),
\end{aligned}
$$

$$
\begin{aligned}
F_l(z)=&\sum_{m=1}^{l-1}\left(\gamma_{m,l+1}(z)\beta_{m,l}(z)-\bar{\gamma}_{m,l}(z)\bar{\beta}_{m,l+1}(z)\right)-\gamma_l(z)\bar{\beta}_{l,l+1}(z)\\
&-2\bar{\gamma}_{l,l+1}(z)\beta_{l+1}(z)
\end{aligned}
$$

$$
\begin{aligned}
&+\sum_{m=l+2}^{n}\left(\gamma_{l,m}(z)\bar{\gamma}_{l,m}(z)\bar{\beta}_{l,l+1}(z)-\gamma_{l,m}(z)\beta_{l+1,m}(z)\right)\\
&-\sum_{m=l+2}^{n}\left(2\bar{\gamma}_{l,m}(z)\gamma_{l+1,m}(z)\beta_{l+1}(z)+\bar{\gamma}_{l,m}(z)\bar{\beta}_{l+1,m}(z)\right)\\
&-\sum_{m=1}^{r}\left(\Psi^{+}_{l,m}(z)\bar{\Psi}^{+}_{l,m}(z)\bar{\beta}_{l,l+1}(z)+\Psi^{+}_{l,m}(z)\Psi_{l+1,m}(z)\right)\\
&-\sum_{m=1}^{r}\left(2\bar{\Psi}^{+}_{l,m}(z)\Psi^{+}_{l+1,m}(z)\beta_{l+1}(z)+\bar{\Psi}^{+}_{l,m}(z)\bar{\Psi}_{l+1,m}(z)\right)\\
&-\Psi^{+}_{l}(z)\Psi_{l+1}(z)-\Psi^{+}_{l}(z)\Psi^{+}_{l+1}(z)\beta_{l+1}(z)-\gamma^{2}_{l,l+1}(z)\beta_{l,l+1}(z)\\
&-\gamma_{l,l+1}(z)\sum_{m=l+2}^{n}\left(\gamma_{l,m}(z)\beta_{l,m}(z)+\bar{\gamma}_{l,m}(z)\bar{\beta}_{l,m}(z)\right)\\
&+\gamma_{l,l+1}(z)\sum_{m=l+2}^{n}\left(\gamma_{l+1,m}(z)\beta_{l+1,m}(z)+\bar{\gamma}_{l+1,m}(z)\bar{\beta}_{l+1,m}(z)\right)\\
&-\gamma_{l,l+1}(z)\sum_{m=1}^{r}\left(\Psi^{+}_{l,m}(z)\Psi_{l,m}(z)+\bar{\Psi}^{+}_{l,m}(z)\bar{\Psi}_{l,m}(z)\right)\\
&+\gamma_{l,l+1}(z)\sum_{m=1}^{r}\left(\Psi^{+}_{l+1,m}(z)\Psi_{l+1,m}(z)+\bar{\Psi}^{+}_{l+1,m}(z)\bar{\Psi}_{l+1,m}(z)\right)\\
&+2\gamma_{l,l+1}(z)\gamma_{l+1}(z)\beta_{l+1}(z)-2\gamma_{l,l+1}(z)\gamma_{l}(z)\beta_{l}(z)\\
&+\gamma_{l,l+1}(z)\Psi^{+}_{l+1}(z)\Psi_{l+1}(z)-\gamma_{l,l+1}(z)\Psi^{+}_{l}(z)\Psi_{l}(z)\\
&+\sqrt{k+2r-2n-1}\,\gamma_{l,l+1}(z)\left(\partial\phi_{l}(z)-\partial\phi_{l+1}(z)\right)\\
&+(-k+2(l-1))\partial\gamma_{l,l+1}(z),\quad 1\le l\le n-1,
\end{aligned}
\tag{11}
$$

Here normal ordering of free fields is implied.

The free field realization of the currents associated with the non-simple roots can be obtained from the OPEs of the simple ones. Moreover, for the case of $n = 0$ our result recovers the free field realization proposed in[44] for $so(2r+1)$ current algebra.

Acknowledgements

The financial support from the National Natural Science Foundation of China (Grant Nos. 11075126 and 11031005) is gratefully acknowledged. I would like to thank my collaborators, S. Kault, X. Liu and Y.-Z. Zhang for their joint works. I also wish to thank the organizers, especially Profs. C.-M Bai, L. Guo and J.-L. Loday, of the conference for giving me the opportunity to present the work.

References

1. O. Andreev, Operator algebra of the $SL(2)$ conformal field theories, *Phys. Lett.* **B 363**, 166-172 (1995).
2. A. Babichenko, Conformal invariance and quantum integrability of sigma models on symmetric superspaces, *Phys. Lett* **B 648**, 254-261 (2007).
3. Z. S. Bassi and A. LeClair, The exact S -matrix for an $osp(2|2)$ disordered system, *Nucl. Phys.* **B 578**, 577-627 (2000).
4. N. Berkovits, M. Bershadsky, T. Hauer, S. Zhukov and B. Zwiebach, Superstring theory on $AdS_2 \times S^2$ as a coset supermanifold, *Nucl. Phys.* **B 567**, 61-86 (2000).
5. N. Berkovits, C. Vafa and E. Witten, Conformal field theory of AdS Background with Ramond-Ramond Flux, *JHEP* **03**, 018 (1999).
6. D. Bernard, (Perturbed) Conformal field theory applied to 2D disordered systems: an introduction, `hep-th/9509137`.
7. D. Bernard and G. Felder, Fock representations and BRST cohomology in $SL(2)$ current algebra, *Commun. Math. Phys.* **127**, 145-168 (1990).
8. M. Bershadsky and H. Ooguri, Hidden $osp(N|2)$ symmetries in superconformal field theories, *Phys. Lett.* **B 229**, 374-378 (1989); Hidden $SL(n)$ symmetry in conformal field theory, *Commun. Math. Phys.* **126**, 49-84 (1989).
9. M. Bershadsky, S. Zhukov and A. Vaintrob, $PSL(n|n)$ sigma model as a conformal field theory, *Nucl. Phys.* **B 559**, 205-234 (1999).
10. M. J. Bhaseen, J.-S. Caux, I. I. Kogan and A. M. Tsveilk, Disordered Dirac fermions: the marriage of three different approaches, *Nucl. Phys.* **B 618**, 465-499 (2001).
11. J. de Boer and L. Feher, An explicit construction of Wakimoto realizations of current algebras, *Mod. Phys. Lett.* **A 11**, 1999-2012 (1996); Wakimoto realizations of current algebras: an explicit construction, *Commun. Math. Phys.* **189**, 759-793 (1997).
12. P. Bouwknegt, J. McCarthy and K. Pilch, Free field approach to two-dimensional conformal field theories, *Prog. Theor. Phys. Suppl.* **102**, 67-135 (1990).
13. P. Bowcock, R-L. K. Koktava and A. Taormina, Wakimoto modules for the affine superalgebra and noncritical $N = 2$ strings, *Phys. Lett.* **B 388**, 303-308 (1996).
14. X.-M. Ding, M. D. Gould, C. J. Mewton and Y.-Z. Zhang, On $osp(2|2)$ conformal field theories, *J. Phys.* **A 36**, 7649-7665 (2003).
15. X.-M. Ding, M. D. Gould and Y.-Z. Zhang, $gl(2|2)$ current superalgebra and non-unitary conformal field theory, *Phys. Lett.* **A 318**, 354-363 (2003).
16. VI. S. Dotsenko and V. A. Fateev, Conformal algebra and multipoint correlation functions in 2D statistical models, *Nucl. Phys.* **B 240**, 312-348 (1984); Four-point correlation functions and the operator algebra in 2D conformal

invariant theories with central charge $C \leq 1$, *Nucl. Phys.* **B 251**, 691-734 (1985).
17. K. Efetov, Supersymmetry and theory of disordered metals, *Adv. Phys.* **32**, 53-127 (1983).
18. V. A. Fateev and A. B. Zamolodchikov, Non-local (parafermion) currents in two-dimensional conformal quantum field theory and self-dual critical points in Z_n-symmetric statistical systems, *Sov. JETP.* **62**, 380-399 (1985).
19. B. Feigin and E. Frenkel, Affine Kac-Moody algebras and semi-infinite flag manifolds, *Commun. Math. Phys.* **128**, 161-189 (1990).
20. M. Flohr, Bits and pieces in logarithmic conformal field theory, *Int. J. Mod. Phys.* **A 18**, 4497-4592 (2003).
21. P. Di Francesco, P. Mathieu and D. Senehal, *Conformal Field Theory* (Springer Press, Berlin, 1997).
22. L. Frappat, P. Sorba and A. Sciarrino, *Dictionary on Lie algebras and superalgebras* (Academic Press, New York, 2000).
23. E. Frenkel, *QFT and Geometric Langlands Program, Langlands Correspondence For Loop Groups An Introduction* (Cambridge University Press, 2007).
24. P. Furlan, A. C. Ganchev, R. Paunov and V. B. Petkova, Solutions of the Knizhnik-Zamolodchikov equation with rational isospins and the reduction to the minimal models, *Nucl. Phys.* **B 394**, 665-706 (1993).
25. M. Gaberdiel, An algebraic approach to logarithmic conformal field theory, *Int. J. Mod. Phys.* **A 18**, 4593-4638 (2003).
26. A. Gerasimov, A. Morozov, M. Olshanetsky, A. Marshakov and S. Shatashvili, Wess-Zumino-Witten model as a free fields, *Int. J. Mod. Phys.* **A 5**, 2495-2589 (1990).
27. P. Goddard, A. Kent and D. Olive, Virasoro algebras and coset space models, *Phys. Lett.* **B 152**, 88-92 (1985).
28. V. Gurarie,Logarithmic operators in conformal field theory, *Nucl. Phys.* **B 410**, 535-549 (1993).
29. S. Guruswamy, A. LeClair and A. W. W. Ludwig, $gl(N|N)$ super-current algebras for disordered Dirac fermions in two dimensions, *Nucl. Phys.* **B 583**, 475-512 (2000).
30. K. Ito and S. Komata, Feigin-Fuchs representations of arbitrary affine Lie algebras, *Mod. Phys. Lett.* **A 6**, 581-587 (1991).
31. V. Kac, Lie superalgebras, *Adv. Math.* **26**, 8-96 (1977).
32. V. Kac, *Infinite-dimensional Lie algebras*, 3rd edition (Cambridge University Press, 1990).
33. D. Kagan and C. A. S. Young, Conformal sigma models on supercoset targets, *Nucl. Phys.* **B 745**, 109-122 (2006).
34. S. M. Khoroshkin and V. N. Tolstoy, UniversalR-matrix for quantized (super)algebras, *Commun. Math. Phys.* **141**, 599-617 (1991).
35. B. Kostant, Verma modules and the existence of quasi-invariant differential operators, *Lecture Notes in Math* **466**, 101-128 (1974).
36. J. Lepowsky and R. L. Wilson, The structure of standard modules I: Universal algebras and the Rogers-Ramanujan identities, *Invent. Math.* **77**, 199-290 (1984); The structure of standard modules II: The case $A_1^{(1)}$, principal gradation, *Invent. Math.* **79**, 417-442 (1985).

37. A. W. W. Ludwig, A free field representation of the osp(2—2) current algebra at level k=-2, and Dirac fermions in a random SU(2) gauge potential, cond-mat/0012189.
38. Z. Maassarani and D. Serban,Non-unitary conformal field theory and logarithmic operators for disordered systems, *Nucl. Phys.* **B 489**, 603-625 (1997).
39. R. R. Metsaev and A. A. Tseytlin, Type IIB superstring action in $AdS_5 \times S^5$ background, *Nucl. Phys.* **B 533**, 109-126 (1998).
40. C. Mudry, C. Chamon and X.-G. Wen, Two-dimensional conformal field theory for disordered systems at criticality, *Nucl. Phys.* **B 466**, 383-443 (1996).
41. J. Rasmussen, Free field realizations of affine current superalgebras, screening currents and primary fields, *Nucl. Phys.* **B 510**, 688-720 (1998).
42. L. Rozansky and H. Saleur, Quantum field theory for the multi-variable Alexander-Conway polynomial, *Nucl. Phys.* **B 376**, 461-509 (1992).
43. M. Wakimoto, Fock representation of the algebra $A_1^{(1)}$, *Commun. Math. Phys.* **104**, 605-609 (1986).
44. W.-L. Yang and Y.-Z. Zhang, On explicit free field realizations of current algebras, *Nucl. Phys.* **B 800**, 527-546 (2008).
45. W.-L. Yang, Y.-Z. Zhang, Free field realization of the $osp(2n|2n)$ current algebra, *Phys. Rev.* **D 78**, 106004 (2008).
46. W.-L. Yang, Y.-Z. Zhang and S. Kault, Differential operator realizations of superalgebras and free field representations of corresponding current algebras, *Nucl. Phys.* **B 823**, 372-402 (2009).
47. W.-L. Yang, Y.-Z. Zhang and X. Liu, $gl(4|4)$ current algebra: free field realization and screening currents, *Phys. Lett.* **B 641**, 329-334 (2006); Free-field representation of super affine $gl(m|n)$ currents at general level, *J. Math. Phys.* **48**, 053514 (2007).
48. Y.-Z. Zhang, X. Liu and W.-L. Yang, Primary field and screening currents of $gl(2|2)$ non-unitary conformal field theory, *Nucl. Phys.* **B 704**, 510-526 (2005).

Free TD Algebras and Rooted Trees

Chenyan Zhou
Office of Public Lessons, Wuhan Technical College of Communications,
Wuhan 430065, P. R. China
E-mail: zhouchy05@lzu.cn

This paper uses rooted trees and forests to give explicit constructions of free noncommutative TD algebras on modules and sets. We also obtain constructions of free TD algebras in terms of Motzkin paths. These highlight the combinatorial nature of TD algebras.

Keywords: TD algebras; free objects; rooted trees; planar rooted trees.

1. Introduction

Let $\mathbf{k}$ be a commutative unitary ring. A TD $\mathbf{k}$-algebra is an associative $\mathbf{k}$-algebra R with a $\mathbf{k}$-linear endomorphism $P : R \to R$ satisfying the **TD relation**:

$$P(x)P(y) = P\big(P(x)y + xP(y) - xP(1)y\big), \ \forall\ x, y \in R. \tag{1}$$

The relation was introduced by Leroux[9] as a relation closely related to Rota-Baxter relation

$$P(x)P(y) = P\big(P(x)y + xP(y) + \lambda xy\big), \ \forall\ x, y \in R.$$

where λ is a fixed element in the base ring $\mathbf{k}$.

In the recent literature further Rota-Baxter type algebras appeared, to wit, the associative Nijenhuis algebra and the TD-algebra. In fact, the latter is a particular case of Rota-Baxter algebras of generalized weight. The TD-algebras only recently entered the scene in the work of Leroux[9] in the context of Loday's work on dendriform algebras, however with motivations from associative Nijenhuis algebras. Free commutative TD algebras have been constructed in[4] by generalized shuffles that are similar to the mixable shuffle construction of free commutative Rota-Baxter algebras.[6] Our goal in this paper is to give an explicit construction of free *noncommutative* TD algebras in terms of rooted trees. In our constructions, we follow closely the constructions of free Rota-Baxter algebras in.[2,3,5]

In Section 2 we will consider the set of planar rooted forests $\mathcal{F}$ and the corresponding free **k**-modules $\mathbf{k}\mathcal{F}$. We equip this module with a TD algebra structure (Theorem 2.1). By decorating angles of the forests in this TD algebra by elements of a module M, we construct in Section 3 the free unitary TD algebra $TD(M)$ on M, in Theorem 3.1. By taking $M = \mathbf{k}X$ for a set X, we obtain the free TD algebra on the set X in Section 3.2 and display a canonical basis in the form of angularly decorated forests (Theorem 3.2). In Section 4 we give another construction of free TD algebras in terms of Motzkin paths.

Notations: In this paper, by a **k**-algebra we mean a unitary algebra unless otherwise stated. The same applies to TD algebras.

Acknowledgements: This paper is based the Master's thesis of the author from Lanzhou University. The author would like to thank his advisors, Professors Li Guo and Yanfeng Luo for guidance.

2. The TD algebra of planar rooted forests

We first obtain a TD algebra structure on planar rooted forests and their various subsets. This allows us to give a uniform construction of free TD algebras in different settings in Section 3.

2.1. *Planar rooted forests*

For the convenience of the reader and for fixing notations, we recall basic concepts and facts of planar rooted trees. For references, see.[1,13]

A **free tree** is an undirected graph that is connected and contains no cycles. A **rooted tree** is a free tree in which a particular vertex has been distinguished as the **root**. Such a distinguished vertex endows the tree with a directed graph structure when the edges of the tree are given the orientation of pointing away from the root. If two vertices of a rooted tree are connected by such an oriented edge, then the vertex on the side of the root is called the **parent** and the vertex on the opposite side of the root is called a **child**. A vertex with no children is called a **leaf**. By our convention, in a tree with only one vertex, this vertex is a leaf, as well as the root. The number of edges in a path connecting two vertices in a rooted tree is called the **length** of the path. The **depth** $\mathrm{d}(T)$ (or **height**) of a rooted tree T is the length of the longest path from its root to its leafs. A **planar rooted tree** is a rooted tree with a fixed embedding into the plane.

There are various ways of drawing planar rooted trees. In our drawing all vertices are represented by a dot and the root is usually at the top of the tree. The first few planar rooted trees are shown below. The tree $\bullet$ with only the root is called the **empty tree**.

Let $\mathcal{T}$ be the set of planar rooted trees and let $\mathcal{F}$ be the free semigroup generated by $\mathcal{T}$ in which the product is given by the concatenation, denoted by $\sqcup$. In this way, each element in $\mathcal{F}$ is a noncommutative product $T_1 \sqcup \cdots \sqcup T_n$ of of trees $T_1, \cdots, T_n \in \mathcal{T}$. Such a product is called a **planar rooted forest**. We also use the abbreviation

$$T^{\sqcup n} = \underbrace{T \sqcup \cdots \sqcup T}_{n \text{ terms}}. \tag{1}$$

We use the (grafting) **bracket** $\lfloor T_1 \sqcup \cdots \sqcup T_n \rfloor$ to denote the tree obtained by **grafting**, that is, by adding a new root together with an edge from the new root to the root of each of the trees $T_1, \cdots, T_n$.

The **depth** of a forest F is the maximal depth $\mathrm{d} = \mathrm{d}(F)$ of trees in F. Furthermore, for a forest $F = T_1 \sqcup \cdots \sqcup T_b$ with trees $T_1, \cdots, T_b$, we define $b = b(F)$ to be the **breadth** of F. Let $\ell(F)$ be the number of leafs of F. Then

$$\ell(F) = \sum_{i=1}^{b} \ell(T_i). \tag{2}$$

We will often use the following recursive structure on forests. For any subset X of $\mathcal{F}$, let $\langle X \rangle$ be the sub-semigroup of $\mathcal{F}$ generated by X. Let $\mathcal{F}_0 = \langle \bullet \rangle$, consisting of forests $\bullet^{\sqcup n}, n \geq 1$. These are also the forests of depth zero. We then recursively define

$$\mathcal{F}_n = \langle \{\bullet\} \cup \lfloor \mathcal{F}_{n-1} \rfloor \rangle. \tag{3}$$

It is clear that $\mathcal{F}_n$ is the set of forests with depth less or equal to n. From this observation, we see that $\mathcal{F}_n$ form a linear ordered direct system: $\mathcal{F}_n \supseteq \mathcal{F}_{n-1}$, and

$$\mathcal{F} = \cup_{n \geq 0} \mathcal{F}_n = \varinjlim \mathcal{F}_n. \tag{4}$$

2.2. TD operator on rooted forests

We note that $\mathbf{k}\mathcal{F}$ with the product $\sqcup$ is in fact the free noncommutative nonunitary $\mathbf{k}$-algebra on the alphabet set $\mathcal{T}$. We are going to define another product $\diamond$ on $\mathbf{k}\mathcal{F}$, making it into a unitary TD algebra.

We define $\diamond$ by giving a set map

$$\diamond : \mathcal{F} \times \mathcal{F} \to \mathbf{k}\mathcal{F}$$

and then extending it bilinearly. For this, we use the depth filtration $\mathcal{F} = \cup_{n\geq 0}\mathcal{F}_n$ in Eq. (4) and apply the induction on $i+j$ to define

$$\diamond : \mathcal{F}_i \times \mathcal{F}_j \to \mathbf{k}\mathcal{F}.$$

When $i+j=0$, we have $\mathcal{F}_i = \mathcal{F}_j = \langle \bullet \rangle$. With the notation in Eq. (1), we define

$$\diamond : \mathcal{F}_0 \times \mathcal{F}_0 \to \mathbf{k}\mathcal{F},\ \bullet^{\sqcup m} \diamond \bullet^{\sqcup n} := \bullet^{\sqcup (m+n-1)}. \tag{5}$$

For given $k \geq 0$, suppose that $\diamond : \mathcal{F}_i \times \mathcal{F}_j \to \mathbf{k}\mathcal{F}$ is defined for $i+j \leq k$. Consider forests F, F' with $\mathrm{d}(F) + \mathrm{d}(F') = k+1$.

First suppose that F and F' are trees. Note that a tree is either $\bullet$ or is of the form $\lfloor \overline{F} \rfloor$ for a forest $\overline{F}$ of smaller depth. Thus it makes sense to define

$$F \diamond F' = \begin{cases} F, & \text{if } F' = \bullet, \\ F', & \text{if } F = \bullet, \\ \lfloor F \diamond \overline{F}' \rfloor + \lfloor \overline{F} \diamond F' \rfloor & \\ -\lfloor (\overline{T}_1 \sqcup \cdots \sqcup \overline{T}_{b-1} \sqcup \lfloor \overline{T}_b \rfloor) \diamond \overline{F}' \rfloor, & \text{if } F \neq \bullet, \text{ and } F' \neq \bullet, \end{cases} \tag{6}$$

Here $\overline{F} = \overline{T}_1 \sqcup \cdots \sqcup \overline{T}_{b-1} \sqcup \overline{T}_b$ is the decomposition of $\overline{F}$ into trees. To see that Eq. (6) and Eq. (8) are well-defined assuming the induction hypothesis, note that in the first term (resp. the second term) on the right hand of the third equation in Eq. (6), the sum of the depth of the product is

$$\mathrm{d}(F) + \mathrm{d}(\overline{F}') \quad (\text{resp. } \mathrm{d}(\overline{F}) + \mathrm{d}(F)) \tag{7}$$

which is less than or equal to k. Also in the third term of the third equation, $(\overline{T}_1 \sqcup \cdots \sqcup \overline{T}_{b-1} \sqcup \lfloor \overline{T}_b \rfloor)$ has depth less or equal that of F. Hence

$$\mathrm{d}(\overline{T}_1 \sqcup \cdots \sqcup \overline{T}_{b-1} \sqcup \lfloor \overline{T}_b \rfloor) + \mathrm{d}(\overline{F}') \leq \mathrm{d}(F) + \mathrm{d}(F') \leq k.$$

Note that when F and F' are trees, $F \diamond F'$ is a tree or a sum of trees.

Next let $F = T_1 \sqcup \cdots \sqcup T_b$ and $F' = T'_1 \sqcup \cdots \sqcup T'_{b'}$ be forests. We then define

$$F \diamond F' = T_1 \sqcup \cdots \sqcup T_{b-1} \sqcup (T_b \diamond T'_1) \sqcup T'_2 \cdots \sqcup T_{b'} \tag{8}$$

where $T_b \diamond T'_1$ is defined by Eq. (6).

By the remark after Eq. (7), $F \diamond F'$ is in $\mathbf{k}\mathcal{F}$. This completes the definition of the set map $\diamond$ on $\mathcal{F} \times \mathcal{F}$.

Before we proceed, we display the following interesting properties of $\diamond$.

Lemma 2.1.

(a) For a trees F, we have

$$F \diamond \overset{\bullet}{\bullet} = \overset{\bullet}{\bullet} \diamond F = \lfloor F \rfloor. \tag{9}$$

(b) If $F = \lfloor \overline{F} \rfloor$ where $\overline{F}$ is also a tree and $F' = \lfloor \overline{F}' \rfloor$ where $\overline{F}'$ is a forest, then

$$F \diamond F' = \lfloor \overline{F} \diamond F' \rfloor. \tag{10}$$

(c) If $F = \lfloor \overline{F} \rfloor$ and $F' = \lfloor \overline{F}' \rfloor$ where $\overline{F}$ and $\overline{F}'$ are forests, then

$$F \diamond F' = \lfloor \lfloor \overline{F} \rfloor \diamond \overline{F}' \rfloor + \lfloor \overline{F} \diamond \lfloor \overline{F}' \rfloor \rfloor - \lfloor (\overline{F} \diamond \overset{\bullet}{\bullet}) \diamond \overline{F}' \rfloor. \tag{11}$$

Proof. (a). By Eq. (6), we have

$$\begin{aligned} \overset{\bullet}{\bullet} \diamond F &= \lfloor \overset{\bullet}{\bullet} \diamond \overline{F} \rfloor + \lfloor \bullet \diamond \lfloor \overline{F} \rfloor \rfloor - \lfloor \lfloor \bullet \rfloor \diamond \overline{F} \rfloor \\ &= \lfloor F \rfloor. \end{aligned}$$

Similarly, we have, $F \diamond \overset{\bullet}{\bullet} = \lfloor F \rfloor$.

(b). Similarly by using Eq.(6), we have

$$\begin{aligned} F \diamond F' &= \lfloor \lfloor \overline{F} \rfloor \diamond \overline{F}' \rfloor + \lfloor \overline{F} \diamond \lfloor \overline{F}' \rfloor \rfloor - \lfloor (\overline{T}_1 \sqcup \cdots \sqcup \overline{T}_{b-1} \sqcup \lfloor \overline{T}_b \rfloor \diamond \overline{F}' \rfloor \\ &= \lfloor \lfloor \overline{F} \rfloor \diamond \overline{F}' \rfloor + \lfloor \overline{F} \diamond \lfloor \overline{F}' \rfloor \rfloor - \lfloor \lfloor \overline{F} \rfloor \diamond \overline{F}' \rfloor \\ &= \lfloor \overline{F} \diamond F' \rfloor. \end{aligned}$$

(c). We only need to verify $\overline{F} \diamond \overset{\bullet}{\bullet} = \overline{T}_1 \sqcup \cdots \sqcup \overline{T}_{b-1} \sqcup \lfloor \overline{T}_b \rfloor$. It is obvious by Eq.(10), since

$$\begin{aligned} \overline{F} \diamond \overset{\bullet}{\bullet} &= (\overline{T}_1 \sqcup \cdots \sqcup \overline{T}_{b-1} \sqcup \overline{T}_b) \diamond \overset{\bullet}{\bullet} \\ &= \overline{T}_1 \sqcup \cdots \sqcup \overline{T}_{b-1} \sqcup (\overline{T}_b \diamond \overset{\bullet}{\bullet}) \\ &= \overline{T}_1 \sqcup \cdots \sqcup \overline{T}_{b-1} \sqcup \lfloor \overline{T}_b \rfloor. \end{aligned}$$

$\square$

As an example, we have

$$\begin{aligned} \lfloor \bullet \sqcup \bullet \rfloor \diamond \lfloor \bullet \rfloor &= \lfloor \bullet \sqcup \bullet \rfloor \diamond \lfloor \bullet \rfloor \\ &= \lfloor (\bullet \sqcup \bullet) \diamond \lfloor \bullet \rfloor \rfloor + \lfloor \lfloor \bullet \sqcup \bullet \rfloor \diamond \bullet \rfloor - \lfloor (\bullet \sqcup \bullet) \diamond \lfloor \bullet \rfloor \diamond \bullet \rfloor \\ &= \lfloor \bullet \sqcup \lfloor \bullet \rfloor \rfloor + \lfloor \lfloor \bullet \sqcup \bullet \rfloor \rfloor - \lfloor \bullet \sqcup \lfloor \bullet \rfloor \rfloor = \lfloor \lfloor \bullet \sqcup \bullet \rfloor \rfloor. \end{aligned} \tag{12}$$

We record the following simple properties of $\diamond$ for later applications.

Lemma 2.2. *Let F, F', F'' be forests.*

(a) $(F \sqcup F') \diamond F'' = F \sqcup (F' \diamond F''), \quad F'' \diamond (F \sqcup F') = (F'' \diamond F) \sqcup F'.$
(b) $\ell(F \diamond F') = \ell(F) + \ell(F') - 1.$

Thus $\mathbf{k}\mathcal{F}$ with the operations $\sqcup$ and $\diamond$ forms a 2-associative algebra in the sense of [11,12] .

Extending $\diamond$ bilinearly, we obtain a binary operation

$$\diamond : \mathbf{k}\mathcal{F} \otimes \mathbf{k}\mathcal{F} \to \mathbf{k}\mathcal{F}.$$

For $F \in \mathcal{F}$, we use the grafting operation to define

$$P_{\mathcal{F}}(F) = \lfloor F \rfloor. \tag{13}$$

Then $P_{\mathcal{F}}$ extends to a linear operator on $\mathbf{k}\mathcal{F}$.

We now state our first main result which will be proved in the next subsection.

Theorem 2.1.

(a) The pair $(\mathbf{k}\mathcal{F}, \diamond)$ is a unitary associative algebra.
(b) The triple $(\mathbf{k}\mathcal{F}, \diamond, P_{\mathcal{F}})$ is a unitary TD algebra.

2.3. *The proof of Theorem 2.1*

Proof. (a). By Eq. (6), $\bullet$ is the identity under the product $\diamond$. So we just need to verify the associativity. For this we only need to verify

$$(F \diamond F') \diamond F'' = F \diamond (F' \diamond F'') \tag{14}$$

for forests $F, F', F'' \in \mathcal{F}$. We will accomplish this by induction on the sum of the depths $n := \mathrm{d}(F) + \mathrm{d}(F') + \mathrm{d}(F'')$. If $n = 0$, then all of F, F', F'' have depth zero and so are in $\mathcal{F}_0 = \langle \bullet \rangle$, the sub-semigroup of $\mathcal{F}$ generated by $\bullet$. Then we have $F = \bullet^{\sqcup i}$, $F' = \bullet^{\sqcup i'}$ and $F'' = \bullet^{\sqcup i''}$, for $i, i', i'' \geq 1$. Then the associativity follows from Eq. (5) since both sides of Eq. (14) is $\bullet^{\sqcup(i+i'+i''-2)}$ in this case.

Let $k \geq 0$. Assume Eq. (14) holds for $n \leq k$ and assume that $F, F', F'' \in \mathcal{F}$ satisfy $n = \mathrm{d}(F) + \mathrm{d}(F') + \mathrm{d}(F'') = k + 1$. We next reduce the breadths of the forests.

Lemma 2.3. *If the associativity*

$$(F \diamond F') \diamond F'' = F \diamond (F' \diamond F'')$$

holds when F, F' and F'' are trees, then it holds when they are forests.

Proof. The proof of this lemma is the same as for Rota-Baxter algebra in.[5] □

To summarize, our proof of the associativity (14) has been reduced to the special case when the forests $F, F', F'' \in \mathcal{F}$ are chosen such that

(a) $n := \mathrm{d}(F) + \mathrm{d}(F') + \mathrm{d}(F'') = k + 1 \geq 1$ with the assumption that the associativity holds when $n \leq k$, and
(b) the forests are of breadth one, that is, they are trees.

If any one of the trees is $\bullet$ then the associativity is clear since $\bullet$ is the identity for the product $\diamond$. So it remains to consider the case when F, F', F'' are all in $\lfloor \mathcal{F} \rfloor$. Then $F = \lfloor \overline{F} \rfloor, F' = \lfloor \overline{F}' \rfloor, F'' = \lfloor \overline{F}'' \rfloor$ with $\overline{F}, \overline{F}', \overline{F}'' \in \mathcal{F}$. To deal with this case, we prove the following general fact on TD operators on not necessarily associative algebras.

Lemma 2.4. *Let R be a unitary $\mathbf{k}$-module with a distinguished element e and a multiplication $\cdot$ that is not necessarily associative. Let $\lfloor\ \rfloor_R : R \to R$ be a $\mathbf{k}$-linear map such that the TD identity holds:*

$$\lfloor x \rfloor_R \cdot \lfloor x' \rfloor_R = \lfloor x \cdot \lfloor x' \rfloor_R \rfloor_R + \lfloor \lfloor x \rfloor_R \cdot x' \rfloor_R - \lfloor (x \cdot \lfloor e \rfloor_R) \cdot x' \rfloor_R, \ \forall x,\, x' \in R. \quad (15)$$

Let x, x' and x'' be in R. If

$$(x \cdot x') \cdot x'' = x \cdot (x' \cdot x''),$$

*then we say that (x, x', x'') is an **associative triple** for the product $\cdot$. Fix $y, y', y'', e \in R$. Suppose all the triples*

$$(y, y', y''), (\lfloor y \rfloor_R, y', y''), (y, \lfloor y' \rfloor_R, y''), (y, y', \lfloor y'' \rfloor_R), (\lfloor y \rfloor_R, y', \lfloor y'' \rfloor_R), \quad (16)$$

$$(\lfloor y \rfloor_R, \lfloor y' \rfloor_R, y''),\ (y, \lfloor y' \rfloor_R, \lfloor y'' \rfloor_R) \quad (17)$$

are associative triples for $\cdot$. In addition, the following equations hold.

$$\begin{aligned}
&((\lfloor y \rfloor \cdot y') \cdot \lfloor e \rfloor_R) \cdot y'' = \lfloor y \rfloor \cdot ((y' \cdot \lfloor e \rfloor_R) \cdot y''),\\
&((y \cdot \lfloor y' \rfloor) \cdot \lfloor e \rfloor_R) \cdot y'' = ((y \cdot \lfloor e \rfloor_R) \cdot (\lfloor y' \rfloor \cdot y''),\\
&((y \cdot \lfloor e \rfloor_R) \cdot y') \cdot \lfloor y'' \rfloor = (y \cdot \lfloor e \rfloor_R) \cdot (y' \cdot \lfloor y'' \rfloor),\\
&(((y \cdot \lfloor e \rfloor_R) \cdot y') \cdot \lfloor e \rfloor_R) \cdot y'' = (y \cdot \lfloor e \rfloor_R) \cdot ((y' \cdot \lfloor e \rfloor_R) \cdot y'')
\end{aligned} \quad (18)$$

Then $(\lfloor y\rfloor_R, \lfloor y'\rfloor_R, \lfloor y''\rfloor_R)$ *is an associative triple for* $\cdot$.

Proof. Using Eq. (15) and bilinearity of the product $\cdot$, we have

$$
\begin{aligned}
&(\lfloor y\rfloor_R\cdot\lfloor y'\rfloor_R)\cdot\lfloor y''\rfloor_R\\
&=\big(\lfloor\lfloor y\rfloor_R\cdot y'\rfloor_R+\lfloor y\cdot\lfloor y'\rfloor_R\rfloor_R-\lfloor(y\cdot\lfloor e\rfloor_R)\cdot y'\rfloor_R\big)\cdot\lfloor y''\rfloor_R\\
&=\lfloor\lfloor y\rfloor_R\cdot y'\rfloor_R\cdot\lfloor y''\rfloor_R+\lfloor y\cdot\lfloor y'\rfloor_R\rfloor_R\cdot\lfloor y''\rfloor_R-\lfloor(y\cdot\lfloor e\rfloor_R)\cdot y'\rfloor_R\cdot\lfloor y''\rfloor_R\\
&=\lfloor\lfloor\lfloor y\rfloor_R\cdot y'\rfloor_R\cdot y''\rfloor_R+\lfloor(\lfloor y\rfloor_R\cdot y')\cdot\lfloor y''\rfloor_R\rfloor_R\\
&\quad-\lfloor((\lfloor y\rfloor_R\cdot y')\cdot\lfloor e\rfloor_R)\cdot y''\rfloor_R+\lfloor\lfloor y\cdot\lfloor y'\rfloor_R\rfloor_R\cdot y''\rfloor_R\\
&\quad+\lfloor(y\cdot\lfloor y'\rfloor_R)\cdot\lfloor y''\rfloor_R\rfloor_R-\lfloor((y\cdot\lfloor y'\rfloor_R)\cdot\lfloor e\rfloor_R))\cdot y''\rfloor_R\\
&\quad-\lfloor\lfloor(y\cdot\lfloor e\rfloor_R)\cdot y'\rfloor_R\cdot y''\rfloor_R-\lfloor((y\cdot\lfloor e\rfloor_R)\cdot y')\cdot\lfloor y''\rfloor_R\rfloor_R\\
&\quad+\lfloor(((y\cdot\lfloor e\rfloor_R)\cdot y')\cdot\lfloor e\rfloor_R)\cdot y''\rfloor_R
\end{aligned}
$$

Applying the associativity of the second triple in Eq. (17) to $(y\cdot\lfloor y'\rfloor_R)\cdot\lfloor y''\rfloor_R$ in the fifth term above and then using Eq. (15) again, we have

$$
\begin{aligned}
&(\lfloor y\rfloor_R\cdot\lfloor y'\rfloor_R)\cdot\lfloor y''\rfloor_R\\
&=\lfloor\lfloor\lfloor y\rfloor_R\cdot y'\rfloor_R\cdot y''\rfloor_R+\lfloor(\lfloor y\rfloor_R\cdot y')\cdot\lfloor y''\rfloor_R\rfloor_R\\
&\quad-\lfloor((\lfloor y\rfloor_R\cdot y')\cdot\lfloor e\rfloor_R)\cdot y''\rfloor_R+\lfloor\lfloor y\cdot\lfloor y'\rfloor_R\rfloor_R\cdot y''\rfloor_R\\
&\quad+\lfloor y\cdot\lfloor\lfloor y'\rfloor_R\cdot y''\rfloor_R\rfloor_R+\lfloor y\cdot\lfloor y'\cdot\lfloor y''\rfloor_R\rfloor_R\rfloor_R\\
&\quad-\lfloor y\cdot\lfloor(y'\cdot\lfloor e\rfloor_R)y''\rfloor_R\rfloor_R-\lfloor((y\cdot\lfloor y'\rfloor_R)\cdot\lfloor e\rfloor_R))\cdot y''\rfloor_R\\
&\quad-\lfloor\lfloor(y\cdot\lfloor e\rfloor_R)\cdot y'\rfloor_R\cdot y''\rfloor_R-\lfloor((y\cdot\lfloor e\rfloor_R)\cdot y')\cdot\lfloor y''\rfloor_R\rfloor_R\\
&\quad+\lfloor(((y\cdot\lfloor e\rfloor_R)\cdot y')\cdot\lfloor e\rfloor_R)\cdot y''\rfloor_R.
\end{aligned}
$$

By a similar calculation, we have

$$
\begin{aligned}
&\lfloor y\rfloor_R\cdot(\lfloor y'\rfloor_R\cdot\lfloor y''\rfloor_R)\\
&=\lfloor\lfloor y\rfloor_R\cdot(\lfloor y'\rfloor_R\cdot y'')\rfloor_R+\lfloor y\cdot\lfloor\lfloor y'\rfloor_R\cdot y''\rfloor_R\rfloor_R\\
&\quad-\lfloor(y\cdot\lfloor e\rfloor_R)\cdot(\lfloor y'\rfloor_R\cdot y'')\rfloor_R+\lfloor\lfloor y\rfloor_R\cdot(y'\cdot\lfloor y''\rfloor_R)\rfloor_R\\
&\quad+\lfloor y\cdot\lfloor y'\cdot\lfloor y''\rfloor_R\rfloor_R\rfloor_R-\lfloor(y\cdot\lfloor e\rfloor_R)\cdot(y'\cdot\lfloor y''\rfloor_R)\rfloor_R\\
&\quad-\lfloor y\cdot\lfloor(y'\cdot\lfloor e\rfloor_R)\cdot y''\rfloor_R\rfloor_R-\lfloor\lfloor y\rfloor_R\cdot((y'\cdot\lfloor e\rfloor_R)\cdot y'')\rfloor_R\\
&\quad+\lfloor(y\cdot\lfloor e\rfloor_R)\cdot((y'\cdot\lfloor e\rfloor_R)\cdot y'')\rfloor_R.
\end{aligned}
$$

We use the associativity of the first triple in Eq. (17) to $\lfloor y\rfloor_R\cdot(\lfloor y'\rfloor_R\cdot y'')$ in the first term above and then using Eq. (15) again, we have

$$
\begin{aligned}
&\lfloor y\rfloor_R\cdot(\lfloor y'\rfloor_R\cdot\lfloor y''\rfloor_R)\\
&=\lfloor\lfloor\lfloor y\rfloor_R\cdot y'\rfloor_R\cdot y''\rfloor_R+\lfloor\lfloor y\cdot\lfloor y'\rfloor_R\rfloor_R\cdot y''\rfloor_R-\lfloor\lfloor(y\cdot\lfloor e\rfloor_R)\cdot y'\rfloor_R\cdot y''\rfloor_R
\end{aligned}
$$

$$
\begin{aligned}
&+\lfloor y\cdot\lfloor\lfloor y'\rfloor_R\cdot y''\rfloor_R\rfloor_R-\lfloor((y\cdot\lfloor e\rfloor_R)\cdot(\lfloor y'\rfloor_R\cdot y''))\rfloor_R\\
&+\lfloor\lfloor y\rfloor_R\cdot(y'\cdot\lfloor y''\rfloor_R)\rfloor_R+\lfloor y\cdot\lfloor y'\cdot\lfloor y''\rfloor_R\rfloor_R\rfloor_R\\
&-\lfloor(y\cdot\lfloor e\rfloor_R)\cdot(y'\cdot\lfloor y''\rfloor_R)\rfloor_R-\lfloor\lfloor y\rfloor_R\cdot((y'\cdot\lfloor e\rfloor_R)\cdot y''))\rfloor_R\\
&-\lfloor y\cdot\lfloor(y'\cdot\lfloor e\rfloor_R)\cdot y''\rfloor_R\rfloor_R+\lfloor(y\cdot\lfloor e\rfloor_R)\cdot((y'\cdot\lfloor e\rfloor_R)\cdot y''))\rfloor_R).
\end{aligned}
$$

Now by the associativity of the triples in Eq. (16) and the relations in Eq. (18), the i-th term in the expansion of $(\lfloor y\rfloor_R\cdot\lfloor y'\rfloor_R)\cdot\lfloor y''\rfloor_R$ matches with the $\sigma(i)$-th term in the expansion of $\lfloor y\rfloor_R\cdot\big(\lfloor y'\rfloor_R\cdot\lfloor y''\rfloor_R\big)$. Here the permutation $\sigma\in\Sigma_{11}$ is

$$
\begin{pmatrix} i \\ \sigma(i)\end{pmatrix}=\begin{pmatrix}1&2&3&4&5&6&7&8&9&10&11\\1&6&9&2&4&7&10&5&3&8&11\end{pmatrix}. \tag{19}
$$

This proves the lemma. □

To continue the proof of Theorem 2.1, we apply Lemma 2.4 to the situation where R is $\mathbf{k}\mathcal{F}$ with the multiplication $\cdot=\diamond$, the operator $\lfloor\ \rfloor_R=\lfloor\ \rfloor$ and the quadruple $(e,y,y',y'')=(\bullet,\overline{F},\overline{F}',\overline{F}'')$. By the induction hypothesis on n , all the triples in Eq. (16) and(17) are associative for $\diamond$. We only need to verify Eq. (18). It corresponds the following equation in $\mathbf{k}\mathcal{F}$.

$$
\begin{aligned}
((\lfloor\overline{F}\rfloor\diamond\overline{F}')\diamond\overset{\bullet}{\bullet})\diamond\overline{F}''&=\lfloor\overline{F}\rfloor\diamond((\overline{F}'\diamond\overset{\bullet}{\bullet})\diamond\overline{F}''),\\
((\overline{F}\diamond\lfloor\overline{F}'\rfloor)\diamond\overset{\bullet}{\bullet})\diamond\overline{F}''&=((\overline{F}\diamond\overset{\bullet}{\bullet})\diamond(\lfloor\overline{F}'\rfloor\diamond\overline{F}''),\\
((\overline{F}\diamond\overset{\bullet}{\bullet})\diamond\overline{F}')\diamond\lfloor\overline{F}''\rfloor&=(\overline{F}\diamond\overset{\bullet}{\bullet})\diamond(\overline{F}'\diamond\lfloor\overline{F}''\rfloor),\\
(((\overline{F}\diamond\overset{\bullet}{\bullet})\diamond\overline{F}')\diamond\overset{\bullet}{\bullet})\diamond\overline{F}''&=(\overline{F}\diamond\overset{\bullet}{\bullet})\diamond((\overline{F}'\diamond\overset{\bullet}{\bullet})\diamond\overline{F}''.
\end{aligned}\tag{20}
$$

Firstly, let us consider the first equation of Eq. (20). Since

$$
\mathrm{d}(F)+\mathrm{d}(\overline{F}')+\mathrm{d}(\overset{\bullet}{\bullet})=\mathrm{d}(F)+(\mathrm{d}(F')-1)+2=\mathrm{d}(F)+\mathrm{d}(F')+1,
$$

$$
\mathrm{d}(F)+\mathrm{d}(\overline{F}'\diamond\overset{\bullet}{\bullet})+\mathrm{d}(\overline{F}'')\le\mathrm{d}(F)+\mathrm{d}(F')+\mathrm{d}(F'')-1
$$

satisfy the induction hypothesis. Then it fulfills the associativity condition. Therefore, we have

$$
\begin{aligned}
((\lfloor\overline{F}\rfloor\diamond\overline{F}')\diamond\overset{\bullet}{\bullet})\diamond\overline{F}''&=(\lfloor\overline{F}\rfloor\diamond(\overline{F}'\diamond\overset{\bullet}{\bullet}))\diamond\overline{F}''\\
&=\lfloor\overline{F}\rfloor\diamond((\overline{F}'\diamond\overset{\bullet}{\bullet})\diamond\overline{F}'').
\end{aligned}
$$

Similarly, we can proof the 2nd, 3rd and 4th equations of Eq. (20). We remark that in the proof of the 2nd equation, the commutativity of $\overset{\bullet}{\bullet}$ and F' has been used, which is a consequence of Lemma 2.1.(a). So by Lemma 2.4, the triple (F, F', F'') is associative for $\diamond$. This completes the induction and therefore the proof of the first part of Theorem 2.1.

(b). We just need to prove that $P_{\mathcal{F}}(F) = \lfloor F \rfloor$ is a TD operator . This is immediate from Eq. (6).

Since the product $\diamond$ satisfy the associativity on $\mathbf{k}\mathcal{F}$, we have $(\overline{F} \diamond \mathbf{I}) \diamond \overline{F}' = \overline{F} \diamond (\mathbf{I} \diamond \overline{F}')$, Therefore it is meaningful to write the form $\overline{F} \diamond \mathbf{I} \diamond \overline{F}'$ instead of $(\overline{F} \diamond \mathbf{I}) \diamond \overline{F}'$. Now we give another equivalent form and corollary of the definition (6) :

Definition 2.1. Let F and F' are both trees, where $F = \lfloor \overline{F} \rfloor$, $F' = \lfloor \overline{F}' \rfloor$. then the product $\diamond$ can be given in the following equivalent way

$$F \diamond F' = \begin{cases} F, & \text{if } F' = \bullet, \\ F', & \text{if } F = \bullet, \\ \lfloor \lfloor \overline{F} \rfloor \diamond \overline{F}' \rfloor + \lfloor \overline{F} \diamond \lfloor \overline{F}' \rfloor \rfloor - \lfloor \overline{F} \diamond \mathbf{I} \diamond \overline{F}' \rfloor, & \text{if } F \neq \bullet,\ F' \neq \bullet. \end{cases} \tag{21}$$

Corollary 2.1. *If* $F = \lfloor \overline{F} \rfloor$ *and* $F' = \lfloor \overline{F}' \rfloor$, *where* $\overline{F}$ *and* $\overline{F}'$ *are also trees, we have*

$$F \diamond F' = \lfloor \lfloor \overline{F} \rfloor \diamond \overline{F}' \rfloor = \lfloor \overline{F} \diamond \lfloor \overline{F}' \rfloor \rfloor. \tag{22}$$

Proof. It is immediately from lemma 2.1(a). □

3. Free TD algebras on a module or a set

We will construct the free unitary TD algebra on a $\mathbf{k}$-module or on a set by expressing elements in the TD algebra in terms of forests from Section 2, in addition with angles decorated by elements from the $\mathbf{k}$-module or set. The free unitary TD algebra will be constructed in Section 3.1. When the $\mathbf{k}$-module is taken to be the free $\mathbf{k}$-module on a set, we obtain the free unitary TD algebra on the set. This will be discussed in Section 3.2.

3.1. *Free TD algebra on a module as decorated forests*

Let M be a non-zero $\mathbf{k}$-module. Let F be in $\mathcal{F}$ with ℓ leafs. We let $M^{\otimes F}$ denote the tensor power $M^{\otimes(\ell-1)}$ labeled by F. In other words,

$$M^{\otimes F} = \{(F; \mathfrak{m}) \mid \mathfrak{m} \in M^{\otimes(\ell-1)}\} \tag{1}$$

with the $\mathbf{k}$-module structure coming from the second component and with the convention that $M^{\otimes 0} = \mathbf{k}$. We can think of $M^{\otimes F}$ as the tensor power of

M with exponent F with the usual tensor power $M^{\otimes n}, n \geq 0$, corresponding to $M^{\otimes F}$ when F is the forest $\bullet^{\sqcup(n+1)}$. We define the **k**-module

$$\mathrm{T}D(M) = \bigoplus_{F \in \mathcal{F}} M^{\otimes F}.$$

and define a product $\overline{\diamond}$ on $\mathrm{T}D(M)$ by using the product $\diamond$ on $\mathcal{F}$ in Section 2.2.

Let $T(M) = \oplus_{n \geq 0} M^{\otimes n}$ be the (unitary) tensor algebra with $M^{\otimes 0} := \mathbf{k}$. Let $\overline{\otimes}$ be its product, so for $\mathfrak{m} \in M^{\otimes n}$ and $\mathfrak{m}' \in M^{\otimes n'}$, we have

$$\mathfrak{m} \overline{\otimes} \mathfrak{m}' = \begin{cases} \mathfrak{m} \otimes \mathfrak{m}' \in M^{\otimes n+n'}, & \text{if } n > 0, n' > 0, \\ \mathfrak{m}\mathfrak{m}' \in M^{\otimes n'}, & \text{if } n = 0, n' > 0, \\ \mathfrak{m}'\mathfrak{m} \in M^{\otimes n}, & \text{if } n > 0, n' = 0, \\ \mathfrak{m}'\mathfrak{m} \in \mathbf{k}, & \text{if } n = n' = 0. \end{cases} \tag{2}$$

where the multiplications in second and the third cases are the scalar product.

Definition 3.1. For tensors $D = (F; \mathfrak{m}) \in M^{\otimes F}$ and $D' = (F'; \mathfrak{m}') \in M^{\otimes F'}$, define

$$D \overline{\diamond} D' = (F \diamond F'; \mathfrak{m} \overline{\otimes} \mathfrak{m}'). \tag{3}$$

The right hand side is well-defined since $\mathfrak{m} \overline{\otimes} \mathfrak{m}'$ has tensor degree

$$\deg(\mathfrak{m} \overline{\otimes} \mathfrak{m}') = \deg(\mathfrak{m}) + \deg(\mathfrak{m}') = \ell(F) - 1 + \ell(F') - 1$$

which equals $\ell(F \diamond F') - 1$ by Lemma 2.2.(b). For example, from Eq. (12) we have

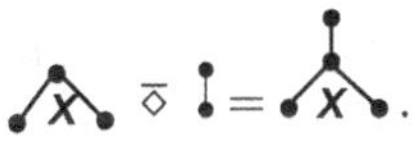

$$X \;\overline{\diamond}\; = X .$$

By Eq. (5) – (8), we have a more explicit expression.

$$D \overline{\diamond} D' = \begin{cases} (\bullet; cc'), & \text{if } D = (\bullet; c), D' = (\bullet; c'), \\ (F; c'\mathfrak{m}), & \text{if } D' = (\bullet, c'), F \neq \bullet, \\ (F'; c\mathfrak{m}'), & \text{if } D = (\bullet, c), F' \neq \bullet, \\ (F \diamond F'; \mathfrak{m} \overline{\otimes} \mathfrak{m}'), & \text{if } F \neq \bullet, F' \neq \bullet. \end{cases} \tag{4}$$

We can give an even more explicit description of $\overline{\diamond}$. For pure tensors $\mathfrak{m}$ and $\mathfrak{m}'$, consider the standard decompositions of $D = (F; \mathfrak{m})$ and $D' = (F'; \mathfrak{m}')$[3]

$$D = (F; \mathfrak{m}) = (T_1; \mathfrak{m}_1) \sqcup_{u_1} (T_2; \mathfrak{m}_2) \sqcup_{u_2} \cdots \sqcup_{u_{b-1}} (T_b; \mathfrak{m}_b),$$

$$D' = (F'; \mathfrak{m}') = (T_1'; \mathfrak{m}_1') \sqcup_{u_1'} (T_2'; \mathfrak{m}_2') \sqcup_{u_2'} \cdots \sqcup_{u_{b'-1}'} (T_{b'}'; \mathfrak{m}_{b'}').$$

Then by Eq. (5) – (8) and Eq. (3) – (4), it is easy to see that the product $\bar{\diamond}$ can be defined by induction on the sum $d + d'$ of the depths $\mathrm{d} = \mathrm{d}(F)$ and $\mathrm{d}' = \mathrm{d}(F')$ as follows: If $\mathrm{d} + \mathrm{d}' = 0$, then $F = \bullet^{\sqcup i}$ and $F' = \bullet^{\sqcup j}$ for $i, j \geq 1$. If $i = 1$, then $D = (F; \mathfrak{m}) = (\bullet; c) = c(\bullet; \mathbf{1})$ and we define $D \,\bar{\diamond}\, D' = cD' = (F'; c\mathfrak{m}')$. Similarly define $D \,\bar{\diamond}\, D'$ if $j = 1$. If $i > 1$ and $j > 1$, then $(F; \mathfrak{m}) = (\bullet; \mathbf{1}) \sqcup_{u_1} \cdots \sqcup_{u_{b-1}} (\bullet; \mathbf{1})$ with $u_1, \cdots, u_{b-1} \in M$. Similarly, $(F'; \mathfrak{m}') = (\bullet; \mathbf{1}) \sqcup_{u'_1} \cdots \sqcup_{u'_{b'-1}} (\bullet; \mathbf{1})$. Then define

$$(F; \mathfrak{m}) \,\bar{\diamond}\, (F'; \mathfrak{m}') = (\bullet; \mathbf{1}) \sqcup_{u_1} \cdots \sqcup_{u_{b-1}} (\bullet; \mathbf{1}) \sqcup_{u'_1} \cdots \sqcup_{u'_{b'-1}} (\bullet; \mathbf{1}).$$

For $k \geq 0$, suppose $D \,\bar{\diamond}\, D'$ has been defined for all $D = (F; \mathfrak{m})$ and $D' = (F'; \mathfrak{m}')$ with $\mathrm{d}(F) + \mathrm{d}(F') \leq k$ and consider D and D' with $\mathrm{d}(F) + \mathrm{d}(F') = k + 1$. Then we define

$$D \,\bar{\diamond}\, D' = (T_1; \mathfrak{m}_1) \sqcup_{u_1} \cdots \sqcup_{u_{b-1}} \left((T_b; \mathfrak{m}_b) \,\bar{\diamond}\, (T'_1; \mathfrak{m}'_1)\right) \sqcup_{u'_1} \cdots \sqcup_{u'_{b'-1}} (T'_{b'}; \mathfrak{m}'_{b'}) \tag{5}$$

where

$$(T_b; \mathfrak{m}_b) \,\bar{\diamond}\, (T'_1; \mathfrak{m}'_1) \tag{6}$$
$$= \begin{cases} (\bullet; \mathbf{1}), & \text{if } T_b = T'_1 = \bullet \text{ (so } \mathfrak{m}_b = \mathfrak{m}'_1 = \mathbf{1}), \\ (T_b, \mathfrak{m}_b), & \text{if } T'_1 = \bullet, T_b \neq \bullet, \\ (T'_1, \mathfrak{m}'_1), & \text{if } T'_1 \neq \bullet, T_b = \bullet, \\ \lfloor (T_b; \mathfrak{m}_b) \,\bar{\diamond}\, (\overline{F}'_1; \mathfrak{m}'_1) \rfloor & \\ + \lfloor (\overline{F}_b; \mathfrak{m}_b) \,\bar{\diamond}\, (T'_1; \mathfrak{m}'_1) \rfloor & \\ - \lfloor (\overline{F}_b; \mathfrak{m}_b) \,\bar{\diamond}\, (\text{ǐ}, \mathbf{1}) \,\bar{\diamond}\, (\overline{F}'_1; \mathfrak{m}'_1) \rfloor, & \text{if } T'_1 = \lfloor \overline{F}'_1 \rfloor \neq \bullet, T_b = \lfloor \overline{F}_b \rfloor \neq \bullet. \end{cases}$$

In the last case, we have applied the induction hypothesis on $\mathrm{d}(F) + \mathrm{d}(F')$ to define the terms in the brackets on the right hand side. Further, for $(F; \mathfrak{m}) \in M^{\otimes F}$, define $\lfloor (F; \mathfrak{m}) \rfloor = (\lfloor F \rfloor; \mathfrak{m})$. This is well-defined since $\ell(F) = \ell(\lfloor F \rfloor)$.

The product $\bar{\diamond}$ is clearly bilinear. So extending it biadditively, we obtain a binary operation

$$\bar{\diamond} : \mathrm{T}\mathcal{D}(M) \otimes \mathrm{T}\mathcal{D}(M) \to \mathrm{T}\mathcal{D}(M).$$

For $(F; \mathfrak{m}) \in (F; M)$, define

$$P_M(F; \mathfrak{m}) = \lfloor (F; \mathfrak{m}) \rfloor = (\lfloor F \rfloor; \mathfrak{m}) \in (\lfloor F \rfloor; M). \tag{7}$$

Thus P_M defines a linear operator on $\mathrm{T}\mathcal{D}(M)$. Alternatively

$$P_M(F; \mathfrak{m}) = (P_{\mathcal{F}}(F); \mathfrak{m})$$

with $P_{\mathcal{F}}$ defined in Eq. (13). Finally, let

$$j_M : M \to \mathrm{T}D(M) \tag{8}$$

denote the **k**-module map sending $a \in M$ to $(\bullet\sqcup\bullet; a)$. The following theorem follows from the same argument as that of Theorem 3.4 in.[3]

Theorem 3.1. *Let M be a* **k***-module.*

(a) The pair $(\mathrm{T}D(M), \bar{\diamond})$ is a unitary associative algebra.
(b) The triple $(\mathrm{T}D(M), \bar{\diamond}, P_M)$ is a unitary TD algebra .
(c) The quadruple $(\mathrm{T}D(M), \bar{\diamond}, P_M, j_M)$ is the free unitary TD algebra on the module M. More precisely, for any unitary TD algebra (R, P) and module morphism $f : M \to R$, there is a unique unitary TD algebra morphism $\bar{f} : \mathrm{T}D(M) \to R$ such that $f = \bar{f} \circ j_M$.

3.2. *Free TD algebra on a set*

We now apply the tree construction of free TD algebra on a module in the last section to obtain a tree construction of a free TD algebra on a set. We accomplish this by displaying a canonical basis of the free TD algebra in terms of forests decorated by the set and express the product and operator in terms of this basis.

Let X be a given set and let $M = \mathbf{k}X$ be the free **k**-module on X. For any $n \geq 1$, the tensor power $M^{\otimes n}$ has a natural basis $X^n = \{(x_1, \cdots, x_n) \mid x_i \in X,\ 1 \leq i \leq n\}$. Accordingly, for any rooted forest $F \in \mathcal{F}$, with $\ell = \ell(F) \geq 2$, the set

$$X^F := \{(F; (x_1, \cdots, x_{\ell-1})) := (F; x_1 \otimes \cdots \otimes x_{\ell-1}) \mid x_i \in X,\ 1 \leq i \leq \ell - 1\}$$

form a basis of $M^{\otimes F}$. In particular, when $\ell(F) = 1$, $M^{\otimes F} = \mathbf{k}\, F$ has a basis $X^F := \{(F; \mathbf{1})\}$. Thus for each $F \in \mathcal{F}$, the **k**-module $M^{\otimes F}$ has a basis

$$X^F := \{(F; \vec{x}) \mid \vec{x} \in X^{\ell(F)-1}\}, \tag{9}$$

with the convention that $X^0 = \{\mathbf{1}\}$. Thus the disjoint union

$$X^{\mathcal{F}} := \coprod_{F \in \mathcal{F}} X^F. \tag{10}$$

forms a basis of

$$\mathrm{T}D(X) := \mathrm{T}D(M).$$

We call $X^{\mathcal{F}}$ the set of **angularly decorated rooted forests with decoration set X.**

For $(F;\vec{x}) \in X^{\mathcal{F}}$, we can give the **standard decomposition**

$$(F;\vec{x}) = (T_1;\vec{x}_1) \sqcup_{u_1} (T_2;\vec{x}_2) \sqcup_{u_2} \cdots \sqcup_{u_{b-1}} (T_b;\vec{x}_b) \quad (11)$$

where $F = T_1 \sqcup \cdots \sqcup T_b$ is the decomposition of F into trees and $\vec{x}$ is the vector concatenation of the elements of $\vec{x}_1, u_1, \vec{x}_2, \cdots, u_{b-1}, \vec{x}_b$ which are not the unit $\mathbf{1}$. As a corollary of Theorem 3.1, we have

Theorem 3.2. *For $D = (F;(x_1,\cdots,x_b))$, $D' = (F';(x'_1,\cdots,x'_{b'}))$ in $X^{\mathcal{F}}$, define*

$$D \,\bar{\diamond}\, D' = \begin{cases} (\bullet;\mathbf{1}), & \text{if } F = F' = \bullet, \\ D, & \text{if } F' = \bullet, F \neq \bullet, \\ D', & \text{if } F = \bullet, F' \neq \bullet, \\ (F \diamond F';(x_1,\cdots,x_b,x'_1,\cdots,x'_{b'})), & \text{if } F \neq \bullet, F' \neq \bullet, \end{cases} \quad (12)$$

where $\diamond$ is defined in Eq. (6) and (8). Define

$$P_X : \mathrm{T}D(X) \to \mathrm{T}D(X), \quad P_X(F;(x_1,\cdots,x_b)) = (\lfloor F \rfloor;(x_1,\cdots,x_b)),$$

and

$$j_X : X \to \mathrm{T}D(X), \quad j_X(x) = (\bullet \sqcup \bullet;(x)), \quad x \in X.$$

Then the quadruple $(\mathrm{T}D(X), \bar{\diamond}, P_X, j_X)$ is the free TD algebra on X.

4. Construction of free TD algebras by Motzkin paths

We have used angularly decorated planar forests to construct the free TD algebras on a set X in Section 3. We shall give a new description of this TD algebra structure on Motzkin paths.

4.1. *Review of the Motzkin path.*

For later references, we review the concept of Motzkin paths. See[5] for further details. A **Motzkin path** is a lattice path in $\mathbb{N}^2$ from $(0,0)$ to $(n,0)$ whose permitted steps are an up diagonal step (or **up step** for short) $(1,1)$, a down diagonal step (or **down step**) $(1,-1)$ and a horizontal step (or **level step**) $(1,0)$. The first few Motzkin paths are

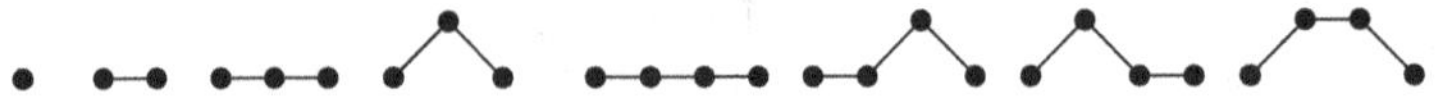

(1)

Let $\mathcal{P}$ be the set of Motzkin paths. For Motzkin paths $\mathfrak{m}$ and $\mathfrak{m}'$, define $\mathfrak{m}\circ\mathfrak{m}$, called the **link product** of $\mathfrak{m}$ and $\mathfrak{m}'$, to be the Motzkin path obtained by joining the last vertex of $\mathfrak{m}$ with the first vertex of $\mathfrak{m}'$. For example,

$\circ$ $=$, $\circ$ $=$

Let $\mathcal{I}$ be the set of **indecomposable** (also called **prime**) Motzkin paths, consisting of Motzkin paths that touch the x-axis only at the two end vertices.

Next for a Motzkin path $\mathfrak{m}$, let $/\mathfrak{m}\backslash$ denote the Motzkin path obtained by raising $\mathfrak{m}$ on the left end by an up step and on the right end by a down step. For example,

$$/ \bullet \backslash = \bullet\!\nearrow\!\bullet\!\searrow\!\bullet, \quad / \bullet\!\!-\!\!\bullet \backslash = \bullet\!\nearrow\!\bullet\!\!-\!\!\bullet\!\searrow\!\bullet$$

This defines an operator $/\ \backslash$ on $\mathcal{P}$, called the **raising operator**.

A Motzkin path is called **peak-free** if it is not $\bullet$ and does not have an up step followed immediately by a down step. For example, $\bullet\!\!-\!\!\bullet$ $\bullet\!\!-\!\!\bullet\!\!-\!\!\bullet$ $\bullet\!\!-\!\!\bullet\!\!-\!\!\bullet\!\!-\!\!\bullet$ $\bullet\!\nearrow\!\bullet\!\!-\!\!\bullet\!\searrow\!\bullet$ and $\bullet\!\!-\!\!\bullet\!\!-\!\!\bullet\!\!-\!\!\bullet\!\!-\!\!\bullet$ in the list (1) are peak-free while the rest are not.

Let X be a set. An **X-decorated (or colored) Motzkin path**[5] is a Motzkin path whose level steps are decorated (colored) by elements in X.

Let $\mathcal{P}(X)$ be the set of X-decorated Motzkin paths , $\mathcal{L}(X)$ be the set of peak-free X-decorated Motzkin paths, and let $\mathcal{V}(X)$ be the set of valley-free Motzkin paths with level steps decorated by X, consisting of Motzkin paths in $\mathcal{P}(X)$ with no down steps followed immediately by a up step. Note that Motzkin paths with no decorations can be identified with X-decorated Motzkin paths where X is a singleton.

4.2. *Free TD algebras in terms of Motzkin paths*

Theorem 4.1. *Let $X^{\mathcal{F}}$ denotes the set of angularly decorated rooted forests with decoration set X which is defined in section 3. There exits a bijective map $\phi_{X^{\mathcal{F}},\mathcal{V}} : X^{\mathcal{F}} \to \mathcal{V}(X)$.*

Proof. See [5]for more details. □

We now transport the TD algebra structure from $\mathbf{k}X^{\mathcal{F}}$ in Eq. (10) to $\mathbf{k}\mathcal{V}(X)$ through the bijection $\phi_{X^{\mathcal{F}},\mathcal{V}}$ and its inverse $\phi_{\mathcal{V},X^{\mathcal{F}}}$ defined in.[5] Note that an indecomposable X-decorated Motzkin path is either $\bullet$ or $\bullet\!\!-\!\!\bullet$ or $/\bar{\mathfrak{m}}\backslash$ for another X-decorated Motzkin path $\bar{\mathfrak{m}}$.

Theorem 4.2. *The bijection $\phi_{X^{\mathcal{F}},\mathcal{V}} : X^{\mathcal{F}} \to \mathcal{V}(X)$ extends to an isomorphism*

$$\phi_{X^{\mathcal{F}},\mathcal{V}} : (\mathbf{k}X^{\mathcal{F}}, \bar{\diamond}, P_X) \to (\mathbf{k}\mathcal{V}(X), \diamond_v, \lfloor\ \rfloor)$$

of TD algebras where the multiplication $\diamond_v$ *on* $\mathbf{k}\mathcal{V}(X)$ *is defined recursively with respect to the height of Motzkin paths and is characterized by the following properties.*

(a) The trivial path $\bullet$ *is the multiplication identity;*
(b) If $\mathfrak{m}$ *and* $\mathfrak{m}'$ *are indecomposable* X*-decorated Motzkin paths not equal to* $\bullet$*, then*

$$\mathfrak{m}\diamond_v \mathfrak{m}' = \begin{cases} \mathfrak{m}\circ\mathfrak{m}', & \text{if } \mathfrak{m} = \bullet\!\overset{x}{-}\!\bullet \ \text{ or } \mathfrak{m}' = \bullet\!\overset{x'}{-}\!\bullet\,, \\ /\bar{\mathfrak{m}}\diamond_v \mathfrak{m}'\backslash + /\mathfrak{m}\diamond\bar{\mathfrak{m}}'\backslash & \\ -\,/\bar{\mathfrak{m}}\diamond_v \bullet\!\nearrow\!\bullet\!\searrow\!\bullet \diamond_v \bar{\mathfrak{m}}'\backslash\,, & \text{if } \mathfrak{m} = /\bar{\mathfrak{m}}\backslash\,, \mathfrak{m}' = /\bar{\mathfrak{m}}'\backslash\,; \end{cases} \tag{2}$$

(c) If $\mathfrak{m} = \mathfrak{m}_1\circ\cdots\circ\mathfrak{m}_p$ *and* $\mathfrak{m}' = \mathfrak{m}'_1\circ\cdots\circ\mathfrak{m}'_{p'}$ *are the decompositions of* $\mathfrak{m},\mathfrak{m}'\in\mathcal{V}(X)$ *into indecomposable paths, then*

$$\mathfrak{m}\diamond_v\mathfrak{m}' = \mathfrak{m}_1\circ\cdots\circ(\mathfrak{m}_p\diamond_v\mathfrak{m}'_1)\circ\cdots\circ\mathfrak{m}_{p'}. \tag{3}$$

Furthermore the map $j_X : X\to X^{\mathcal{F}}$ in Eq. (8) is translated to

$$j_X : X\to\mathcal{P}(X), \quad j_X(x) = \bullet\!\overset{x}{-}\!\bullet\,. \tag{4}$$

Then by Theorem 3.1 and Theorem 4.2 we have

Corollary 4.1. *The quadruple* $(\mathbf{k}\mathcal{V}(X),\diamond_v,/\ \backslash, j_X)$ *(resp.* $(\mathbf{k}(\mathcal{V}(X)\cap\mathcal{L}(X)),\diamond_v,/\ \backslash, j_X)$*) is the freeTD algebra (resp. free nonunitaryTD algebra) on* X*.*

References

1. R. Diestel, "Graph Theory", Third edition, Springer-Verlag, 2005.
2. K. Ebrahimi-Fard and L. Guo, Rota–Baxter algebras and dendriform algebras, *J. Pure and Appl. Algebra,* **212** (2008), 320-339, arXiv: math.RA/0503647.
3. K. Ebrahimi-Fard and L. Guo, Free Rota-Baxter algebras and rooted trees, to appear in *J. Algebra and Its Applications*, **7** (2008), no.2, 167-194.
4. K. Ebrahimi-Fard and P. Leroux, Generalized shuffles related to Nijenhuis and TD-algebras, *Comm. Alg.* **37** (2009) 3064-3094.
5. L. Guo, Operated semigroups, Motzkin paths and rooted trees, to appear in *J. Algebraic Combinatorics*, **29** (2009), no.1, 35-62.
6. L. Guo, W. Keigher, Baxter algebras and shuffle products, *Adv. Math.*, **150** (2000), 117-149.
7. L. Guo, W. Keigher, On differential Rota-Baxter algebras, *J. Pure and Appl. Algebra,* **212** (2008), 522-540, arXiv: math.RA/0703780.

8. L. Guo and B. Zhang, Renormalization of multiple zeta values, to appear in *J. Algebra*, **319** (2008), 3770-3809.
9. P. Leroux, Construction of Nijenhuis operators and dendriform trialgebras, *Int. J. Math. Math. Sci.* **49-52** (2006), 2595-2615.
10. J.-L. Loday and M. Ronco, Trialgebras and families of polytopes, in "Homotopy Theory: Relations with Algebraic Geometry, Group Cohomology, and Algebraic K-theory" Contemporary Mathematics, 346, (2004).
11. J.-L. Loday and M. Ronco, On the structure of cofree Hopf algebras, to appear in *J. reine angew. Math*, **592** (2006), 123-155.
12. T. Pirashvili, Sets with two associative operations, *C. E. J. M.* **2** (2003), p. 169-183.
13. E. W. Weisstein. "Tree." From MathWorld–A Wolfram Web Resource. http://mathworld.wolfram.com/Tree.html

Encyclopedia of Types of Algebras 2010

G. W. Zinbiel
Zinbiel Institute of Mathematics (France)
E-mail: gw.zinbiel@free.fr

This is a cornucopia of types of algebras with some of their properties from the operadic point of view.

Introduction

The following is a list of some types of algebras together with their properties under an operadic and homological point of view. In this version we restrict ourselves to types of algebras which are encoded by an algebraic operad, cf. [44].

We keep the information to one page per type and we provide one reference in full as Ariadne's thread (so that one can print only one page). More references are listed by the end of the paper. We work over a fixed field $\mathbb{K}$ though in many instances everything makes sense and holds over a commutative ground ring ($\mathbb{Z}$ for instance). The category of vector spaces over $\mathbb{K}$ is denoted by `Vect`. All tensor products are over $\mathbb{K}$ unless otherwise stated.

The items of a standard page (which is to be found at the end of this introduction) are as follows. Sometimes a given type appears under different names in the literature. The choice made in `Name` is, most of the time, the most common one (up to a few exceptions). The other possibilities appear under the item `Alternative`.

The presentation given in `Definit.` is the most common one (*Lie* excepted). When others are used in the literature they are given in `Alternative`. The item `oper.` gives the generating operations. The item `sym.` gives their symmetry properties, if any. The item `rel.` gives the relation(s). They are supposed to hold for any value of the variables $x, y, z, \ldots$. If, in the presentation, only binary operations appear, then the type is said to be *binary*. Analogously, there are *ternary, k-ary, multi-ary* types.

If, in the presentation, the relations involve only the composition of two operations at a time (hence 3 variables in the binary case, 5 variables in the ternary case), then the type is said to be *quadratic.*

For a given type of algebras $\mathcal{P}$ the category of $\mathcal{P}$-algebras is denoted by $\mathcal{P}$-alg. For each type there is defined a notion of *free algebra.* By definition the free algebra of type $\mathcal{P}$ over the vector space V is an algebra denoted by $\mathcal{P}(V)$ satisfying the following universal condition:

for any algebra A of type $\mathcal{P}$ and any linear map $\phi : V \to A$ there is a unique $\mathcal{P}$-algebra morphism $\tilde{\phi} : \mathcal{P}(V) \to A$ which lifts ϕ. In other words the forgetful functor $\mathcal{P}$-$\mathsf{alg} \longrightarrow \mathtt{Vect}$ admits a left adjoint $\mathcal{P} : \mathtt{Vect} \longrightarrow \mathcal{P}$-$\mathsf{alg}$. In all the cases mentioned here the relations involved in the presentation of the given type are (or can be made) multilinear. Hence the functor $\mathcal{P}(V)$ is of the form (at least in characteristic zero),

$$\mathcal{P}(V) = \bigoplus_{n\geq 1} \mathcal{P}(n) \otimes_{\mathbb{S}_n} V^{\otimes n} \ ,$$

where $\mathcal{P}(n)$ is some $\mathbb{S}_n$-module. The $\mathbb{S}_n$-module $\mathcal{P}(n)$ is called the space of n-ary operations since for any algebra A there is a map

$$\mathcal{P}(n) \otimes_{\mathbb{S}_n} A^{\otimes n} \to A.$$

The functor $\mathcal{P} : \mathtt{Vect} \to \mathtt{Vect}$ inherits a monoid structure from the properties of the free algebra. Hence there exist transformations of functors $\iota : \mathrm{Id} \to \mathcal{P}$ and $\gamma : \mathcal{P} \circ \mathcal{P} \to \mathcal{P}$ such that γ is associative and unital. The monoid $(\mathcal{P}, \gamma, \iota)$ is called a *symmetric operad.*

The symmetric operad $\mathcal{P}$ can also be described as a family of $\mathbb{S}_n$-modules $\mathcal{P}(n)$ together with maps

$$\gamma(i_1, \ldots, i_k) : \mathcal{P}(k) \otimes \mathcal{P}(i_1) \otimes \cdots \otimes \mathcal{P}(i_k) \longrightarrow \mathcal{P}(i_1 + \cdots + i_k)$$

satisfying some compatibility with the action of the symmetric group and satisfying the associativity property.

If $\mathbb{S}_n$ is acting freely on $\mathcal{P}(n)$, then $\mathcal{P}(n) = \mathcal{P}_n \otimes \mathbb{K}[\mathbb{S}_n]$ where $\mathcal{P}_n$ is some vector space, and $\mathbb{K}[\mathbb{S}_n]$ is the regular representation. If, moreover, the maps $\gamma(i_1, \ldots, i_n)$ are induced by maps

$$\gamma_{i_1,\ldots,i_k} : \mathcal{P}_k \otimes \mathcal{P}_{i_1} \otimes \cdots \otimes \mathcal{P}_{i_k} \longrightarrow \mathcal{P}_{i_1+\cdots+i_k},$$

then the operad $\mathcal{P}$ comes from a *nonsymmetric operad* (abbreviated ns operad), still denoted by $\mathcal{P}$ in general.

For more terminology and details about algebraic operads we refer to [46] or [44].

The generating series of the operad $\mathcal{P}$ is defined as

$$f^{\mathcal{P}}(t) := \sum_{n\geq 1} \frac{\dim \mathcal{P}(n)}{n!} \, t^n,$$

in the binary case. When dealing with a nonsymmetric operad it becomes

$$f^{\mathcal{P}}(t) := \sum_{n\geq 1} \dim \mathcal{P}_n \, t^n.$$

The Koszul duality theory of associative algebras has been extended to binary quadratic operads by Ginzburg and Kapranov, cf. [22], then to quadratic operads by Fresse, cf. [18]. A conceptual treatment of this theory, together with applications, is given in [44]. So, to any quadratic operad $\mathcal{P}$, there is associated a quadratic **Koszul dual operad** denoted $\mathcal{P}^!$. It is often a challenge to find a presentation of $\mathcal{P}^!$ out of a presentation of $\mathcal{P}$. One of the main results of the Koszul duality theory of operads is to show the existence of a natural differential map on the composite $\mathcal{P}^{!*} \circ \mathcal{P}$ giving rise to the *Koszul complex*. If it is acyclic, then $\mathcal{P}$ is said to be *Koszul*. One can show that, if $\mathcal{P}$ is Koszul, then so is $\mathcal{P}^!$. In this case the generating series are inverse to each other for composition, up to sign, that is:

$$f^{\mathcal{P}^!}(-f^{\mathcal{P}}(t)) = -t.$$

Recall that if $f^{\mathcal{P}^!}(t) = t + \sum_{n\geq 2} a_n t^n$ and $g^{\mathcal{P}}(t) = t + \sum_{n\geq 2} b_n t^n$, then

$$\begin{aligned}
b_2 &= a_2, \\
b_3 &= -a_3 + 2a_2^2, \\
b_4 &= a_4 - 5a_3a_2 + 5a_2^3, \\
b_5 &= -a_5 + 3a_3^2 + 6a_2a_4 - 21a_2^2a_3 + 14a_2^4.
\end{aligned}$$

In the k-ary case one introduces the skew-generating series

$$g^{\mathcal{P}}(t) := \sum_{n\geq 1} (-1)^k \frac{\dim \mathcal{P}((k-1)n+1)}{n!} \, t^{((k-1)n+1)}.$$

If the operad $\mathcal{P}$ is Koszul, then by [59] the following formula holds:

$$f^{\mathcal{P}^!}(-g^{\mathcal{P}}(t)) = -t.$$

The items `Free alg.`, `rep.` $\mathcal{P}(n)$ or $\mathcal{P}_n$, $\dim \mathcal{P}(n)$ or $\dim \mathcal{P}_n$, and `Gen.series` speak for themselves.

Koszulity of an operad implies the existence of a small chain complex to compute the (co)homology of a $\mathcal{P}$-algebra. When possible, the information on it is given in the item `Chain-cplx`. Moreover it permits us to construct the notion of $\mathcal{P}$-algebra up to homotopy, whose associated operad, which is a differential graded operad, is denoted by $\mathcal{P}_\infty$. The importance of this notion is due to the "Homotopy Transfer Theorem", see [44] section 10.3.

The item `Properties` lists the main features of the operad. *Set-theoretic* means that there is a set operad $\mathcal{P}_{Set}$ (monoid in the category of $\mathbb{S}$-Sets) such that $\mathcal{P} = \mathbb{K}[\mathcal{P}_{Set}]$. Usually this property can be read on the presentation of the operad: no algebraic sums. Quasi-regular means that $\mathcal{P}(n)$ is a sum of regular representations, but the operad is not necessarily coming from a ns operad.

In the item `Relationsh.` we list some of the ways to obtain this operad under some natural constructions like tensor product (Hadamard product) or Manin products (white ○ or black ●), denoted □ and ■ in the nonsymmetric framework, cf. [60] or [44] for instance. We also list some of the most common functors to other types of algebras. Keep in mind that a functor $\mathcal{P} \to \mathcal{Q}$ induces a functor $\mathcal{Q}\text{-alg} \to \mathcal{P}\text{-alg}$ on the categories of algebras.

Though we describe only algebras without unit, for some types there is a possibility of introducing an element 1 which is either a unit or a partial unit for some of the operations, see the discussion in [36]. We indicate it in the item `Unit`.

For binary operads the *opposite type* consists in defining new operations by $x \cdot y = yx$, etc. If the new type is isomorphic to the former one, then the operad is said to be *self-opposite*. When it is not the case, we mention whether the given type is called *right* or *left* in the item `Comment`.

In some cases the structure can be "integrated". For instance Lie algebras are integrated into Lie groups (Lie third problem). If so, we indicate it in the item `Comment`.

In the item `Ref.` we indicate a reference where information on the operad and/or on the (co)homology theory can be obtained. It is not necessarily the first paper in which this type of algebras first appeared. For the "three graces", that is the operads As, Com, Lie, the classical books by Cartan and Eilenberg "Homological Algebra" and by MacLane "Homology" are standard references.

Notation. We use the notation $\mathbb{S}_n$ for the symmetric group. Trees are very much in use in the description of operads. We use the following notation:

– PBT_n is the set of planar binary rooted trees with $n-1$ internal vertices (and hence n leaves). The number of elements in PBT_{n+1} is the Catalan number $c_n = \frac{1}{n+1}\binom{2n}{n}$.

–PT_n is the set of planar rooted trees with n leaves, whose vertices have valency greater than $1+2$ (one root, at least 2 inputs). So we have $PBT_n \subset PT_{n+1}$. The number of elements in PT_n is the super Catalan number, also called Schröder number, denoted C_n.

A planar binary rooted tree t is completely determined by its right part t^r and its left part t^l. More precisely t is the grafting of t^l and t^r: $t = t^l \vee t^r$.

Comments. Many thanks to Walter Moreira for setting up a software which computes the first dimensions of the operad from its presentation.

We remind the reader that we can replace the symmetric monoidal category `Vect` by many other symmetric monoidal categories. So there are notions of graded algebras, differential graded algebras, twisted algebras, and so forth. In the graded cases the Koszul sign rule is in order. Observe that there are also operads where the operations may have different degree (operad encoding Gerstenhaber algebras for instance).

We end this paper with a tableau of integer sequences appearing in this document.

This list of types of algebras is not as encyclopedic as the title suggests. We put only the types which are defined by a finite number of generating operations and whose relations are multilinear. You will not find the "restricted types" (like divided power algebras), nor bialgebras. Moreover we only put those which have been used some way or another. We plan to update this encyclopedia every now and then.

Please report any error or comment or possible addition to:

`gw.zinbiel@free.fr`

Here is the list of the types included so far (with letter K indicating that they are Koszul dual to each other):

sample	As	self-dual
Com	Lie	K
$Pois$	none	self-dual
$Leib$	$Zinb$	K
$Dend$	$Dias$	K
$PreLie$	$Perm$	K
$Dipt$	$Dipt^!$	K
$2as$	$2as^!$	K
$Tridend$	$Trias$	K
$PostLie$	$ComTrias$	K
CTD	$CTD^!$	K
$L\text{-}dend$	$Ennea$	
$Gerst$	BV	
Mag	Nil_2	K
$ComMag$	$ComMag^!$	K
$Quad$	$Quad^!$	K
Dup	$Dup^!$	K
$As^{(2)}$	$As^{\langle 2\rangle}$	
$Lie\text{-}adm$	$PreLiePerm$	
$Altern$	$Param1rel$	
$MagFine$	$GenMag$	
NAP	$Moufang$	
$Malcev$	$Novikov$	
$DoubleLie$	$DiPreLie$	
$Akivis$	$Sabinin$	
$Jordan\ triples$	$t\text{-}As^{(3)}$	
$p\text{-}As^{(3)}$	LTS	
$Lie\text{-}Yamaguti$	$Comtrans$	
$Interchange$	$HyperCom$	
A_∞	C_∞	
L_∞	$Dend_\infty$	
$\mathcal{P}_\infty$	$Brace$	
MB	$n\text{-}Lie$	
$n\text{-}Leib$	$\mathcal{X}^{\pm}$	
your own		

An index is to be found at the end of the paper.

A page for personal notes (in fact, to ensure that an operad and its Koszul dual fit on opposite pages).

1. Type of algebras

Name	**Most common terminology**
Notation	our favorite notation for the operad (generic notation: $\mathcal{P}$)
Def.oper.	list of the generating operations
sym.	their symmetry if any
rel.	the relation(s)
Free alg.	the free algebra as a functor in V
rep. $\mathcal{P}(n)$	$\mathbb{S}_n$-representation $\mathcal{P}(n)$ and/or the space $\mathcal{P}_n$ if nonsymmetric
$\dim \mathcal{P}(n)$	the series (if close formula available), the list of the 7 first numbers beginning at $n = 1$
Gen.series	close formula for $f^{\mathcal{P}}(t) = \sum_{n\geq 1} \frac{\dim \mathcal{P}(n)}{n!} t^n$ when available
Dual operad	the Koszul dual operad
Chain-cplx	Explicitation of the chain complex, if not too complicated
Properties	among: nonsymmetric, binary, quadratic, set-theoretic, ternary, multi-ary, cubic, Koszul.
Alternative	alternative terminology, and/or notation, and/or presentation
Relationsh.	some of the relationships with other operads, either under some construction like symmetrizing, Hadamard product, Manin products, or under the existence of functors
Unit	whether one can assume the existence of a unit (or partial unit)
Comment	whatever needs to be said which does not fit into the other items
Ref.	a reference, usually dealing with the homology of the $\mathcal{P}$-algebras (not necessarily containing all the results of this page)

`Name`	**Associative algebra**
`Notation`	As (as nonsymmetric operad) Ass (as symmetric operad)

`Def.oper.`
`sym.`
`rel.`
	xy, operadically: μ so that $\mu(x,y) = xy$
	
	$(xy)z = x(yz)$, operadically $\mu \circ_1 \mu = \mu \circ_2 \mu$ (associativity)

`Free alg.`	$As(V) = \overline{T}(V) = \oplus_{n\geq 1} V^{\otimes n}$ tensor algebra (noncommutative polynomials) $(x_1 \dots x_p)(x_{p+1} \dots x_{p+q}) = x_1 \dots x_{p+q}$ (concatenation)
`rep.` $\mathcal{P}(n)$	$Ass(n) = \mathbb{K}[\mathbb{S}_n]$ (regular representation), $As_n = \mathbb{K}$
$\dim \mathcal{P}(n)$	$1, 2, 6, 24, 120, 720, 5040, \dots, n!, \dots$
`Gen.series`	$f^{As}(t) = \frac{t}{1-t}$
`Dual operad`	$As^! = As$
`Chain-cplx`	non-unital Hochschild complex, $C_n^{As}(A) = A^{\otimes n}$ $b'(a_1, \dots, a_n) := \sum_{i=1}^{i=n-1} (-1)^{i-1}(a_1, \dots, a_i a_{i+1}, \dots, a_n)$ important variation: cyclic homology
`Properties`	ns, binary, quadratic, set-theoretic, Koszul, self-dual.
`Alternative`	associative algebra is often simply called *algebra*. Can be presented with commutative operation $x \cdot y := xy + yx$ and anti-symmetric operation $[x,y] = xy - yx$ satisfying $\begin{cases} [x \cdot y, z] = x \cdot [y,z] + [x,z] \cdot y, \\ (x \cdot y) \cdot z - x \cdot (y \cdot z) = [y,[x,z]] \,. \end{cases}$ (Livernet and Loday, unpublished)
`Relationsh.`	Ass-alg $\to$ Lie-alg, $[x,y] = xy - yx$, Com-alg $\to$ Ass-alg (inclusion), and many others
`Unit`	$1x = x = x1$
`Comment`	one the "three graces"

Name	**Commutative algebra**
Notation	*Com*
Def.oper.	xy
sym.	$xy = yx$ (commutativity)
rel.	$(xy)z = x(yz)$ (associativity)
Free alg.	$Com(V) = \overline{S}(V)$ (polynomials) if $V = \mathbb{K}x_1 \oplus \cdots \oplus \mathbb{K}x_n$, then $\mathbb{K}1 \oplus Com(V) = \mathbb{K}[x_1, \ldots, x_n]$
rep. $\mathcal{P}(n)$	$Com(n) = \mathbb{K}$ (trivial representation)
$\dim \mathcal{P}(n)$	$1, 1, 1, 1, 1, 1, 1, \ldots, 1, \ldots$
Gen.series	$f^{Com}(t) = \exp(t) - 1$
Dual operad	$Com^! = Lie$
Chain-cplx	Harrison complex in char. 0, André-Quillen cplx in general
Properties	binary, quadratic, set-theoretic, Koszul.
Alternative	sometimes called associative and commutative algebra other notation *Comm*
Relationsh.	Com-alg $\to$ Ass-alg, $Zinb$-alg $\to$ Com-alg
Unit	$1x = x = x1$
Comment	one of the “three graces”

Name	**Lie algebra**
Notation	*Lie*
Def.oper.	$[x,y]$ (bracket)
sym.	$[x,y] = -[y,x]$ (anti-symmetry)
rel.	$[[x,y],z] = [x,[y,z]] + [[x,z],y]$ (Leibniz relation)
Free alg.	$Lie(V)$ = subspace of the tensor algebra $T(V)$ generated by V under the bracket
rep. $\mathcal{P}(n)$	$Lie(n) = \mathrm{Ind}_{C_n}^{\mathbb{S}_n}(\sqrt[n]{1})$
$\dim \mathcal{P}(n)$	$1, 1, 2, 6, 24, 120, 720, \ldots, (n-1)!, \ldots$
Gen.series	$f^{Lie}(t) = -\log(1-t)$
Dual operad	$Lie^! = Com$
Chain-cplx	Chevalley-Eilenberg complex $C_n^{Lie}(\mathfrak{g}) = \Lambda^n \mathfrak{g}$ $d(x_1 \wedge \ldots \wedge x_n) :=$ $\sum_{1 \le i < j \le n} (-1)^j (x_1 \wedge \ldots \wedge [x_i, x_j] \wedge \ldots \wedge \widehat{x_j} \wedge \ldots \wedge x_n)$
Properties	binary, quadratic, Koszul.
Alternative	The relation is more commonly written as the Jacobi identity: $[x,[y,z]] + [y,[z,x]] + [z,[x,y]] = 0$
Relationsh.	Ass-alg $\to Lie$-alg, Lie-alg $\to Leib$-alg, $PreLie$-alg $\to Lie$-alg
Unit	no
Comment	one of the "three graces". Named after Sophus Lie. Integration: Lie groups.

Name	**Poisson algebra**
Notation	*Pois*
Def.oper.	$xy, \{x,y\}$
sym.	$xy = yx, \quad \{x,y\} = -\{y,x\}$
rel.	$\begin{cases} \{\{x,y\},z\} = \{x,\{y,z\}\} + \{\{x,z\},y\}, \\ \{xy,z\} = x\{y,z\} + \{x,z\}y, \\ (xy)z - x(yz) = 0 \ . \end{cases}$
Free alg.	$Pois(V) \cong \overline{T}(V)$ (tensor module, iso as Schur functors)
rep. $\mathcal{P}(n)$	$Pois(n) \cong \mathbb{K}[\mathbb{S}_n]$ (regular representation)
$\dim \mathcal{P}(n)$	$1, 2!, 3!, 4!, 5!, 6!, 7!, \ldots, n!, \ldots$
Gen.series	$f^{Pois}(t) = \frac{t}{1-t}$
Dual operad	$Pois^! = Pois$
Chain-cplx	Isomorphic to the total complex of a certain bicomplex constructed from the action of the Eulerian idempotents
Properties	binary, quadratic, quasi-regular, Koszul, self-dual.
Alternative	Can be presented with one operation $x * y$ with no symmetry satisfying the relation $(x*y)*z = x*(y*z) + \frac{1}{3}\big(+ x*(z*y) - z*(x*y) - y*(x*z) + y*(z*x)\big)$
Relationsh.	$Pois$-alg $\rightleftarrows$ Lie-alg, $Pois$-alg $\rightleftarrows$ Com-alg,
Unit	$1x = x = x1, [1,x] = 0 = [x,1]$
Comment	Named after Siméon Poisson.
Ref.	[18] B.Fresse, *Théorie des opérades de Koszul et homologie des algèbres de Poisson,* Ann. Math. Blaise Pascal 13 (2006), 237–312.

This page is inserted so that, in the following part, an operad and its dual appear on page $2n$ and $2n+1$ respectively. Since *Pois* is self-dual there is no point to write a page about its dual.

Let us take the opportunity to mention that if A is a $\mathcal{P}$-algebra and B is a $\mathcal{P}^!$-algebra, then the tensor product $A \otimes B$ inherits naturally a structure of Lie algebra. If $\mathcal{P}$ is nonsymmetric, then so is $\mathcal{P}^!$, and $A \otimes B$ is in fact an associative algebra.

In some cases (like the Leibniz case for instance), $A \otimes B$ is a pre-Lie algebra.

Name	**Leibniz algebra**
Notation	*Leib*
Def.oper.	$[x, y]$
sym.	
rel.	$[[x, y], z] = [x, [y, z]] + [[x, z], y]$ (Leibniz relation)
Free alg.	$Leib(V) \cong \overline{T}(V)$ (reduced tensor module, iso as Schur functors)
rep. $\mathcal{P}(n)$	$Leib(n) = \mathbb{K}[\mathbb{S}_n]$ (regular representation)
$\dim \mathcal{P}(n)$	1, 2!, 3!, 4!, 5!, 6!, 7!, ..., n!, ...
Gen.series	$f^{Leib}(t) = \frac{t}{1-t}$
Dual operad	$Leib^! = Zinb$
Chain-cplx	$C_n^{Leib}(\mathfrak{g}) = \mathfrak{g}^{\otimes n}$ $d(x_1, \ldots, x_n)$ $= \sum_{1 \leq i < j \leq n} (-1)^j (x_1, \ldots, [x_i, x_j], \ldots, \widehat{x_j}, \ldots, x_n)$
Properties	binary, quadratic, quasi-regular, Koszul.
Alternative	Sometimes improperly called Loday algebra.
Relationsh.	$Leib = Perm \bigcirc Lie$ (Manin white product), see [60] Lie-alg $\to$ $Leib$-alg, $Dias$-alg $\to$ $Leib$-alg, $Dend$-alg $\to$ $Leib$-alg
Unit	no
Comment	Named after G.W. Leibniz. This is the *left* Leibniz algebra. The opposite type is called *right* Leibniz algebra. Integration : "coquecigrues" ! see for instance [13]
Ref.	[33] J.-L. Loday, *Une version non commutative des algèbres de Lie: les algèbres de Leibniz.* Enseign. Math. (2) 39 (1993), no. 3-4, 269–293.

Name	**Zinbiel algebra**
Notation	$Zinb$
Def.oper.	$x \cdot y$
sym.	
rel.	$(x \cdot y) \cdot z = x \cdot (y \cdot z) + x \cdot (z \cdot y)$ (Zinbiel relation)
Free alg.	$Zinb(V) = \overline{T}(V), \quad \cdot =$ halfshuffle $x_1 \dots x_p \cdot x_{p+1} \dots x_{p+q} = x_1 \mathrm{sh}_{p-1,q}(x_2 \dots x_p, x_{p+1} \dots x_{p+q})$
rep. $\mathcal{P}(n)$	$Zinb(n) \cong \mathbb{K}[\mathbb{S}_n]$ (regular representation)
$\dim \mathcal{P}(n)$	1, 2!, 3!, 4!, 5!, 6!, 7!, ..., n!, ...
Gen.series	$f^{Zinb}(t) = \frac{t}{1-t}$
Dual operad	$Zinb^! = Leib$
Chain-cplx	known, see Ref.
Properties	binary, quadratic, quasi-regular, Koszul.
Alternative	$Zinb = ComDend$ (commutative dendriform algebra), previously called *dual Leibniz algebra.*
Relationsh.	$Zinb$-alg $\to$ Com-alg, $\quad xy = x \cdot y + y \cdot x$, $Zinb$-alg $\to$ $Dend$-alg, $\quad x \prec y = x \cdot y = y \succ x$ $Zinb = PreLie \bullet Com$, see [60].
Unit	$1 \cdot x = 0, \quad x \cdot 1 = x$
Comment	symmetrization of the dot product gives, not only a commutative alg., but in fact a *divided power algebra.* Named after G.W. Zinbiel. This is right Zinbiel algebra.
Ref.	[34] J.-L. Loday, *Cup-product for Leibniz cohomology and dual Leibniz algebras.* Math. Scand. 77 (1995), no. 2, 189–196.

Name	**Dendriform algebra**
Notation	*Dend*
Def.oper. sym.	$x \prec y,\ x \succ y$ (left and right operation)
rel.	$\begin{cases} (x \prec y) \prec z = x \prec (y \prec z) + x \prec (y \succ z), \\ (x \succ y) \prec z = x \succ (y \prec z), \\ (x \prec y) \succ z + (x \succ y) \succ z = x \succ (y \succ z). \end{cases}$
Free alg.	$Dend(V) = \bigoplus_{n\geq 1} \mathbb{K}[PBT_{n+1}] \otimes V^{\otimes n}$, for pb trees s and t: $s \prec t := s^l \vee (s^r * t)$, and $s \succ t := (s * t^l) \vee t^r$ where $x * y := x \prec y + x \succ y$.
$\mathcal{P}_n$	$Dend_n = \mathbb{K}[PBT_{n+1}]$
$\dim \mathcal{P}_n$	$1, 2, 5, 14, 42, 132, 429, \ldots, c_n, \ldots$ where $c_n = \frac{1}{n+1}\binom{2n}{n}$ is the Catalan number
Gen.series	$f^{Dend}(t) = \frac{1-2t-\sqrt{1-4t}}{2t} = y, \quad y^2 - (1-2t)y + t = 0$
Dual operad	$Dend^! = Dias$
Chain-cplx	Isomorphic to the total complex of a certain explicit bicomplex
Properties	ns, binary, quadratic, Koszul.
Alternative	Handy to introduce $x * y := x \prec y + x \succ y$ which is associative.
Relationsh.	$Dend$-alg $\to As$-alg, $\quad x * y := x \prec y + y \succ x$, $Zinb$-alg $\to Dend$-alg, $\quad x \prec y := x \cdot y =: y \succ x$ $Dend$-alg $\to PreLie$-alg, $\quad x \circ y := x \prec y - y \succ x$ $Dend$-alg $\to Brace$-alg, see Ronco [52] $Dend = PreLie \bullet As$, see [60]
Unit	$1 \prec x = 0,\ x \prec 1 = x, \quad 1 \succ x = x,\ x \succ 1 = 0.$
Comment	dendro = tree in greek. There exist many variations.
Ref.	[35] J.-L. Loday, *Dialgebras,* Springer Lecture Notes in Math. 1763 (2001), 7-66.

Name	**Diassociative algebra**
Notation	*Dias*
Def.oper. sym.	$x \dashv y,\ x \vdash y$ (left and right operation)
rel.	$\begin{cases} (x \dashv y) \dashv z = x \dashv (y \dashv z), \\ (x \dashv y) \dashv z = x \dashv (y \vdash z), \\ (x \vdash y) \dashv z = x \vdash (y \dashv z), \\ (x \dashv y) \vdash z = x \vdash (y \vdash z), \\ (x \vdash y) \vdash z = x \vdash (y \vdash z). \end{cases}$
Free alg.	$Dias(V) = \bigoplus_{n\geq 1}(\underbrace{V^{\otimes n} \oplus \cdots \oplus V^{\otimes n}}_{n \text{ copies}})$ noncommutative polynomials with one variable marked
$\mathcal{P}_n$	$Dias_n = \mathbb{K}^n$
$\dim \mathcal{P}_n$	$1, 2, 3, 4, 5, 6, 7, \ldots, n, \ldots,$
Gen.series	$f^{Dias}(t) = \frac{t}{(1-t)^2}$
Dual operad	$Dias^! = Dend$
Chain-cplx	see Ref.
Properties	ns, binary, quadratic, set-theoretic, Koszul.
Alternative	previously called associative dialgebras or dialgebras.
Relationsh.	$As\text{-alg} \to Dias\text{-alg}, \quad x \dashv y := xy =: x \vdash y$ $Dias\text{-alg} \to Leib\text{-alg}, \quad [x, y] := x \dashv y - x \vdash y$ $Dias = Perm \circ As = Perm \underset{\mathrm{H}}{\otimes} As$
Unit	Bar-unit: $x \dashv 1 = x = 1 \vdash x,\ 1 \dashv x = 0 = x \vdash 1$
Comment	
Ref.	[35] J.-L. Loday, *Dialgebras*, Springer Lecture Notes in Math. 1763 (2001), 7-66.

Name	**Pre-Lie algebra**
Notation	$PreLie$
Def.oper.	$\{x,y\}$
sym.	
rel.	$\{\{x,y\},z\} - \{x,\{y,z\}\} = \{\{x,z\},y\} - \{x,\{z,y\}\}$
Free alg.	$PreLie(V) = \{$rooted trees labeled by elements of $V\}$
rep. $\mathcal{P}(n)$	$PreLie(n) = \mathbb{K}[\{$rooted trees, vertices labeled by $1,\ldots,n\}]$
$\dim \mathcal{P}(n)$	$1, 2, 9, 64, 625, 1296, 117649, \ldots, n^{n-1}, \ldots$
Gen.series	$f^{PreLie}(t) = y$ which satisfies $y = t\ \exp(y)$
Dual operad	$PreLie^! = Perm$
Chain-cplx	see Ref.
Properties	binary, quadratic, Koszul.
Alternative	The relation is $as(x,y,z) = as(x,z,y)$.
Relationsh.	$PreLie$-alg $\to Lie$-alg, $[x,y] := \{x,y\} - \{y,x\}$ $Dend$-alg $\to PreLie$-alg, $\{x,y\} := x \prec y - y \succ x$ $Brace$-alg $\to PreLie$-alg, forgetful functor
Unit	$\{1,x\} = x = \{x,1\}$
Comment	This is *right* pre-Lie algebra, also called *right-symmetric algebra*, or *Vinberg algebra*. The opposite type is left symmetric, first appeared in [20,61]. Symmetric brace algebra equivalent to pre-Lie algebra [49]
Ref.	[11] F. Chapoton, M. Livernet, *Pre-Lie algebras and the rooted trees operad.* Internat. Math. Res. Notices 2001, no. 8, 395–408.

Name	**Perm algebra**
Notation	$Perm$
Def.oper.	xy
sym.	
rel.	$(xy)z = x(yz) = x(zy)$
Free alg.	$Perm(V) = V \otimes S(V)$
rep. $\mathcal{P}(n)$	$Perm(n) = \mathbb{K}^n$
$\dim \mathcal{P}(n)$	$1, 2, 3, 4, 5, 6, 7, \ldots, n, \ldots$
Gen.series	$f^{Perm}(t) = t \exp(t)$
Dual operad	$Perm^! = PreLie$
Chain-cplx	
Properties	binary, quadratic, set-theoretic, Koszul.
Alternative	$Perm = ComDias$
Relationsh.	$Perm$-alg $\to$ Ass-alg Com-alg $\to$ $Perm$-alg, NAP-alg $\to$ $Perm$-alg, $Perm$-alg $\to$ $Dias$-alg
Unit	no, unless it is a commutative algebra
Comment	
Ref.	[10] F. Chapoton, *Un endofoncteur de la catégorie des opérades.* Dialgebras & related operads, 105–110, Lect. Notes in Math., 1763, Springer, 2001.

Name	**Dipterous algebra**
Notation	*Dipt*
Def.oper.	$x * y,\ x \prec y$
sym.	
rel.	$\begin{cases} (x * y) * z = x * (y * z) & \text{(associativity)} \\ (x \prec y) \prec z = x \prec (y * z) & \text{(dipterous relation)} \end{cases}$
Free alg.	$Dipt(V) = \bigoplus_{n\geq 1}(\mathbb{K}[PT_n] \oplus \mathbb{K}[PT_n]) \otimes V^{\otimes n}$, for $n = 1$ the two copies of PT_1 are identified
$\mathcal{P}_n$	$Dipt_n = \mathbb{K}[PT_n] \oplus \mathbb{K}[PT_n], n \geq 2$
$\dim \mathcal{P}_n$	$1, 2, 6, 22, 90, 394, 1806, \ldots, 2C_n, \ldots$ where C_n is the Schröder number: $\sum_{n\geq 1} C_n t^n = \frac{1+t-\sqrt{1-6t+t^2}}{4}$
Gen.series	$f^{Dipt}(t) = \frac{1-t-\sqrt{1-6t+t^2}}{2}$
Dual operad	$Dipt^!$
Chain-cplx	Isomorphic to the total complex of a certain explicit bicomplex
Properties	ns, binary, quadratic, Koszul.
Alternative	“diptère” in French
Relationsh.	*Dend*-alg $\to$ *Dipt*-alg, $x * y := x \prec y + y \succ x$, a variation: replace the dipterous relation by $(x \prec y) * z + (x * y) \prec z = x \prec (y * z) + x * (y \prec z)$ to get *Hoch*-algebras, see [31]. Same properties.
Unit	$1 \prec x = 0,\ x \prec 1 = x, \quad 1 * x = x = x * 1.$
Comment	dipterous = 2-fold in greek (free algebra has two planar trees copies)
Ref.	[41] J.-L. Loday, M. Ronco, *Algèbres de Hopf colibres*, C. R. Math. Acad. Sci. Paris 337, Ser. I (2003), 153 -158.

Name	**Dual dipterous algebra**
Notation	$Dipt^!$
Def.oper.	$x \dashv y,\ x * y$
sym.	
rel.	$\begin{cases} (x*y)*z = x*(y*z), \\ (x \dashv y) \dashv z = x \dashv (y*z), \\ (x*y) \dashv z = 0, \\ (x \dashv y) * z = 0, \\ 0 = x*(y \dashv z) \\ 0 = x \dashv (y \dashv z). \end{cases}$
Free alg.	$Dipt^!(V) = \overline{T}(V) \oplus \overline{T}(V)$
$\mathcal{P}_n$	$Dipt^!_n = \mathbb{K}^2, n \geq 2$
$\dim \mathcal{P}_n$	$1, 2, 2, 2, 2, 2, 2, \ldots, 2, \ldots,$
Gen.series	$f^{Dipt^!}(t) = \frac{t+t^2}{1-t}$
Dual operad	$Dipt^{!!} = Dipt$
Chain-cplx	see Ref.
Properties	ns, binary, quadratic, Koszul.
Alternative	
Relationsh.	As-alg $\to Dipt^!$-alg, $\quad x \dashv y := xy =: x \vdash y$
Unit	no
Comment	
Ref.	[41] J.-L. Loday, M. Ronco, *Algèbres de Hopf colibres*, C. R. Math Acad. Sci. Paris 337, Ser. I (2003), 153 -158.

Name	**Two-associative algebra**
Notation	$2as$
Def.oper.	$x * y,\ x \cdot y,$
sym.	
rel.	$\begin{cases} (x * y) * z = x * (y * z), \\ (x \cdot y) \cdot z = x \cdot (y \cdot z). \end{cases}$
Free alg.	$2as(V) = \bigoplus_{n\geq 1}(\mathbb{K}[T_n] \oplus \mathbb{K}[T_n]) \otimes V^{\otimes n}$, where $T_n =$ planar trees for $n = 1$ the two copies of T_1 are identified
$\mathcal{P}_n$	$Dipt_n = \mathbb{K}[T_n] \oplus \mathbb{K}[T_n], n \geq 2$
$\dim \mathcal{P}_n$	$1, 2, 6, 22, 90, 394, 1806, \ldots, 2C_n, \ldots$ where C_n is the Schröder number: $\sum_{n\geq 1} C_n t^n = \frac{1+t-\sqrt{1-6t+t^2}}{4}$
Gen.series	$f^{2as}(t) = \frac{1-t-\sqrt{1-6t+t^2}}{2}$
Dual operad	$2as^!$
Chain-cplx	Isomorphic to the total complex of a certain explicit bicomplex
Properties	ns, binary, quadratic, set-theoretic, Koszul.
Alternative	
Relationsh.	$2as$-alg $\to$ Dup-alg, $2as$-alg $\to$ B_∞-alg
Unit	$1 \cdot x = x = x \cdot 1, \quad 1 * x = x = x * 1.$
Comment	
Ref.	[43] J.-L. Loday, M. Ronco, *On the structure of cofree Hopf algebras*, J. reine angew. Math. 592 (2006), 123–155.

Name	**Dual 2-associative algebra**
Notation	$2as^!$
Def.oper.	$x \cdot y,\ x * y$
sym.	
rel.	$\begin{cases} (x*y)*z = x*(y*z), \\ (x\cdot y)\cdot z = x\cdot(y\cdot z), \\ (x\cdot y)*z = 0, \\ (x*y)\cdot z = 0, \\ 0 = x*(y\cdot z) \\ 0 = x\cdot(y*z). \end{cases}$
Free alg.	$2as^!(V) = V \oplus \bigoplus_{n\geq 2} (V^{\otimes n} \oplus V^{\otimes n})$
$\mathcal{P}_n$	$Dipt^!_n = \mathbb{K} \oplus \mathbb{K}, n \geq 2.$
$\dim \mathcal{P}_n$	$1, 2, 2, 2, 2, 2, 2, \ldots, 2, \ldots$
Gen.series	$f^{2as^!}(t) = \frac{t+t^2}{1-t}$
Dual operad	$(2as^!)^! = 2as$
Chain-cplx	see Ref.
Properties	ns, binary, quadratic, Koszul.
Alternative	
Relationsh.	
Unit	no
Comment	
Ref.	[43] J.-L. Loday, M. Ronco, *On the structure of cofree Hopf algebras*, J. reine angew. Math. 592 (2006), 123–155.

`Name` **Tridendriform algebra**

`Notation` *Tridend*

`Def.oper.` $x \prec y,\ x \succ y,\ x \cdot y$

`sym.`

`rel.`

$$\begin{cases} (x \prec y) \prec z = x \prec (y * z)\ , \\ (x \succ y) \prec z = x \succ (y \prec z)\ , \\ (x * y) \succ z = x \succ (y \succ z)\ , \\ (x \succ y) \cdot z = x \succ (y \cdot z)\ , \\ (x \prec y) \cdot z = x \cdot (y \succ z)\ , \\ (x \cdot y) \prec z = x \cdot (y \prec z)\ , \\ (x \cdot y) \cdot z = x \cdot (y \cdot z)\ . \end{cases}$$

where $x * y := x \prec y + x \succ y + x \cdot y$.
One relation for each cell of the triangle.

`Free alg.` Planar rooted trees with variables in between the leaves

$\mathcal{P}_n$ $Tridend_n = \mathbb{K}[PT_n]$

$\dim \mathcal{P}_n$ $1, 3, 11, 45, 197, 903, \ldots, C_n, \ldots$
where C_n is the Schröder (or super Catalan) number

`Gen.series` $f^{Tridend}(t) = \frac{-1+3t+\sqrt{1-6t+t^2}}{4t}$

`Dual operad` $Triend^! = Trias$

`Chain-cplx` Isomorphic to the total complex of a certain explicit tricomplex

`Properties` ns, binary, quadratic, Koszul.

`Alternative` sometimes called dendriform trialgebra

`Relationsh.` $Tridend\text{-alg} \to As\text{-alg}, \quad x * y := x \prec y + y \succ x + x \cdot y,$
$ComTridend\text{-alg} \to Tridend\text{-alg}, \quad x \prec y := x \cdot y =: y \succ x$
$Tridend = PostLie \bullet Ass$

`Unit` $1 \prec x = 0,\ x \prec 1 = x, \quad 1 \succ x = x,\ x \succ 1 = 0,$
$1 \cdot x = 0 = x \cdot 1.$

`Comment` There exist several variations (see [12] for instance).

`Ref.` [42] J.-L. Loday and M. Ronco, *Trialgebras and families of polytopes*, Contemp. Math. (AMS) 346 (2004), 369–398.

Name	**Triassociative algebra**
Notation	$Trias$
Def.oper.	$x \dashv y,\ x \vdash y,\ x \perp y$ (no symmetry)
rel.	$\begin{cases} (x \dashv y) \dashv z = x \dashv (y \dashv z), \\ (x \dashv y) \dashv z = x \dashv (y \vdash z), \\ (x \vdash y) \dashv z = x \vdash (y \dashv z), \\ (x \dashv y) \vdash z = x \vdash (y \vdash z), \\ (x \vdash y) \vdash z = x \vdash (y \vdash z), \\ (x \dashv y) \dashv z = x \dashv (y \perp z), \\ (x \perp y) \dashv z = x \perp (y \dashv z), \\ (x \dashv y) \perp z = x \perp (y \vdash z), \\ (x \vdash y) \perp z = x \vdash (y \perp z), \\ (x \perp y) \vdash z = x \vdash (y \vdash z), \\ (x \perp y) \perp z = x \perp (y \perp z). \end{cases}$ one relation for each cell of the pentagon.
Free alg.	noncommutative polynomials with several variables marked
$\mathcal{P}_n$	$Trias_n = \mathbb{K}^{2^n - 1}$
$\dim \mathcal{P}_n$	$1, 3, 7, 15, 31, 63, 127, \ldots, (2^n - 1), \ldots$
Gen.series	$f^{Trias}(t) = \frac{t}{(1-t)(1-2t)}$
Dual operad	$Trias^! = Tridend$
Properties	ns, binary, quadratic, set-theoretic, Koszul.
Alternative	Also called *associative trialgebra*, or for short, *trialgebra*.
Relationsh.	$As\text{-alg} \to Trias\text{-alg}, \quad x \dashv y = x \vdash y = x \perp y = xy$
Unit	Bar-unit: $x \dashv 1 = x = 1 \vdash x,\ 1 \dashv x = 0 = x \vdash 1,$ $1 \perp x = 0 = x \perp 1$
Comment	Relations easy to understand in terms of planar trees
Ref.	[42] J.-L. Loday and M. Ronco, *Trialgebras and families of polytopes*, Contemp. Math. (AMS) 346 (2004), 369–398.

Name	**PostLie algebra**
Notation	$PostLie$
Def.oper.	$x \circ y,\ [x, y]$
sym.	$[x, y] = -[y, x]$
rel.	$[x, [y, z]] + [y, [z, x]] + [z, [x, y]] = 0$ $(x \circ y) \circ z - x \circ (y \circ z) - (x \circ z) \circ y + x \circ (z \circ y) = x \circ [y, z]$ $[x, y] \circ z = [x \circ z, y] + [x, y \circ z]$
Free alg.	$PostLie(V) \cong Lie(Mag(V))$
rep. $\mathcal{P}(n)$	
$\dim \mathcal{P}(n)$	$1, 3, 20, 210, 3024, \ldots$
Gen.series	$f(t) = -\log\left(\frac{1+\sqrt{1-4t}}{2}\right)$
Dual operad	$PostLie^{!} = ComTrias$
Chain-cplx	See Ref.
Properties	binary, quadratic, Koszul.
Alternative	
Relationsh.	$PostLie$-alg $\to$??-alg, $xy = x \circ y$??? $PostLie$-alg $\to Lie$-alg, $\{x, y\} = x \circ y - y \circ x + [x, y]$ $PreLie$-alg $\to PostLie$-alg, $[x, y] = 0$
Unit	
Comment	
Ref.	[59] Vallette B., *Homology of generalized partition posets*, J. Pure Appl. Algebra 208 (2007), no. 2, 699–725.

Name	**Commutative triassociative algebra**
Notation	$ComTrias$
Def.oper.	$x \dashv y, x \perp y$
sym.	$x \perp y = y \perp x$
rel.	$\begin{cases} (x \dashv y) \dashv z = x \dashv (y \dashv z), \\ (x \dashv y) \dashv z = x \dashv (z \dashv y), \\ (x \dashv y) \dashv z = x \dashv (y \perp z), \\ (x \perp y) \dashv z = x \perp (y \dashv z), \\ (x \perp y) \perp z = x \perp (y \perp z). \end{cases}$
Free alg.	
rep. $\mathcal{P}(n)$	
$\dim \mathcal{P}(n)$	
Gen.series	$f(t) =$
Dual operad	$ComTrias^! = PostLie$
Chain-cplx	
Properties	binary, quadratic, Koszul.
Alternative	Triassociative with the following symmetry: $x \dashv y = y \vdash x$ and $x \perp y = y \perp x$
Relationsh.	$ComTrias$-alg $\to$ $Perm$-alg, forgetful functor $ComTrias$-alg $\to$ $Trias$-alg
Unit	$x \dashv 1 = x, 1 \dashv x = 0, 1 \perp x = 0$
Comment	
Ref.	[59] Vallette B., *Homology of generalized partition posets*, J. Pure Appl. Algebra 208 (2007), no. 2, 699–725.

Name	**Commutative tridendriform algebra**
Notation	CTD
Def.oper.	$x \prec y,\ x \cdot y$
sym.	$x \cdot y = y \cdot x$
rel.	$\begin{cases} (x \prec y) \prec z = x \prec (y \prec z) + x \prec (z \prec y) \\ \qquad\qquad + x \prec (y \cdot z), \\ (x \cdot y) \prec z = x \cdot (y \prec z)\ , \\ (x \prec z) \cdot y = x \cdot (y \prec z)\ , \\ (x \prec z) \cdot y = (x \prec z) \cdot y\ , \\ (x \cdot y) \cdot z = x \cdot (y \cdot z)\ . \end{cases}$
Free alg.	$CTD(V)$ =quasi-shuffle algebra on V $= QSym(V)$
rep. $\mathcal{P}(n)$	
$\dim \mathcal{P}(n)$	$1, 3, 13, 75, 541, 4683, \ldots$
Gen.series	$f^{CTD}(t) = \frac{\exp(t)-1}{2-\exp(t)}$
Dual operad	$CTD^! =$ see next page
Chain-cplx	
Properties	binary, quadratic, Koszul.
Alternative	Handy to introduce $x * y := x \prec y + y \prec x + x \cdot y$ (assoc. and comm.) equivalently: tridendriform with symmetry: $x \prec y = y \succ x,\ x \cdot y = y \cdot x$
Relationsh.	CTD-alg $\to$ $Trident$-alg CTD-alg $\to$ Com-alg CTD-alg $\to$ $Zinb$-alg
Unit	$1 \prec x = 0,\ x \prec 1 = x, \quad 1 \cdot x = 0 = x \cdot 1.$
Comment	
Ref.	[39] J.-L. Loday, *On the algebra of quasi-shuffles*, Manuscripta Mathematica 123 (1), (2007), 79–93.

Name	**Dual CTD algebra**
Notation	$CTD^!$
Def.oper.	$x \dashv y,\ [x,y]$
sym.	$[x,y] = -[y,x]$
rel.	to be done
Free alg.	
rep. $\mathcal{P}(n)$	
$\dim \mathcal{P}(n)$	$1, 3, 14, 90, 744, \ldots$
Gen.series	$f^{CTD^!}(t) =$
Dual operad	$(CTD^!)^! = CTD$
Chain-cplx	
Properties	binary, quadratic, Koszul.
Alternative	
Relationsh.	$Trias$-alg $\to CTD^!$-alg Lie-alg $\to CTD^!$-alg $Leib$-alg $\to CTD^!$-alg
Unit	
Comment	
Ref.	

Name	**L-dendriform algebra**
Notation	*L-dend*
Def.oper.	$x \triangleright y$ and $x \triangleleft y$
sym.	
rel.	$x \triangleright (y \triangleright z) - (x \bullet y) \triangleright z = y \triangleright (x \triangleright z) - (y \bullet x) \triangleright z$ $x \triangleright (y \triangleleft z) - (x \triangleright y) \triangleleft z = y \triangleleft (x \bullet z) - (y \triangleleft x) \triangleleft z$ where $x \bullet y := x \triangleright y + x \triangleleft y$.
Free alg.	
rep. $\mathcal{P}(n)$	
$\dim \mathcal{P}(n)$	
Gen.series	$f(t) =$
Dual operad	
Chain-cplx	
Properties	$L\text{-}dend = PreLie \bullet PreLie$
Alternative	
Relationsh.	$Dend$-alg $\to$ L-$dend$-alg via $x \triangleright y := x \succ y$, $x \triangleleft y := x \prec y$ L-$dend$-alg $\to$ $PreLie$-alg, $(A, \triangleright, \triangleleft) \mapsto (A, \bullet)$ $preLie \bullet preLie = L\text{-}dend$
Unit	
Comment	Various variations like L-$quad$-alg, see [4,5]
Ref.	[3] C. Bai, L. Liu, X. Ni, *Some results on L-dendriform algebras*, J. Geom. Phys. 60 (2010), no. 6-8, 940–950.

`Name`	**Ennea algebra**
`Notation`	*Ennea*
`Def.oper.`	9 binary operations
`sym.`	
`rel.`	49 relations, see Ref. (tridendriform splitting of *As* applied twice)
`Free alg.`	
$\mathcal{P}_n$	
$\dim \mathcal{P}_n$	$1, 9, 113, ?, ?, \ldots$
`Gen.series`	$f(t) =$
`Dual operad`	
`Chain-cplx`	
`Properties`	ns, binary, quadratic.
`Alternative`	
`Relationsh.`	several relations with tridendriform, for instance: $Ennea = Tridend \blacksquare Tridend$
`Unit`	
`Comment`	
`Ref.`	[30] Leroux, P., *Ennea-algebras.* J. Algebra 281 (2004), no. 1, 287–302.

Name	**Gerstenhaber algebra**
	underlying objects: graded vector spaces
Notation	*Gerst*
Def.oper.	$m =$ binary operation of degree 0, $c =$ binary operation of degree 1
sym.	m symmetric, c antisymmetric
rel.	$c \circ_1 c + (c \circ_1 c)^{(123)} + (c \circ_1 c)^{(321)} = 0,$ $c \circ_1 m - m \circ_2 c - (m \circ_1 c)^{(23)} = 0,$ $m \circ_1 m - m \circ_2 m = 0\ .$
Free alg.	
rep. $\mathcal{P}(n)$	
$\dim \mathcal{P}(n)$	
Gen.series	$f(t) =$
Dual operad	
Chain-cplx	
Properties	binary, quadratic, Koszul.
Alternative	
Relationsh.	
Unit	
Comment	To get the relations in terms of elements, do not forget to apply the Koszul sign rule. Last relation is associativity of m.
Ref.	[20] M. Gerstenhaber, *The cohomology structure of an associative ring*, Ann. of Math. (2), 78 (1963), 267–288.

Name **Batalin-Vilkovisky algebra**

underlying objects: graded vector spaces

Notation BV

Def.oper. Δ unary degree 1, m binary degree 0, c binary degree 1

sym. m symmetric, c antisymmetric

rel.

$$m \circ_1 m - m \circ_2 m = 0,$$
$$\Delta^2 = 0,$$
$$c = \Delta \circ_1 m + m \circ_1 \Delta + m \circ_2 \Delta,$$
$$c \circ_1 c + (c \circ_1 c)^{(123)} + (c \circ_1 c)^{(321)} = 0,$$
$$c \circ_1 m - m \circ_2 c - (m \circ_1 c)^{(23)} = 0,$$
$$\Delta \circ_1 c + c \circ_1 \Delta + c \circ_2 \Delta = 0 \ .$$

Free alg.

rep. $\mathcal{P}(n)$

$\dim \mathcal{P}(n)$

Gen.series $f(t) =$

Dual operad

Chain-cplx

Properties unary and binary, inhomogeneous quadratic, Koszul.

Alternative Generated by Δ and m only.

Relationsh. BV-alg $\to$ $Gerst$-alg

Unit

Comment To get the relations in terms of elements, do not forget to apply the Koszul sign rule.

Ref. [29] J.-L. Koszul, *Crochet de Schouten-Nijenhuis et cohomologie*, Astérisque (1985), Numéro Hors Série, 257–271. See also [19].

Name	**Magmatic algebra**
Notation	Mag
Def.oper.	xy
sym.	
rel.	
Free alg.	basis: any parenthesizing of words, or planar binary trees, with product: $st = s \vee t$
$\mathcal{P}_n$	$Mag_n = \mathbb{K}[PBT_n]$ (planar binary trees with n leaves)
$\dim \mathcal{P}_n$	$1, 1, 2, 5, 14, 42, 132, \ldots, c_{n-1}, \ldots$ where $c_n = \frac{1}{n+1}\binom{2n}{n}$ (Catalan number)
Gen.series	$f^{Mag}(t) = (1/2)(1 - \sqrt{1-4t})$
Dual operad	$Mag^! = Nil_2$
Chain-cplx	
Properties	ns, binary, quadratic, set-theoretic, Koszul.
Alternative	most often called *nonassociative algebra.*
Relationsh.	many "inclusions" (all types of alg. with only one Gen. op.)
Unit	$1x = x = x1$
Comment	
Ref.	

`Name`	**2-Nilpotent algebra**
`Notation`	Nil_2
`Def.oper.`	xy
`sym.`	
`rel.`	$(xy)z = 0 = x(yz)$
`Free alg.`	$Nil_2(V) = V \oplus V^{\otimes 2}$
$\mathcal{P}_n$	$(Nil_2)_2 = \mathbb{K},\ (Nil_2)_n = 0$ for $n \geq 3$
$\dim \mathcal{P}_n$	$1, 1, 0, 0, 0, 0, 0, \ldots$
`Gen.series`	$f^{\mathcal{P}}(t) = t + t^2$
`Dual operad`	${Nil_2}^! = Mag$
`Chain-cplx`	
`Properties`	ns, binary, quadratic, Koszul.
`Alternative`	
`Relationsh.`	
`Unit`	no
`Comment`	
`Ref.`	

Name	**Commutative magmatic algebra**
Notation	$ComMag$
Def.oper.	$x \cdot y$
sym.	$x \cdot y = y \cdot x$
rel.	none
Free alg.	any parenthesizing of commutative words
rep. $\mathcal{P}(n)$	$ComMag(n) = \mathbb{K}[shBT(n)]$ $shBT(n) = \{\text{shuffle binary trees with } n \text{ leaves}\}$
$\dim \mathcal{P}(n)$	$1, 1, 3, 15, 105, \ldots, (2n-3)!!, \ldots,$ $\dim ComMag(n) = (2n-3)!! = 1 \times 3 \times \cdots \times (2n-3)$
Gen.series	$f^{ComMag}(t) = 1 - \sqrt{1-2t}$
Dual operad	$ComMag^!$-alg: $[x, y]$ antisymmetric, $[[x, y], z] = 0$.
Chain-cplx	
Properties	binary, quadratic, set-theoretic, Koszul.
Alternative	
Relationsh.	$ComMag \rightarrowtail PreLie, x \cdot y := \{x, y\} + \{y, x\}$, cf. [6]
Unit	$1x = x = x1$
Comment	
Ref.	

Name	**Anti-symmetric nilpotent algebra**
Notation	$ComMag^!$
Def.oper.	$[x,y]$
sym.	$[x,y] = -[y,x]$
rel.	$[[x,y],z] = 0$
Free alg.	$V \oplus \Lambda^2 V$
rep. $\mathcal{P}(n)$	
$\dim \mathcal{P}(n)$	$1, 1, 0, \ldots, 0, \ldots$
Gen.series	$f^{ComMag^!}(t) = t + \frac{t^2}{2}$
Dual operad	$ComMag$
Chain-cplx	
Properties	binary, quadratic, set-theoretic, Koszul.
Alternative	
Relationsh.	Lie-alg $\to ComMag^!$-alg
Unit	no
Comment	
Ref.	

Name **Quadri-algebra**

Notation *Quad*

Def.oper. rel. $x \nwarrow y,\ x \nearrow y,\ x \searrow y,\ x \swarrow y$ called NW, NE, SE, SW oper.

$$(x \nwarrow y) \nwarrow z = x \nwarrow (y \star z) \quad (x \nearrow y) \nwarrow z = x \nearrow (y \prec z) \quad (x \wedge y) \nearrow z = x \nearrow (y \succ z)$$

$$(x \swarrow y) \nwarrow z = x \swarrow (y \wedge z) \quad (x \searrow y) \nwarrow z = x \searrow (y \nwarrow z) \quad (x \vee y) \nearrow z = x \searrow (y \nearrow z)$$

$$(x \prec y) \swarrow z = x \swarrow (y \vee z) \quad (x \succ y) \swarrow z = x \searrow (y \swarrow z) \quad (x \star y) \searrow z = x \searrow (y \searrow z)$$

where

$$x \succ y := x \nearrow y + x \searrow y, \quad x \prec y := x \nwarrow y + x \swarrow y$$
$$x \vee y := x \searrow y + x \swarrow y, \quad x \wedge y := x \nearrow y + x \nwarrow y$$
$$x \star y := x \searrow y + x \nearrow y + x \nwarrow y + x \swarrow y$$
$$= x \succ y + x \prec y = x \vee y + x \wedge y$$

Free alg.

$\mathcal{P}_n$

$\dim \mathcal{P}_n$ 1, 4, 23, 156, 1162, 9192, ...

$$\dim \mathcal{P}_n = \frac{1}{n} \sum_{j=n}^{2n-1} \binom{3n}{n+1+j} \binom{j-1}{j-n}$$

Gen.series $f^{Quad}(t) =$

Dual operad $Quad^!$

Properties ns, binary, quadratic, Koszul.

Alternative $Quad = Dend \blacksquare Dend = PreLie \bullet Dend$

Relationsh. Related to dendriform in several ways

Unit partial unit (like in dendriform)

Comment There exist several variations like $\mathcal{P} \blacksquare \mathcal{P}$ (cf. [4,30])

Ref. [1] M. Aguiar, J.-L. Loday, *Quadri-algebras*, J. Pure Applied Algebra 191 (2004), 205–221.

Name	**Dual quadri-algebra**
Notation	$Quad^!$
Def.oper.	$x \nwarrow y,\ x \nearrow y,\ x \searrow y,\ x \swarrow y$
sym.	
rel.	To be done
Free alg.	
$\mathcal{P}_n$	
$\dim \mathcal{P}_n$	$1, 4, 9, 16, 25, \ldots, n^2, \ldots$
Gen.series	$f^{Quad^!}(t) = \frac{t(1+t)}{(1-t)^3}$
Dual operad	
Chain-cplx	
Properties	ns, binary, quadratic, Koszul.
Alternative	
Relationsh.	$Quad^! = Dias \,\square\, Dias = Perm \circ Dias$ (cf. [60])
Unit	
Comment	
Ref.	[60] B.Vallette, *Manin products, Koszul duality, Loday algebras and Deligne conjecture.* J. Reine Angew. Math. 620 (2008), 105–164.

Name	**Duplicial algebra**
Notation	*Dup*
Def.oper.	$x \prec y,\ x \succ y$
sym.	
rel.	$\begin{cases}(x \prec y) \prec z = x \prec (y \prec z),\\ (x \succ y) \prec z = x \succ (y \prec z),\\ (x \succ y) \succ z = x \succ (y \succ z).\end{cases}$
Free alg.	$Dup(V) = \oplus_{n\geq 1}\mathbb{K}[PBT_{n+1}] \otimes V^{\otimes n}$, for p.b. trees s and t: $x \succ y$ the *over* operation is grafting of x on the leftmost leaf of y; $x \prec y$ the *under* operation is grafting of y on the rightmost leaf of x
$\mathcal{P}_n$	$Dup_n = \mathbb{K}[PBT_{n+1}]$
$\dim \mathcal{P}_n$	$1, 2, 5, 14, 42, 132, 429, \ldots, c_n, \ldots$ where $c_n = \frac{1}{n+1}\binom{2n}{n}$ is the Catalan number
Gen.series	$f^{Dup}(t) = \frac{1-2t-\sqrt{1-4t}}{2t} = y, \quad y^2 - (1-2t)y + t = 0$
Dual operad	$Dup^!$
Chain-cplx	Isomorphic to the total complex of a certain explicit bicomplex
Properties	ns, binary, quadratic, set-theoretic, Koszul.
Alternative	
Relationsh.	$2as$-alg $\to$ Dup-alg $\to$ As^2-alg
Unit	
Comment	The associator of $xy := x \succ y - x \prec y$ is $as(x, y, z) = x \prec (y \succ z) - (x \prec y) \succ z$. Appeared first in [8]
Ref.	[40] J.-L. Loday, *Generalized bialgebras and triples of operads*, Astérisque (2008), no 320, x+116 p.

Name **Dual duplicial algebra**

Notation $Dup^!$

Def.oper. $x \prec y,\ x \succ y$
sym.

rel.
$$\begin{cases} (x \prec y) \prec z = x \prec (y \prec z), \\ (x \prec y) \succ z = 0, \\ (x \succ y) \prec z = x \succ (y \prec z), \\ \qquad\qquad 0 = x \prec (y \succ z), \\ (x \succ y) \succ z = x \succ (y \succ z). \end{cases}$$

Free alg. $Dup^!(V) = \oplus_{n\geq 1}\, n\, V^{\otimes n}$
noncommutative polynomials with one variable marked

$\mathcal{P}_n$ $Dup^!_n = \mathbb{K}^n$

$\dim \mathcal{P}_n$ $1, 2, 3, 4, 5, 6, 7, \ldots, n, \ldots,$

Gen.series $f^{Dup^!}(t) = \frac{t}{(1-t)^2}$

Dual operad $Dup^{!!} = Dup$

Chain-cplx see ref.

Properties ns, binary, quadratic, Koszul.

Alternative

Relationsh.
Unit

Comment

Ref.

Name	**$As^{(2)}$-algebra**
Notation	$As^{(2)}$
Def.oper.	$x * y, x \cdot y$
sym.	
rel.	$(x \circ_1 y) \circ_2 z = x \circ_1 (y \circ_2 z)$ for $\circ_i = *$ or $\cdot$ (4 relations)
Free alg.	
$\mathcal{P}_n$	$As^{(2)}_n = \mathbb{K}[\{0,1\}^{n-1}]$
$\dim \mathcal{P}_n$	2^{n-1}
Gen.series	$f^{As^{(2)}}(t) = \frac{t}{1-2t}$
Dual operad	$As^{(2)!} = As^{(2)}$
Chain-cplx	
Properties	ns, binary, quadratic, set-theoretic.
Alternative	
Relationsh.	$2as$-alg $\to$ Dup-alg $\to$ $As^{(2)}$-alg
Unit	
Comment	Variations: $As^{(k)}$
Ref.	

`Name` **$As^{\langle 2\rangle}$-algebra**

`Notation` $As^{\langle 2\rangle}$, compatible products algebra

`Def.oper.` $x * y, x \cdot y$

`sym.`

`rel.`
$$(x * y) * z = x * (y * z)$$
$$(x * y) \cdot z + (x \cdot y) * z = x * (y \cdot z) + x \cdot (y * z)$$
$$(x \cdot y) \cdot z = x \cdot (y \cdot z)$$

`Free alg.` similar to dendriform and duplicial

`rep.` $\mathcal{P}_n$ $As_n^{\langle 2\rangle} = \mathbb{K}[PBT_{n+1}]$

$\dim \mathcal{P}_n$ $c_n = \frac{1}{n+1}\binom{2n}{n}$

`Gen.series` $f^{As^{\langle 2\rangle}}(t) = \frac{1-2t-\sqrt{1-4t}}{2t}$

`Dual operad` An $(As^{\langle 2\rangle})^!$-algebra has 2 generating binary operations and 5 relations:
$(x \circ_i y) \circ_j z = x \circ_i (y \circ_j z)$ and
$x \circ_1 (y \circ_2 z) = x \circ_2 (y \circ_1 z)$.

`Chain-cplx`

`Properties` ns, binary, quadratic, Koszul.

`Alternative`

`Relationsh.` $(As^{\langle 2\rangle})^!$-alg $\to As^{(2)}$-alg $\to As^{\langle 2\rangle}$-alg

`Unit`

`Comment` equivalently $\lambda\, x * y + \mu\, x \cdot y$ is associative for any λ, μ
Variations: $As^{\langle k\rangle}$, $Hoch$-alg.
Do not confuse $As^{(k)}$ and $As^{\langle k\rangle}$.

`Ref.` [15] Dotsenko, V., *Compatible associative products and trees.* Algebra Number Theory 3 (2009), no. 5, 567–586.
See also [24], [48].

Name	**Lie-admissible algebra**
Notation	*Lie-adm*
Def.oper.	xy
sym.	
rel.	$[x,y] = xy - yx$ is a Lie bracket, that is $\sum_\sigma \mathrm{sgn}(\sigma)\sigma\big((xy)z - x(yz)\big) = 0$
Free alg.	
$\mathcal{P}_n$	$Lie\text{-}adm(n) = ?$
$\dim \mathcal{P}_n$	$1, 2, 11, ?, ?, \ldots$
Gen.series	$f^{Lie\text{-}adm}(t) = ?$
Dual operad	$Lie\text{-}adm^!$
Chain-cplx	
Properties	ns, binary, quadratic, Koszul ??.
Alternative	
Relationsh.	As-alg $\to$ $PreLie$-alg $\to$ $Lie\text{-}adm$-alg
Unit	
Comment	
Ref.	

Name **PreLiePerm algebra**

Notation $PreLiePerm$

Def.oper. $x \prec y,\ x \succ y,\ x * y$

sym.

rel.
$$\begin{cases} (x \prec y) \prec z = x \prec (y * z), \\ (x \prec y) \prec z = x \prec (z * y), \\ (x \succ y) \prec z = x \succ (y \prec z), \\ (x \succ y) \prec z = x \succ (z \succ y), \\ (x * y) \succ z = x \succ (y \succ z), \end{cases}$$

Free alg.

rep. $\mathcal{P}(n)$

$\dim \mathcal{P}(n)$ $1, 6,$

Gen.series

Dual operad $PermPreLie^! = Perm \circ PreLie = Perm \otimes PreLie$

Chain-cplx

Properties binary, quadratic, set-theoretic.

Alternative

Relationsh. $PreLiePerm = PreLie \bullet Perm$, see [60], p. 132.
$Zinb$-alg $\to$ $PreLiePerm$-alg $\to$ $Dend$-alg

Unit no

Comment

Ref. [58] B.Vallette, *Homology of generalized partition posets,* J. Pure Appl. Algebra 208 (2007), no. 2, 69–725.

Name	**Alternative algebra**
Notation	$Altern$
Def.oper.	xy
sym.	
rel.	$(xy)z - x(yz) = -(yx)z + y(xz)$, $(xy)z - x(yz) = -(xz)y + x(zy)$,
Free alg.	
rep. $\mathcal{P}(n)$	
$\dim \mathcal{P}(n)$	$1, 2, 7, 32, 175, ??$
Gen.series	$f(t) =$
Dual operad	$(xy)z = x(yz)$, $xyz + yxz + zxy + xzy + yzx + zyx = 0$ $\dim Altern^!(n) = 1, 2, 5, 12, 15, \ldots$
Chain-cplx	
Properties	binary, quadratic, nonKoszul [16].
Alternative	Equivalent presentation: the associator $as(x, y, z)$ is skew-symmetric: $\sigma \cdot as(x, y, z) = \text{sgn}(\sigma) as(x, y, z)$
Relationsh.	Ass-alg $\to$ $Altern$-alg
Unit	$1x = x = x1$
Comment	The octonions are an example of alternative algebra Integration: Moufang loops $\dim \mathcal{P}(n)$ computed by W. Moreira
Ref.	[54] Shestakov, I. P., *Moufang loops and alternative algebras.* Proc. Amer. Math. Soc. 132 (2004), no. 2, 313–316.

Name	**Parametrized-one-relation algebra**
Notation	$Param1rel$
Def.oper.	xy
sym.	none
rel.	$(xy)z = \sum_{\sigma \in \mathbb{S}_3} a_\sigma \sigma \cdot x(yz)$ where $a_\sigma \in \mathbb{K}$
Free alg.	
rep. $\mathcal{P}(n)$	
$\dim \mathcal{P}(n)$	
Gen.series	$f(t) =$
Dual operad	$x(yz) = \sum_{\sigma \in \mathbb{S}_3} \operatorname{sgn}(\sigma) a_\sigma \sigma^{-1} \cdot (xy)z$
Chain-cplx	
Properties	
Alternative	
Relationsh.	Many classical examples are particular case: $As, Leib, Zinb, Pois$
Unit	
Comment	Problem: for which families of parameters $\{a_\sigma\}_{\sigma \in \mathbb{S}_3}$ is the operad a Koszul operad?
Ref.	

Name	**Magmatic-Fine algebra**
Notation	$MagFine$
Def.oper.	$(x_1, \ldots, x_n)_i^n$ for $1 \le i \le n-2, n \ge 3$
sym.	
rel.	
Free alg.	Described in terms of some coloured planar rooted trees
$\mathcal{P}_n$	vector space indexed as said above
$\dim \mathcal{P}_n$	$1, 0, 1, 2, 6, 18, 57, \ldots, F_{n-1}, \ldots$ where F_n = Fine number
Gen.series	$f^{MagFine}(t) = \frac{1+2t-\sqrt{1-4t}}{2(2+t)}$
Dual operad	$MagFine^!$ same generating operations, any composition is trivial $\dim MagFine^!_n = n-2$ $f^{MagFine^!}(t) = t + \frac{t^3}{(1-t)^2}$
Chain-cplx	
Properties	ns, multi-ary, quadratic, Koszul.
Alternative	
Relationsh.	
Unit	
Comment	
Ref.	[26] Holtkamp, R., Loday, J.-L., Ronco, M., *Coassociative magmatic bialgebras and the Fine numbers,* J. Alg. Comb. 28 (2008), 97–114.

Name	**Generic magmatic algebra**
Notation	$GenMag$
Def.oper.	a_n generating operations of arity n, $a_1 = 1$
sym.	
rel.	
Free alg.	
$\mathcal{P}_n$	based on trees
$\dim \mathcal{P}_n$	
Gen.series	$f^{GenMag}(t) = \sum_n b_n t^n$, $b_n =$ polynomial in $a_1, \ldots, a_n$
Dual operad	same generating operations, any composition is trivial $\dim GenMag^!_n = a_n$, $f^{GenMag^!}(t) = \sum_n a_n t^n$
Chain-cplx	
Properties	ns, multi-ary, quadratic, Koszul.
Alternative	
Relationsh.	For $MagFine^!$ $a_n = n - 2$.
Unit	
Comment	Give a nice proof of the Lagrange inversion formula for a generic power series (computation of the polynomial b_n)
Ref.	[37] Loday, J.-L., *Inversion of integral series enumerating planar trees.* Séminaire lotharingien Comb. 53 (2005), exposé B53d, 16pp.

Name	**Nonassociative permutative algebra**
Notation	NAP
Def.oper.	xy
sym.	
rel.	$(xy)z = (xz)y$
Free alg.	$NAP(V)$ can be described in terms of rooted trees
rep. $\mathcal{P}(n)$	$NAP(n) = PreLie(n)$ as $\mathbb{S}_n$-modules
$\dim \mathcal{P}(n)$	$1, 2, 9, 64, 625, \ldots, n^{n-1}, \ldots$
Gen.series	$f^{NAP}(t) = y$ which satisfies $y = t\ \exp(y)$
Dual operad	$NAP^!$
Chain-cplx	
Properties	binary, quadratic, set-theoretic, Koszul.
Alternative	
Relationsh.	$Perm\text{-}\mathsf{alg} \to NAP\text{-}\mathsf{alg}$
Unit	no
Comment	This is right NAP algebra
Ref.	[32] Livernet, M., *A rigidity theorem for pre-Lie algebras*, J. Pure Appl. Algebra 207 (2006), no. 1, 1–18.

Name	**Moufang algebra**
Notation	$Moufang$
Def.oper.	xy
sym.	
rel.	$x(yz) + z(yx) = (xy)z + (zy)x$, $((xy)z)t + ((zy)x)t = x(y(zt)) + z(y(xt))$, $t(x(yz) + z(yx)) = ((tx)y)z + ((tz)y)x$, $(xy)(tz) + (zy)(tx) = (x(yt))z + (z(yt))x$.
Free alg.	
rep. $\mathcal{P}(n)$	
$\dim \mathcal{P}(n)$	$1, 2, 7, 40, ??$
Gen.series	$f(t) =$
Dual operad	
Chain-cplx	
Properties	binary.
Alternative	Relation may be written in terms of the Jacobiator
Relationsh.	$Altern$-alg $\xrightarrow{-} Moufang$-alg $\to NCJordan$-alg
Unit	
Comment	Integration: Moufang loops. From this presentation there is an obvious definition of "nonantisymmetric Malcev algebra".
Ref.	[51] Shestakov, I., Pérez-Izquierdo, J.M., *An envelope for Malcev algebras.* J. Alg. 272 (2004), 379–393.

Name	**Malcev algebra**
Notation	$Malcev$
Def.oper.	xy
sym.	$xy = -yx$
rel.	$((xy)z)t + (x(yz))t + x((yz)t) + x(y(zt)) +$ $+((ty)z)x + (t(yz))x + t((yz)x) + t(y(zx)) =$ $(xy)(zt) + (ty)(zx)$
Free alg.	
rep. $\mathcal{P}(n)$	
$\dim \mathcal{P}(n)$	$1, 1, 3, 9, ??$
Gen.series	$f(t) =$
Dual operad	
Chain-cplx	
Properties	cubic.
Alternative	
Relationsh.	$Altern\text{-alg} \xrightarrow{-} Malcev\text{-alg}$, $Lie\text{-alg} \to Malcev\text{-alg}$
Unit	
Comment	
Ref.	[51] Shestakov, I., Pérez-Izquierdo, J.M., *An envelope for Malcev algebras.* J. Alg. 272 (2004), 379–393.

Name	**Novikov algebra**
Notation	$Novikov$
Def.oper.	xy
sym.	
rel.	$(xy)z - x(yz) = (xz)y - x(zy)$ $x(yz) = y(xz)$
Free alg.	
rep. $\mathcal{P}(n)$	
$\dim \mathcal{P}(n)$	$1, 2, ??$
Gen.series	$f(t) =$
Dual operad	
Chain-cplx	
Properties	binary, quadratic.
Alternative	Novikov is pre-Lie + $x(yz) = y(xz)$
Relationsh.	$Novikov\text{-alg} \to PreLie\text{-alg}$
Unit	
Comment	
Ref.	

Name	**Double Lie algebra**
Notation	*DoubleLie*
Def.oper.	$[x, y], \{x, y\}$
sym.	$[x, y] = -[y, x], \{x, y\} = -\{y, x\}$
rel.	Any linear combination is a Lie bracket
Free alg.	
rep. $\mathcal{P}(n)$	See Ref.
$\dim \mathcal{P}(n)$	
Gen.series	$f(t) =$
Dual operad	
Chain-cplx	
Properties	binary, quadratic.
Alternative	
Relationsh.	
Unit	
Comment	
Ref.	[14] Dotsenko V., Khoroshkin A., *Character formulas for the operad of two compatible brackets and for the bihamiltonian operad*, Functional Analysis and Its Applications, 41 (2007), no.1, 1-17.

Name	**DipreLie algebra**
Notation	*DipreLie*
Def.oper.	$x \circ y,\ x \bullet y$
sym.	
rel.	$(x \circ y) \circ z - x \circ (y \circ z) = (x \circ z) \circ y - x \circ (z \circ y)$ $(x \bullet y) \bullet z - x \bullet (y \bullet z) = (x \bullet z) \bullet y - x \bullet (z \bullet y)$ $(x \circ y) \bullet z - x \circ (y \bullet z) = (x \bullet z) \circ y - x \bullet (z \circ y)$
Free alg.	
rep. $\mathcal{P}(n)$	
$\dim \mathcal{P}(n)$	
Gen.series	$f(t) =$
Dual operad	
Chain-cplx	
Properties	binary, quadratic.
Alternative	
Relationsh.	relationship with the Jacobian conjecture (T. Maszczsyk)
Unit	
Comment	
Ref.	T. Maszczyk, unpublished.

Name	**Akivis algebra**
Notation	*Akivis*
Def.oper.	$[x, y], (x, y, z)$
sym.	$[x, y] = -[y, x]$
rel.	$[[x, y], z] + [[y, z], x] + [[z, x], y] =$ $(x, y, z) + (y, z, x) + (z, x, y) - (x, z, y) - (y, x, z) - (z, y, x)$ (Akivis relation)
Free alg.	
rep. $\mathcal{P}(n)$	
$\dim \mathcal{P}(n)$	$1, 1, 8, \ldots,$
Gen.series	$f(t) =$
Dual operad	
Chain-cplx	
Properties	binary and ternary, quadratic.
Alternative	relation also called "nonassociative Jacobi identity"
Relationsh.	$Akivis$-alg $\to$ $Sabinin$-alg, Mag-alg $\to$ $Akivis$-alg, $[x, y] = xy - yx$, $(x, y, z) = (xy)z - x(yz)$
Unit	
Comment	
Ref.	[2] Akivis, M. A. *The local algebras of a multidimensional three-web.* (Russian) Sibirsk. Mat. Zh. 17 (1976), no. 1, 5-11, 237. See also [7,55]

Name	**Sabinin algebra**
Notation	*Sabinin*
Def.oper.	$\langle x_1,\ldots,x_m;y,z\rangle, m\geq 0$ $\Phi(x_1,\ldots,x_m;y_1,\ldots,y_n),\quad m\geq 1, n\geq 2,$
sym.	$\langle x_1,\ldots,x_m;y,z\rangle = -\langle x_1,\ldots,x_m;z,y\rangle$, and for $\omega\in\mathbb{S}_m, \theta\in\mathbb{S}_n$ $\Phi(x_1,\ldots,x_m;y_1,\ldots,y_n)=\Phi(\omega(x_1,\ldots,x_m);\theta(y_1,\ldots,y_n))$,
rel.	$\langle x_1,\ldots,x_r,u,v,x_{r+1},\ldots,x_m;y,z\rangle$ $-\langle x_1,\ldots,x_r,v,u,x_{r+1},\ldots,x_m;y,z\rangle$ $+\sum_{k=0}^{r}\sum_{\sigma}\langle x_{\sigma(1)},\ldots,x_{\sigma(k)};$ $\langle x_{\sigma(k+1)},\ldots,x_{\sigma(r)};u,v\rangle,x_{r+1},\ldots,x_m;y,z\rangle$ where σ is a $(k,r-k)$-shuffle $K_{u,v,w}\big[\langle x_1,\ldots,x_r;y,z\rangle+$ $\sum_{k=0}^{r}\sum_{\sigma}\langle x_{\sigma(1)},\ldots,x_{\sigma(k)};\langle x_{\sigma(k+1)},\ldots,x_{\sigma(r)};v,w\rangle,u\rangle\big]=0$ where $K_{u,v,w}$ is the sum over all cyclic permutations
Free alg.	
rep. $\mathcal{P}(n)$	
$\dim\mathcal{P}(n)$	$1,1,8,78,1104,\ldots$
Gen.series	$f^{Sab}(t)=\log(1+(1/2)(1-\sqrt{1-4t}))$
Dual operad	
Chain-cplx	
Properties	quadratic, ns.
Alternative	There exists a more compact form of the relations which uses the tensor algebra over the Sabinin algebra
Relationsh.	*Mag*-alg $\longrightarrow$ *Sabinin*-alg, $\langle y,z\rangle = yz-zy$, $\langle x;y,z\rangle = ??$ *Akivis*-alg $\longrightarrow$ *Sabinin*-alg, $\langle y,z\rangle=-[y,z]$, $\langle x;y,z\rangle=(x,z,y)-(x,y,z)$, $\langle x_1,\ldots,x_m;y,z\rangle=0, m\geq 2$
Unit	
Comment	Integration: local analytic loops
Ref.	[50] D. Pérez-Izquierdo, *Algebras, hyperalgebras, nonassociative bialgebras and loops.* Adv. Math. 208 (2007), 834–876.

Name	**Jordan triple systems**
Notation	*Jordantriples* or JT
Def.oper.	(xyz) or (x,y,z)
sym.	$(xyz) = (zyx)$
rel.	$(xy(ztu)) = ((xyz)tu) - (z(txy)u) + (zt(xyu))$
Free alg.	
rep. $\mathcal{P}(2n-1)$	
$\dim \mathcal{P}(2n-1)$	$1, 3, 50, ??$
Gen.series	$f(t) =$
Dual operad	
Chain-cplx	
Properties	ternary, quadratic, Koszul ?.
Alternative	
Relationsh.	
Unit	
Comment	Remark that the quadratic relation, as written here, has a Leibniz flavor. Dimension computed by Walter Moreira
Ref.	[47] E. Neher, Jordan triple systems by the grid approach, Lecture Notes in Mathematics, 1280, Springer-Verlag, 1987.

Name	**Totally associative ternary algebra**
Notation	$t\text{-}As^{\langle 3\rangle}$
Def.oper.	(xyz)
sym.	
rel.	$((xyz)uv) = (x(yzu)v) = (xy(zuv))$
Free alg.	
$\mathcal{P}_{2n-1}$	
$\dim \mathcal{P}_{2n-1}$	$1, 1, 1, \ldots, 1, \ldots$
Gen.series	$f^{t\text{-}As^3}(t) = \frac{t}{1-t^2}$
Dual operad	$t\text{-}As^{\langle 3\rangle\,!} = p\text{-}As^{\langle 3\rangle}$
Chain-cplx	
Properties	ns, ternary, quadratic, set-theoretic, Koszul ?.
Alternative	
Relationsh.	$As\text{-alg} \to t\text{-}As^{\langle 3\rangle}$
Unit	
Comment	
Ref.	[23] Gnedbaye, A.V., *Opérades des algèbres* $(k+1)$*-aires.* Operads: Proceedings of Renaissance Conferences, 83–113, Contemp. Math., 202, Amer. Math. Soc., Providence, RI, 1997.

Name	**Partially associative ternary algebra**
Notation	$p\text{-}As^{\langle 3\rangle}$
Def.oper.	(xyz)
sym.	
rel.	$((xyz)uv) + (x(yzu)v) + (xy(zuv)) = 0$
Free alg.	
$\mathcal{P}_{2n-1}$	
$\dim \mathcal{P}_{2n-1}$	
Gen.series	$f(t) =$
Dual operad	$p\text{-}As^{\langle 3\rangle\,!} = t\text{-}As^{\langle 3\rangle}$
Chain-cplx	
Properties	ns, ternary, quadratic, Koszul ?.
Alternative	
Relationsh.	
Unit	
Comment	
Ref.	[23] Gnedbaye, A.V., *Opérades des algèbres* $(k+1)$*-aires.* Operads: Proceedings of Renaissance Conferences, 83–113, Contemp. Math., 202, Amer. Math. Soc., Providence, RI, 1997.

Name	**Lie triple systems**
Notation	LTS
Def.oper.	$[xyz]$
sym.	$[xyz] = -[yxz]$,
	$[xyz] + [yzx] + [zxy] = 0.$
rel.	$[xy[ztu]] = [[xyz]tu] - [z[txy]u] + [zt[xyu]]$
Free alg.	
rep. $\mathcal{P}(n)$	
$\dim \mathcal{P}(n)$	
Gen.series	$f(t) =$
Dual operad	
Chain-cplx	
Properties	ternary, quadratic, Koszul ?.
Alternative	the relation admits many different versions due to the symmetry
Relationsh.	Lie-alg $\to$ LTS-alg, $[xyz] = [[xy]z]$
Comment	Appreciate the Leibniz presentation Integration: symmetric spaces
Ref.	[45] Loos O., Symmetric spaces. I. General theory. W. A. Benjamin, Inc., New York-Amsterdam (1969) viii+198 pp.

Name	**Lie-Yamaguti algebra**
Notation	$Lie-Yamaguti$ or LY
Def.oper.	$x\cdot y, \quad [x,y,z]$
sym.	$x\cdot y=-y\cdot x, \quad [x,y,z]=-[y,x,z]$
	$[x,y,z]+[y,z,x]+[z,x,y]+(x\cdot y)\cdot z+(y\cdot z)\cdot x$
	$+(z\cdot x)\cdot y=0$
rel.	$\sum_{cyclic}[x\cdot y,z,t]=0$
	$[x,y,u\cdot v]=u\cdot[x,y,v]+[x,y,u]\cdot v$
	$[x,y,[z,t,u]]=[[x,y,z],t,u]-[z,[t,x,y],u]+[z,t,[x,y,u]]$
Free alg.	
rep. $\mathcal{P}(n)$	
$\dim\mathcal{P}(n)$	
Gen.series	$f(t)=$
Dual operad	
Chain-cplx	
Properties	binary and ternary, quadratic, Koszul ?.
Alternative	Generalized Lie triple systems
Relationsh.	$LTS\text{-alg}\to LY\text{-alg}, x\cdot y=0, [x,y,z]=[xyz]$
Unit	
Comment	
Ref.	[28] M. K. Kinyon, A. Weinstein, *Leibniz algebras, Courant algebroids, and multiplications on reductive homogeneous spaces.* Amer. J. Math. 123 (2001), no. 3, 525–550.

Name **Comtrans algebras**

Notation *Comtrans*

Def.oper. $[x,y,z]$ and $\langle x,y,z\rangle$ (2 ternary operations)

sym. $[x,y,z]+[y,x,z]=0$,
$\langle x,y,z\rangle+\langle y,z,x\rangle+\langle z,x,y\rangle=0$,
$[x,y,z]+[z,y,x]=\langle x,y,z\rangle+\langle z,y,x\rangle$.

rel.

Free alg.

rep. $\mathcal{P}_{2n-1}$

$\dim \mathcal{P}_{2n-1}$

Gen.series $f(t)=$

Dual operad

Chain-cplx

Properties ternary.

Alternative

Relationsh.

Unit

Comment

Ref.

[56] J.D.H. Smith, *Multilinear algebras and Lie's Theorem for formal n-loops*, Arch. Math. 51 (1988), 169–177.

Name	**Interchange algebra**
Notation	*Interchange*
Def.oper.	$x \cdot y, \quad x * y$
sym.	
rel.	$(x \cdot y) * (z \cdot t) = (x * z) \cdot (y * t)$
Free alg.	
rep. $\mathcal{P}(n)$	
$\dim \mathcal{P}(n)$	
Gen.series	$f(t) =$
Dual operad	
Chain-cplx	
Properties	binary, cubic, set-theoretic.
Alternative	
Relationsh.	Strongly related with the notions of 2-category and 2-group
Unit	if a unit for both, then $* = \cdot$ and they are commutative (Eckmann-Hilton trick)
Comment	Many variations depending on the hypotheses on $*$ and $\cdot$ Most common $\cdot$ and $*$ are associative.
Ref.	

Name	**Hypercommutative algebra**
	underlying objects: graded vector spaces
Notation	*HyperCom*
Def.oper.	$(x_1, \dots, x_n)$ n-ary operation of degree $2(n-2)$ for $n \geq 2$
sym.	totally symmetric
rel.	$\sum_{S_1 \sqcup S_2 = \{1,\dots,n\}} ((a, b, x_{S_1}), c, x_{S_2}) = \sum_{S_1 \sqcup S_2 = \{1,\dots,n\}} (-1)^{\lvert c\rvert \lvert x_{S_1}\rvert} (a, (b, x_{S_1}, c), x_{S_2})$, for any $n \geq 0$.
Free alg.	
rep. $\mathcal{P}(n)$	
$\dim \mathcal{P}(n)$	
Gen.series	$f(t) =$
Dual operad	Gravity algebra, see Ref below
Chain-cplx	
Properties	
Alternative	
Relationsh.	
Unit	
Comment	
Ref.	[21] E. Getzler, *Operads and moduli spaces of genus* 0 *Riemann surfaces*, The moduli space of curves (Texel Island, 1994), Progr. Math., 129, (1995), 199–230.

Name	**Associative algebra up to homotopy**
	operad with underlying space in dgVect
Notation	A_∞
Def.oper.	m_n for $n \geq 2$ (operation of arity n and degree $n-2$)
sym.	
rel.	$\partial(m_n) = \sum_{\substack{n=p+q+r \\ k=p+1+r \\ k>1, q>1}} (-1)^{p+qr} m_k \circ (\mathrm{id}^{\otimes p} \otimes m_q \otimes \mathrm{id}^{\otimes r})$.
Free alg.	
$\mathcal{P}_n$	$(A_\infty)_n = \mathbb{K}[PT_n]$ isomorphic to $C_\bullet(\mathcal{K}^{n-2})$ as chain complex where $\mathcal{K}^n$ is the Stasheff polytope of dimension n
$\dim \mathcal{P}_n$	
Gen.series	$f(t) =$
Dual operad	
Chain-cplx	
Properties	ns, multi-ary, quadratic, minimal model for As
Alternative	Cobar construction on $As^{¡}$: $A_\infty = As_\infty := \Omega As^{¡}$
Relationsh.	Many, see the literature.
Unit	Good question !
Comment	There are two levels of morphisms between A_∞-algebras: the morphisms and the ∞-morphisms, see [44] for instance.
Ref.	[57] J. Stasheff, *Homotopy associativity of H-spaces. I, II.* TAMS 108 (1963), 275-292 ; ibid. 108 (1963), 293312.

Name **Commutative algebra up to homotopy**

operad with underlying space in dgVect

Notation C_∞

Def.oper. m_n for $n \geq 2$ (operation of arity n and degree $n-2$) which vanishes on the sum of $(p, n-p)$-shuffles, $1 \leq p \leq n-1$.

sym.

rel. $\partial(m_n) = \sum_{\substack{n=p+q+r \\ k=p+1+r \\ k>1, q>1}} (-1)^{p+qr} m_k \circ (\mathrm{id}^{\otimes p} \otimes m_q \otimes \mathrm{id}^{\otimes r})$.

Free alg.

$\mathcal{P}(n)$

$\dim \mathcal{P}(n)$

Gen.series $f(t) =$

Dual operad

Chain-cplx

Properties multi-ary, quadratic, minimal model for Com

Alternative Cobar construction on $Com^{¡}$: $C_\infty = Com_\infty := \Omega Com^{¡}$

Relationsh.

Unit

Comment

Ref. [27] T. Kadeishvili, *The category of differential coalgebras and the category of $A(\infty)$-algebras.* Proc. Tbilisi Math.Inst. 77 (1985), 50-70.

Name	**Lie algebra up to homotopy**
	operad with underlying space in dgVect
Notation	L_∞
Def.oper.	ℓ_n, n-ary operation of degree $n-2$, for all $n \geq 2$
sym.	
rel.	$\partial_A(\ell_n) = \sum_{\substack{p+q=n+1 \\ p,q>1}} \sum_{\sigma \in Sh_{p,q}^{-1}} \mathrm{sgn}(\sigma)(-1)^{(p-1)q}(\ell_p \circ_1 \ell_q)^\sigma$,
Free alg.	
rep. $\mathcal{P}(n)$	
$\dim \mathcal{P}(n)$	
Gen.series	$f(t) =$
Dual operad	
Chain-cplx	
Properties	multi-ary, quadratic, minimal model for *Lie*
Alternative	Cobar construction on $Lie^{¡}$: $L_\infty = Lie_\infty := \Omega Lie^{¡}$
Relationsh.	
Unit	
Comment	
Ref.	[25] V. Hinich, and V. Schechtman, *Homotopy Lie algebras.* I. M. Gel'fand Seminar, 128, Adv. Soviet Math., 16, Part 2, Amer. Math. Soc., Providence, RI, 1993.

`Name`	**Dendriform algebra up to homotopy**
	operad with underlying space in `dgVect`
`Notation`	$Dend_\infty$
`Def.oper.`	$m_{n,i}$ is an n-ary operation, $1 \leq i \leq n$, for all $n \geq 2$
`sym.`	none
`rel.`	$\partial(m_{n,i}) = \sum (-1)^{p+qr} m_{p+1+r,\ell}(\underbrace{\mathrm{id}, \cdots, \mathrm{id}}_{p}, m_{q,j}, \underbrace{\mathrm{id}, \cdots, \mathrm{id}}_{r})$ sum extended to all the quintuples p, q, r, ℓ, j satisfying: $p \geq 0, q \geq 2, r \geq 0, p+q+r = n, 1 \leq \ell \leq p+1+q, 1 \leq j \leq q$ and either one of the following: $i = q + \ell$, when $1 \leq p+1 \leq \ell - 1$, $i = \ell - 1 + j$, when $p + 1 = \ell$, $i = \ell$, when $\ell + 1 \leq p + 1$.
`Free alg.`	
`rep.` $\mathcal{P}_n$	
$\dim \mathcal{P}_n$	
`Gen.series`	$f(t) =$
`Dual operad`	
`Chain-cplx`	
`Properties`	ns, multi-ary, quadratic, minimal model for *Dend*
`Alternative`	
`Relationsh.`	
`Unit`	
`Comment`	
`Ref.`	See for instance [44]

Name	**$\mathcal{P}$-algebra up to homotopy**
	operad with underlying space in dgVect
Notation	$\mathcal{P}_\infty$
	the operad $\mathcal{P}$ is supposed to be quadratic and Koszul
Def.oper. sym. rel.	
Free alg.	
rep. $\mathcal{P}(n)$	
$\dim \mathcal{P}(n)$	
Gen.series	$f(t) =$
Dual operad	
Chain-cplx	
Properties	quadratic, ns if $\mathcal{P}(n) = \mathcal{P}_n \otimes \mathbb{K}[\mathbb{S}_n]$.
Alternative	Cobar construction on $\mathcal{P}^{¡}$: $\mathcal{P}_\infty := \Omega\, \mathcal{P}^{¡}$
Relationsh.	
Unit	
Comment	
Ref.	See for instance [44]

Name	**Brace algebra**
Notation	*Brace*
Def.oper.	$\{x_0; x_1, \ldots, x_n\}$ for $n \geq 0$
sym.	$\{x; \emptyset\} = x$
rel.	$\{\{x; y_1, \ldots, y_n\}; z_1, \ldots, z_m\} =$ $\sum\{x; \ldots, \{y_1; \ldots\}, \ldots \quad \ldots, \{y_n; \ldots, \}, \ldots\}$. the dots are filled with the variables z_i's (in order).
Free alg.	
rep. $\mathcal{P}(n)$	$Brace(n) = \mathbb{K}[PBT_{n+1}] \otimes \mathbb{K}[\mathbb{S}_n]$
$\dim \mathcal{P}(n)$	$1, 1 \times 2!, 2 \times 3!, 5 \times 4!, 14 \times 5!, 42 \times 6!, 132 \times 7!, \ldots,$ $c_{n-1} \times n!, \ldots$
Gen.series	$f(t) =$
Dual operad	
Chain-cplx	
Properties	multi-ary, quadratic, quasi-regular.
Alternative	
Relationsh.	$Brace$-alg $\to MB$-alg $Brace$-alg $\to PreLie$-alg, ($\{-; -\}$ is a pre-Lie product) If A is a brace algebra, then $T^c(A)$ is a cofree Hopf algebra
Unit	
Comment	There exists a notion of brace algebra with differential useful in algebraic topology
Ref.	[53] Ronco, M. *Eulerian idempotents and Milnor-Moore theorem for certain non-cocommutative Hopf algebras.* J. Algebra 254 (2002), no. 1, 152–172.

Name	**Multi-brace algebra**
Notation	MB
Def.oper.	$(x_1, \ldots, x_p; y_1, \ldots, y_q\}$ for $p \geq 1, q \geq 1$
sym.	
rel.	$\mathcal{R}_{ijk}$, see Ref.
Free alg.	
rep. $\mathcal{P}(n)$	$MB(n) = (\mathbb{K}[PT_n] \oplus \mathbb{K}[PT_n]) \otimes \mathbb{K}[\mathbb{S}_n], n \geq 2$
$\dim \mathcal{P}(n)$	$1, 1 \times 2!, 6 \times 3!, 22 \times 4!, 90 \times 5!, ?? \times 6!, ??? \times 7!, \ldots,$ $2C_n \times n!, \ldots,$ $C_n =$ Schröder number (super Catalan)
Gen.series	$f(t) =$
Dual operad	
Chain-cplx	
Properties	multi-ary, quadratic, quasi-regular.
Alternative	Used to be denoted by B_∞ or $\mathbf{B}_\infty$ confusing notation with respect to algebras up to homotopy
Relationsh.	$Brace$-alg $\to MB$-alg $Brace$-alg $\to PreLie$-alg, ($\{-;-\}$ is a pre-Lie product) If A is a brace algebra, then $T^c(A)$ is a cofree Hopf algebra (and vice-versa)
Unit	
Comment	There exists a notion of MB-infinity algebra with differentials useful in algebraic topology (and called B_∞-algebra)
Ref.	[43] Loday, J.-L., and Ronco, M. *On the structure of cofree Hopf algebras* J. reine angew. Math. 592 (2006) 123–155.

Name	**n-Lie algebras**
Notation	n-*Lie*
Def.oper.	$[x_1, \ldots, x_n]$ (n-ary operation)
sym.	fully anti-symmetric
rel.	Leibniz type identity, see next page
Free alg.	
rep. $\mathcal{P}(n)$	
$\dim \mathcal{P}(n)$	
Gen.series	$f(t) =$
Dual operad	
Chain-cplx	
Properties	n-ary
Alternative	
Relationsh.	
Unit	
Comment	
Ref.	[17] Filippov, V. T. n-Lie algebras. (Russian) Sibirsk. Mat. Zh. 26 (1985), no. 6, 126140, 191. English translation: Siberian Math. J. 26 (1985), no. 6, 879–891.

Name	**n-Leibniz algebras**
Notation	n-*Leib*
Def.oper.	$[x_1,\ldots,x_n]$ (n-ary operation)
sym.	
rel.	$[[x_1,\ldots,x_n],y_1,\ldots,y_{n-1}] = \sum_{i=1}^{n}[x_1,\ldots,x_{i-1},[x_i,y_1,\ldots,y_{n-1}],x_{i+1},\ldots,x_n]$
Free alg.	
rep. $\mathcal{P}(n)$	
$\dim \mathcal{P}(n)$	
Gen.series	$f(t) =$
Dual operad	
Chain-cplx	
Properties	n-ary
Alternative	
Relationsh.	
Unit	
Comment	
Ref.	[9] Casas, J. M.; Loday, J.-L.; Pirashvili, T., *Leibniz n-algebras.* Forum Math. 14 (2002), no. 2, 189–207.

Name **$\mathcal{X}^{\pm}$-algebra**

Notation $\mathcal{X}^{\pm}$-alg

Def.oper. $x \nwarrow y,\ x \nearrow y,\ x \searrow y,\ x \swarrow y.$

sym.

rel.

$$
\begin{array}{l}
(\nwarrow)\nwarrow = \nwarrow(\nwarrow) + \nwarrow(\swarrow),\ (\swarrow)\nwarrow = \swarrow(\nwarrow),\ (\nwarrow)\swarrow + (\swarrow)\swarrow = \swarrow(\swarrow),\\
(\nwarrow)\nwarrow = \nwarrow(\searrow) + \nwarrow(\nearrow),\ (\swarrow)\nwarrow = \swarrow(\nearrow),\ (\nwarrow)\swarrow + (\swarrow)\swarrow = \swarrow(\searrow),\\
(\nearrow)\nwarrow = \nearrow(\nwarrow) + \nearrow(\swarrow),\ (\searrow)\nwarrow = \searrow(\nwarrow),\ (\nearrow)\swarrow + (\searrow)\swarrow = \searrow(\swarrow),\\
(\nwarrow)\nearrow = \nearrow(\nearrow) + \nearrow(\searrow),\ (\swarrow)\nearrow = \searrow(\nearrow),\ (\nwarrow)\searrow + (\swarrow)\searrow = \searrow(\searrow),\\
(\nearrow)\nearrow = \nearrow(\nearrow) + \nearrow(\searrow),\ (\searrow)\nearrow = \searrow(\nearrow),\ (\nearrow)\searrow + (\searrow)\searrow = \searrow(\searrow).
\end{array}
$$

$$(\nearrow)\searrow - (\nwarrow)\searrow = +\nwarrow(\swarrow) - \nwarrow(\searrow)\ , \qquad (16+)$$
$$(\nearrow)\searrow - (\nwarrow)\searrow = -\nwarrow(\swarrow) + \nwarrow(\searrow)\ . \qquad (16-)$$

Free alg.

$\mathcal{P}_n$

$\dim \mathcal{P}_n$

Gen.series $f(t) =$

Dual operad Both sef-dual.

Chain-cplx

Properties ns, binary, quadratic.

Alternative

Relationsh. *Dend* ■ *Dias* $\twoheadrightarrow \mathcal{X}^{\pm} \twoheadrightarrow$ *Dend* □ *Dias*
Fit into the "operadic butterfly" diagram, see the reference.

Unit

Comment

Ref. [38] Loday J.-L., *Completing the operadic butterfly*, Georgian Math. Journal 13 (2006), no 4. 741–749.

`Name`	*put your own type of* **algebras**
`Notation`	

`Def.oper.`
`sym.`
`rel.`	

`Free alg.`	
`rep.` $\mathcal{P}(n)$	
$\dim \mathcal{P}(n)$	
`Gen.series`	$f(t) =$
`Dual operad`	
`Chain-cplx`	
`Properties`	
`Alternative`	
`Relationsh.`	
`Unit`	
`Comment`	
`Ref.`	

Integer sequences which appear in this paper, up to some shift and up to multiplication by $n!$ or $(n-1)!$.

1	1	1	1	1	...	1	...	Com, As
1	2	0	0	0	...	0	...	Nil
1	2	2	2	2	...	2	...	$Dual2as$
1	2	3	4	5	...	n	...	$Dias, Perm$
1	2	5	12	15	...	??	...	$Altern^!$
1	2	5	14	42	...	c_{n-1}	...	$Mag, Dend, brace, Dup$
1	2	6	18	57	...	f_{n+2}	...	$MagFine$
1	2	6	22	90	...	$2C_n$	...	$Dipt, 2as, brace$
1	2	6	24	120	...	$n!$	...	$As, Lie, Leib, Zinb$
1	2	7			...	??	...	Lie-adm
1	2	7	32	175	...	??	...	$Altern$
1	2	7	40		...	??	...	$Moufang$
1	2	9	64	625	...	n^{n-1}	...	$PreLie, NAP$
1	2	10	26	76	...	??	...	$Parastat$
1	3	7	15	31	...	(2^n-1)	...	$Trias$
1	3	9			...	??	...	$Malcev$
1	3	11	45	197	...	C_n	...	$TriDend$
1	3	13	75	541	...	??	...	CTD
1	3	16	125	6^5	...	$(n+1)^{n-1}$	...	$Park$
1	3	20	210	3024	...	$a(n)$	...	$PostLie$
1	3	50			...	??	...	$Jordan\ triples$
1	4	9	16	25	...	$n^2n!$	...	$Quadri^!$
1	4	23	156	1162	...	??	...	$Quadri$
1	4	23	181		...	??	...	see $PreLie$
1	4	27	256		...	n^n	...	$PreLiePerm$
1	8				...	??	...	$Akivis$
1	8	78	1104		...	??	...	$Sabinin$

References

1. Marcelo Aguiar and Jean-Louis Loday. Quadri-algebras. *J. Pure Appl. Algebra*, 191(3):205–221, 2004.
2. M. A. Akivis. The local algebras of a multidimensional three-web. *Sibirsk. Mat. Ž.*, 17(1):5–11, 237, 1976.
3. Chengming Bai, Ligong Liu, and Xiang Ni. Some results on L-dendriform algebras. *J. Geom. Phys.*, 60(6-8):940–950, 2010.
4. C.M. Bai, O. Bellier, L. Guo, and X. Ni. Splitting of operations, Manin products and Rota-Baxter operators. `arXiv:1106.6080`, 2011.
5. C.M. Bai, L. Guo, and X. Ni. L-quadri-algebras. 2010.

6. Nantel Bergeron and Jean-Louis Loday. The symmetric operation in a free pre-Lie algebra is magmatic. *Proc. Amer. Math. Soc.*, 139(5):1585–1597, 2011.
7. Murray R. Bremner, Irvin R. Hentzel, and Luiz A. Peresi. Dimension formulas for the free nonassociative algebra. *Comm. Algebra*, 33(11):4063–4081, 2005.
8. Christian Brouder and Alessandra Frabetti. QED Hopf algebras on planar binary trees. *J. Algebra*, 267(1):298–322, 2003.
9. J. M. Casas, J.-L. Loday, and T. Pirashvili. Leibniz n-algebras. *Forum Math.*, 14(2):189–207, 2002.
10. F. Chapoton. Un endofoncteur de la catégorie des opérades. In *Dialgebras and related operads*, volume 1763 of *Lecture Notes in Math.*, pages 105–110. Springer, Berlin, 2001.
11. F. Chapoton and M. Livernet. Pre-Lie algebras and the rooted trees operad. *Internat. Math. Res. Notices*, (8):395–408, 2001.
12. Frédéric Chapoton. Opérades différentielles graduées sur les simplexes et les permutoèdres. *Bull. Soc. Math. France*, 130(2):233–251, 2002.
13. S. Covez. The local integration of Leibniz algebras. *arXiv: 1011.4112*, November 2010.
14. V. V. Dotsenko and A. S. Khoroshkin. Character formulas for the operad of a pair of compatible brackets and for the bi-Hamiltonian operad. *Funktsional. Anal. i Prilozhen.*, 41(1):1–22, 96, 2007.
15. Vladimir Dotsenko. Compatible associative products and trees. *Algebra Number Theory*, 3(5):567–586, 2009.
16. A. Dzhumadil′daev and P. Zusmanovich. The alternative operad is not Koszul. *arXiv:0906.1272*, 2009.
17. V. T. Filippov. n-Lie algebras. *Sibirsk. Mat. Zh.*, 26(6):126–140, 191, 1985.
18. B. Fresse. Théorie des opérades de Koszul et homologie des algèbres de Poisson. *Ann. Math. Blaise Pascal*, 13(2):237–312, 2006.
19. I. Galvez-Carrillo, A. Tonks, and B. Vallette. Homotopy Batalin-Vilkovisky algebras. *Journal of Noncommutative Geometry*, to appear:49 pp, 2011.
20. M. Gerstenhaber. The cohomology structure of an associative ring. *Ann. of Math. (2)*, 78:267–288, 1963.
21. E. Getzler. Operads and moduli spaces of genus 0 Riemann surfaces. In *The moduli space of curves (Texel Island, 1994)*, volume 129 of *Progr. Math.*, pages 199–230. Birkhäuser Boston, Boston, MA, 1995.
22. V. Ginzburg and M. Kapranov. Koszul duality for operads. *Duke Math. J.*, 76(1):203–272, 1994.
23. Allahtan Victor Gnedbaye. Opérades des algèbres $(k+1)$-aires. In *Operads: Proceedings of Renaissance Conferences (Hartford, CT/Luminy, 1995)*, volume 202 of *Contemp. Math.*, pages 83–113. Amer. Math. Soc., Providence, RI, 1997.
24. A. B. Goncharov. Galois symmetries of fundamental groupoids and noncommutative geometry. *Duke Math. J.*, 128(2):209–284, 2005.
25. Vladimir Hinich and Vadim Schechtman. Homotopy Lie algebras. In *I. M. Gel′fand Seminar*, volume 16 of *Adv. Soviet Math.*, pages 1–28. 93.
26. Ralf Holtkamp, Jean-Louis Loday, and María Ronco. Coassociative magmatic bialgebras and the Fine numbers. *J. Algebraic Combin.*, 28(1):97–114, 2008.

27. T. V. Kadeishvili. The category of differential coalgebras and the category of $A(\infty)$-algebras. *Trudy Tbiliss. Mat. Inst. Razmadze Akad. Nauk Gruzin. SSR*, 77:50–70, 1985.
28. Michael K. Kinyon and Alan Weinstein. Leibniz algebras, Courant algebroids, and multiplications on reductive homogeneous spaces. *Amer. J. Math.*, 123(3):525–550, 2001.
29. Jean-Louis Koszul. Crochet de Schouten-Nijenhuis et cohomologie. *Astérisque*, (Numero Hors Serie):257–271, 1985. The mathematical heritage of Élie Cartan (Lyon, 1984).
30. Philippe Leroux. Ennea-algebras. *J. Algebra*, 281(1):287–302, 2004.
31. Philippe Leroux. Hochschild two-cocycles and a good triple of operads. *Int. Electron. J. Algebra*, 2011.
32. Muriel Livernet. A rigidity theorem for pre-Lie algebras. *J. Pure Appl. Algebra*, 207(1):1–18, 2006.
33. Jean-Louis Loday. Une version non commutative des algèbres de Lie: les algèbres de Leibniz. *Enseign. Math. (2)*, 39(3-4):269–293, 1993.
34. Jean-Louis Loday. Cup-product for Leibniz cohomology and dual Leibniz algebras. *Math. Scand.*, 77(2):189–196, 1995.
35. Jean-Louis Loday. Dialgebras. In *Dialgebras and related operads*, volume 1763 of *Lecture Notes in Math.*, pages 7–66. Springer, Berlin, 2001.
36. Jean-Louis Loday. Scindement d'associativité et algèbres de Hopf. In *Actes des Journées Mathématiques à la Mémoire de Jean Leray*, volume 9 of *Sémin. Congr.*, pages 155–172. Soc. Math. France, Paris, 2004.
37. Jean-Louis Loday. Inversion of integral series enumerating planar trees. *Sém. Lothar. Combin.*, 53:Art. B53d, 16 pp. (electronic), 2005.
38. Jean-Louis Loday. Completing the operadic butterfly. *Georgian Math. J.*, 13(4):741–749, 2006.
39. Jean-Louis Loday. On the algebra of quasi-shuffles. *Manuscripta Math.*, 123(1):79–93, 2007.
40. Jean-Louis Loday. Generalized bialgebras and triples of operads. *Astérisque*, (320):x+116, 2008.
41. Jean-Louis Loday and María Ronco. Algèbres de Hopf colibres. *C. R. Math. Acad. Sci. Paris*, 337(3):153–158, 2003.
42. Jean-Louis Loday and María Ronco. Trialgebras and families of polytopes. In *Homotopy theory: relations with algebraic geometry, group cohomology, and algebraic K-theory*, volume 346 of *Contemp. Math.*, pages 369–398. Amer. Math. Soc., Providence, RI, 2004.
43. Jean-Louis Loday and María Ronco. On the structure of cofree Hopf algebras. *J. Reine Angew. Math.*, 592:123–155, 2006.
44. Jean-Louis Loday and Bruno Vallette. *Algebraic operads*. 2011.
45. Ottmar Loos. *Symmetric spaces. I: General theory*. W. A. Benjamin, Inc., New York-Amsterdam, 1969.
46. Martin Markl, Steve Shnider, and Jim Stasheff. *Operads in algebra, topology and physics*, volume 96 of *Mathematical Surveys and Monographs*. American Mathematical Society, Providence, RI, 2002.
47. Erhard Neher. *Jordan triple systems by the grid approach*, volume 1280 of *Lecture Notes in Mathematics*. Springer-Verlag, Berlin, 1987.

48. A. V. Odesskii and V. V. Sokolov. Integrable matrix equations related to pairs of compatible associative algebras. *J. Phys. A*, 39(40):12447–12456, 2006.
49. J.-M. Oudom and D. Guin. On the Lie enveloping algebra of a pre-Lie algebra. *J. K-Theory*, 2(1):147–167, 2008.
50. José M. Pérez-Izquierdo. Algebras, hyperalgebras, nonassociative bialgebras and loops. *Adv. Math.*, 208(2):834–876, 2007.
51. José M. Pérez-Izquierdo and Ivan P. Shestakov. An envelope for Malcev algebras. *J. Algebra*, 272(1):379–393, 2004.
52. María Ronco. Primitive elements in a free dendriform algebra. In *New trends in Hopf algebra theory (La Falda, 1999)*, volume 267 of *Contemp. Math.*, pages 245–263. Amer. Math. Soc., Providence, RI, 2000.
53. María Ronco. Eulerian idempotents and Milnor-Moore theorem for certain non-cocommutative Hopf algebras. *J. Algebra*, 254(1):152–172, 2002.
54. Ivan P. Shestakov. Moufang loops and alternative algebras. *Proc. Amer. Math. Soc.*, 132(2):313–316 (electronic), 2004.
55. Ivan P. Shestakov and Ualbai U. Umirbaev. Free Akivis algebras, primitive elements, and hyperalgebras. *J. Algebra*, 250(2):533–548, 2002.
56. J. D. H. Smith. Multilinear algebras and Lie's theorem for formal n-loops. *Arch. Math. (Basel)*, 51(2):169–177, 1988.
57. James Dillon Stasheff. Homotopy associativity of H-spaces. I, II. *Trans. Amer. Math. Soc. 108 (1963), 275-292; ibid.*, 108:293–312, 1963.
58. Bruno Vallette. Homology of generalized partition posets. *J. Pure Appl. Algebra*, 208(2):699–725, 2007.
59. Bruno Vallette. A Koszul duality for props. *Trans. of Amer. Math. Soc.*, 359:4865–4993, 2007.
60. Bruno Vallette. Manin products, Koszul duality, Loday algebras and Deligne conjecture. *J. Reine Angew. Math.*, 620:105–164, 2008.
61. È. B. Vinberg. The theory of homogeneous convex cones. *Trudy Moskov. Mat. Obšč.*, 12:303–358, 1963.

Algebra Index

A_∞, 282
$As^{(2)}$-algebra, 258
$As^{\langle 2\rangle}$-algebra, 259
C_∞, 283
$Dend_\infty$, 285
L_∞, 284
$\mathcal{P}_\infty$, 286
$\mathcal{X}^\pm$-algebra, 291
n-Leibniz, 290
n-Lie, 289
$\mathbf{B}_\infty$, 288

Akivis, 272
algebra, 224
alternative, 262
anti-symmetric nilpotent, 253
associative, 225

Batalin-Vilkovisky, 249
brace, 287

commutative, 226
commutative triassociative, 243
commutative tridendriform, 244
CTD algebra, 244

dendriform, 232
diassociative, 233
dipreLie, 271
dipterous, 236
double Lie, 270
dual commutative tridendriform, 245
dual CTD, 245
dual dipterous, 237
dual duplicial, 257
dual quadri-, 255
dual two associative, 239
duplicial, 256

generic magmatic, 265
Gerstenhaber, 248

hypercommutative, 281

interchange, 280
IT-, 280

Jordan triples, 274

L-dendriform, 246
Leibniz, 230
Lie, 227
Lie triple systems, 277
Lie-admissible, 260
Lie-Yamaguti, 278

magmatic, 250, 252
magmatic-Fine, 264
Malcev, 268
Moufang, 267

NAP-, 266
nilpotent, 251
nonassociative permutative, 266
Novikov, 269

parametrized-one-relation, 263
partially associative ternary, 276
perm, 235
Poisson, 228
post-Lie, 242
pre-Lie, 234
PreLiePerm, 261

quadri-, 254

Sabinin, 273

totally associative ternary, 275
triassociative, 241
tridendriform, 240
two-associative, 238

Vinberg, 234

Zinbiel, 231

Author Index

Bokut, L. A., 1

Chen, Yuqun, 1
Curien, Pierre-Louis, 25

Hu, Naihong, 51

Jimènez, Daniel, 69

Li, Yu, 1
Liu, Dong, 51
Liu, Yun, 89
Livernet, Muriel, 107
Loday, Jean-Louis, 139

Makhlouf, Abdenacer, 147

Ronco, María, 69

Yang, Wen-Li, 173

Zhou, Chenyan, 199
Zhu, Linsheng, 51
Zinbiel, G. W., 217